# Genomics of Plants and Fungi

# MYCOLOGY SERIES

*Editor*

## J. W. Bennett

Professor
Department of Cell and Molecular Biology
Tulane University
New Orleans, Louisiana

*Founding Editor*

## Paul A. Lemke

16. *Pathogenic Fungi in Humans and Animals: Second Edition,* edited by Dexter H. Howard
17. *Fungi in Ecosystem Processes,* John Dighton
18. *Genomics of Plants and Fungi,* edited by Rolf A. Prade and Hans J. Bohnert

# Genomics of Plants and Fungi

edited by
## Rolf A. Prade
*Oklahoma State University*
*Stillwater, Oklahoma, U.S.A.*

## Hans J. Bohnert
*University of Illinois at Urbana-Champaign*
*Urbana, Illinois, U.S.A.*

MARCEL DEKKER, INC.          NEW YORK · BASEL

**Library of Congress Cataloging-in-Publication Data**
A catalog record for this book is available from the Library of Congress.

**ISBN: 0-8247-4125-0**

This book is printed on acid-free paper.

**Headquarters**
Marcel Dekker, Inc.
270 Madison Avenue, New York, NY 10016
tel: 212-696-9000; fax: 212-685-4540

**Eastern Hemisphere Distribution**
Marcel Dekker AG
Hutgasse 4, Postfach 812, CH-4001 Basel, Switzerland
tel: 41-61-260-6300; fax: 41-61-260-6333

**World Wide Web**
http://www.dekker.com

The publisher offers discounts on this book when ordered in bulk quantities. For more information, write to Special Sales/Professional Marketing at the headquarters address above.

Current printing (last digit):
10 9 8 7 6 5 4 3 2 1

**PRINTED IN THE UNITED STATES OF AMERICA**

# Preface

## GENOMICS: THE LINK BETWEEN GENETICS AND PHYSIOLOGY

Access to the interpretation of entire-genome sequence information for numerous organisms representing Archaea, Eubacteria, Cyanobacteria, and a number of Eukaryote model organisms such as *Saccharomyces cerevisiae* (baker's yeast), *Caenorabiditis elegans* (worm), *Drosophila melanogaster* (fruit fly), and *Arabidopsis thaliana* (mustard cress) have brought fundamental changes to biology— the comprehension of life as a whole instead of its smallest model unit. Additional genomes are currently in large-scale sequencing pipelines, and projections are that in the next few years, a wealth of whole-genome sequences spanning most taxonomic boundaries of living organisms will become publicly available. Thus, what not long ago appeared to be fiction or wishful thinking is quickly moving into the realm of reality, making it possible to study entire genomes. Scientists will be able to determine structural organization, monitor expression of the genetic code, and distinguish biochemical function, cellular components, and structural components defining the "wholesome" phenotype—the functioning not only of a living cell but also of the integration of all cells in an organism.

Often, the reconstruction of entire metabolic pathways becomes possible through dynamic retrieval of biochemical functions, stored in the form of the universal genetic information code, comparative overviews of diverse biological systems, life histories, morphogenetic processes, and inference with respect to natural interactions with the environment. This scenario is not typical, however, because even in prokaryotes and other comparatively lower complexity systems we are faced with a large number of open questions that often present themselves through the detection of functionally unknown or novel coding regions. In addition, examples of functional ambiguity—coding regions for more than one function, which had held a position of infrequent oddities—are becoming more numerous. As an immediate result of the fast-paced progress that occurred during the last decade, it became apparent that our knowledge of the complexity and number of genes was woefully incomplete. With new protein-coding regions appearing in every organism whose genomic DNA sequence has been determined, realization came that the function(s) of more than half the genes, encoded by the *Homo sapiens* or *A. thaliana* genome, remain completely unknown. Moreover, assignments of gene function continue to be missed, although the accuracy of predictive algorithms continues to improve rapidly, even in cases where established biochemical data exist for well-defined coding regions. Some of the complications of annotating well-characterized functions to genes are associated with nonuniversal biological processes, such as the generation of multiple functionally diverse proteins through alternate mRNA splicing, single proteins with multiple functions depending on spatiotemporal expression, and a significant and growing number of RNAs that are neither protein coding nor ribosomal RNA nor transfer RNA, yet that seem to have important additional structural or regulatory roles.

Direct interpretation of genome DNA sequence information alone has brought recognition of how much more complementary gene function analysis is still needed and simultaneously, has provided new avenues by which to do so. Genome sequences are catalysts for the generation of hypotheses. Figures 1 and 2 illustrate the concepts underlying the term "genomics." Initially, these concepts were geared toward high-throughput detection of which genes and which transcripts make up the complexity of an organism. In a comparative fashion, patterns of commonalities and uniqueness were established. In subdisciplines, labeled as proteomics and metabolomics, the complexity, cellular localization, dynamics of protein associations, and their interactions with substrates and products are being determined at the whole-cell level. Thus, all disciplines derived from the original concept of genomics—considering the genomic information as a whole—are based on finding commonalities and differences among organisms, spatiotemporal cell stages, physiological conditions, and interactions with the environment between or within organisms of a given species.

In a way, the underlying idea is similar to the collection of the existing knowledge that was undertaken some 200 years ago by scholars, encyclopedists who reasoned that understanding would arise from the compilation and synthesis of all existing knowledge. Genomics is similar in its attempt to gather data points, but it is not at all similar to the encyclopedists' reasoning because, so far, genomics is generating not only the data but also the new tools to understand them.

In this book, we have assembled contributions from dedicated practitioners who have been responsible for shaping the field of fungal and plant genomics. The individual chapters are meant to provide technical insight and to highlight

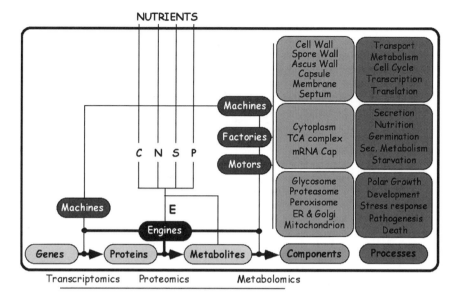

**Figure 1** Schematic representation of cellular organization. *Genes* are the genetic informational units encoding one or more proteins (polypeptide). *Proteins* are the products of gene translation. *Metabolites* are the products of protein function. *Components* are one or more proteins or protein products assembled into a subcellular component, a complex or functional unit that exerts a specific function. *Processes* are combinations of proteins, protein products, and cellular components into a network that exerts a new function. *Engines* are biological processes that provide metabolites—carbon, nitrogen, phosphorus, and sulfur as well as chemical energy required to drive biological processes. *Motors* are cellular components that provide energy and motion to protein complexes, cellular components, or biological processes (e.g., vesicle movement, polar growth, or chromosome division). *Machines* are cellular components exerting constitutive functions such as transcription, translation, and secretion. *Factories* are cellular components exerting customized products used by other cellular components or biological processes. Transcriptomics, proteomics, and metabolomics are whole-genome analysis technologies that approach the transcription, function, and products of the genotype.

how genomics-oriented studies may be used to bring new understanding to established models of fungal development, to analyze and solve problems associated with multiple copies of genes and proteins with seemingly identical functions, and to exemplify how genomics information may be used for industrial applications.

The sea change in biology has had two foundations. Foremost have been new technologies that have harnessed computing power in data acquisition and

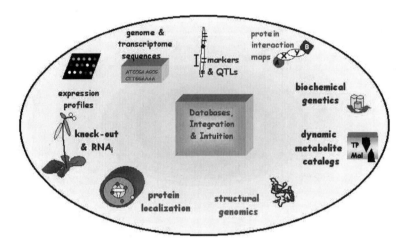

**Figure 2** Components of genomics—the various approaches that use DNA sequence information to identify protein function in terms of where and when proteins are expressed, their interacting partners, the phenotype of gene disruption, biochemical function and three-dimensional structure, the spatial and temporal consequences of protein action, and the relative importance of gene function for performance of the organism.

integration. These technologies are also a crucial component in every instrument in the laboratory. Of equal importance has been the bold thought shown by a number of foresighted individuals whose visionary anticipation of the emerging computational power allowed them to forge new tools to attack complex problems. For those involved in experimentation, "kid in the candy store syndrome" best describes the last decade. There is no indication that this state of mind will end soon.

We wish to thank Joan Bennett (Tulane University) and Anita Lekhwani (Marcel Dekker, Inc.) for the opportunity and encouragement to produce this volume, and the National Science Foundation, from which our work has received funding.

*Rolf A. Prade*
*Hans J. Bohnert*

# Contents

## Plants

## Bioinformatics

# Contributors

**Kiichi Adachi, Ph.D.**   Mitsui Global Strategic Studies Institute, Tokyo, Japan

**Keith D. Allen, Ph.D.**   Department of Computational Biology, Paradigm Genetics, Inc., Research Triangle Park, North Carolina, U.S.A.

**Michael J. Anderson, D.Phil.**   School of Medicine, University of Manchester, Manchester, England

**Jonathan Arnold, Ph.D.**   Department of Genetics, University of Georgia, Athens, Georgia, U.S.A.

**Robert Ascenzi, Ph.D.**   Department of Biochemistry, BASF Plant Science L.L.C., Research Triangle Park, North Carolina, U.S.A.

**Douglas C. Boyes, Ph.D.**   Department of Agricultural Biotechnology, Paradigm Genetics, Inc., Research Triangle Park, North Carolina, U.S.A.

**Jayne L. Brookman, Ph.D.**   Research Division of Biochemistry, School of Biological Sciences, University of Manchester, Manchester, England

**Robert L. Burnap, Ph.D.**   Department of Microbiology and Molecular Genetics, Oklahoma State University, Stillwater, Oklahoma, U.S.A.

**Huey-wen Chuang, Ph.D.**   Department of Biology, Texas A&M University, College Station, Texas, U.S.A.

**John C. Cushman, Ph.D.**   Department of Biochemistry, University of Nevada–Reno, Reno, Nevada, U.S.A.

**Keith R. Davis, Ph.D.**   Department of Agricultural Biotechnology, Paradigm Genetics, Inc., Research Triangle Park, North Carolina, U.S.A.

**David W. Denning, F.R.C.P., F.R.C.Path.**   Section of Infectious Diseases, School of Medicine, University of Manchester, and Wythenshawe Hospital, Manchester, England

**Todd M. DeZwaan, Ph.D.**   Department of Assay Development, Paradigm Genetics, Inc., Research Triangle Park, North Carolina, U.S.A.

**Manuel Duval, Ph.D.**   Department of Biology, Texas A&M University, College Station, Texas, U.S.A.

**Reinhard Fischer, Ph.D.**   Department of Biochemistry, Max-Planck-Institute for Terrestrial Microbiology and Philipps-University of Marburg, Marburg, Germany

**David Hall, Ph.D.**   Research Computer Division, Research Triangle Institute, Research Triangle Park, North Carolina, U.S.A.

**Lisbeth Hamer, Ph.D.**   Department of Microbiology, North Carolina State University, Raleigh, North Carolina, U.S.A.

**Erwin Heberle-Bors, Ph.D.**   Institute of Microbiology and Genetics, Vienna Biocenter, University of Vienna, Vienna, Austria

**Tzung-Fu Hsieh, Ph.D.**   Department of Biology, Texas A&M University, College Station, Texas, U.S.A.

**Susanne Kjemtrup, Ph.D.**   Department of Agricultural Biotechnology, Paradigm Genetics, Inc., Research Triangle Park, North Carolina, U.S.A.

**Andreas Klöti, Ph.D.**   Technology Transfer Office, ETH Zurich, Zurich, Switzerland

**Krzysztof J. Kochut, Ph.D.**   Department of Computer Science, University of Georgia, Athens, Georgia, U.S.A.

**Csaba Koncz, Ph.D.**   Department of Plant Development Biology, Max-Planck-Institute for Plant Breeding Research, Cologne, Germany

**Christian P. Kubicek, Ph.D.**   Institute for Chemical Engineering, Vienna University of Technology, Vienna, Austria

**Ursula Kües, Ph.D.**   ETH Zurich, Zurich, Switzerland, and Institute for Forest Botany, Georg-August-University Göttingen, Göttingen, Germany

**Sanjoy Mahanty, Ph.D.**   Inspire Pharmaceuticals, Inc., Research Triangle Park, North Carolina, U.S.A.

**George A. Marzluf, Ph.D.**   Department of Biochemistry, The Ohio State University, Columbus, Ohio, U.S.A.

**Balázs Melikant, M.Sc.**   Institute of Microbiology and Genetics, Vienna Biocenter, University of Vienna, Vienna, Austria

**John A. Miller, Ph.D.**   Department of Computer Science, University of Georgia, Athens, Georgia, U.S.A.

**Bradley L. Postier, B.S.**   Department of Microbiology and Molecular Genetics, Oklahoma State University, Stillwater, Oklahoma, U.S.A.

**Lakshman Ramamurthy, Ph.D.**   Department of Bioinformatics, GlaxoSmithKline, Inc., Research Triangle Park, North Carolina, U.S.A.

**Mulpuri V. Rao, Ph.D.**   Department of Agricultural Biotechnology, Paradigm Genetics, Inc., Research Triangle Park, North Carolina, U.S.A.

**Amit P. Sheth, Ph.D.**   Department of Computer Science, University of Georgia, Athens, Georgia, U.S.A.

**László Szabados, Ph.D.**   Institute of Plant Biology, Biological Research Center, Szeged, Hungary

**Matthew M. Tanzer, Ph.D.**   Department of Biochemistry, Paradigm Genetics, Inc., Research Triangle Park, North Carolina, U.S.A.

**Terry L. Thomas, Ph.D.**   Department of Biology, Texas A&M University, College Station, Texas, U.S.A.

**Michael J. Weise, Ph.D.**   Life Sciences Scientific Services, Accelys, Inc., San Diego, California, U.S.A.

**Cathal Wilson, Ph.D.**   Institute of Microbiology and Genetics, Vienna Biocenter, University of Vienna, Vienna, Austria

# 1
# *Aspergillus*

**Michael J. Anderson and Jayne L. Brookman**
*University of Manchester, Manchester, England*

**David W. Denning**
*University of Manchester and Wythenshawe Hospital, Manchester, England*

## I. INTRODUCTION

Aspergillosis, the name given to all diseases caused in animals by fungi in the genus *Aspergillus*, includes allergic, superficial, saprophytic, and invasive disease. More than 180 species of *Aspergillus* are recognized (1), but only a few cause disease with any regularity: *A. fumigatus, A. flavus, A. terreus,* and *A. niger* group species. Aspergilli are common saprophytes in the environment, especially in composting facilities. The pathogenic Aspergilli are found the world over, but soil isolation rates do increase towards the equator. Most aerobiology studies have been done in Europe and some have shown a seasonal variation in airborne *Aspergillus* counts. *Aspergillus* species comprise from 1% to 6% of the total air flora outside and, if speciated, *A. fumigatus* comprises from 4% to 41% of the *Aspergillus* total (2). The usual concentration of *Aspergillus* conidia in outside air is 1 to 30 conidia/m$^3$, but can rise to as high as $7 \times 10^7$/m$^3$ inside a barn, after hay or straw disturbance. In hospitals, conidia concentrations in air also vary typically from 0 to $1 \times 10^2$/m$^3$ with much variation in the same site (3–5). Inside human dwellings, *Aspergillus* species may be found in high concentrations in potted plants (50 conidia/g soil) (6), damp cellars, dusty crawl spaces, and condiments, especially pepper (10$^6$ conidia/g) (2) and ground spices.

Human disease has been increasing over the 5 decades since invasive aspergillosis was first described in the immunocompromised patient (7). Invasive aspergillosis (IA) is the most common life-threatening invasive mold infection

1

worldwide. It usually complicates treatments and diseases associated with immunosuppression, including allogeneic bone marrow transplantation, lung and liver transplantation, the treatment of acute leukemia, late-stage acquired immunodeficiency syndrome (AIDS) and a variety of other diseases treated with corticosteroids (8). Invasive aspergillosis rarely affects nonimmunocompromised patients. The incidence of invasive aspergillosis was calculated to have risen 14-fold in the 15-year period up to the end of 1992, as seen in autopsy data from one major teaching hospital in Frankfurt, Germany, with 5% of autopsied patients having invasive aspergillosis in the last year of this survey (9). In a national autopsy survey in Japan from 1969 to 1994, invasive aspergillosis increased from 0.4% to 1.3% in all autopsies (10). Another autopsy series in a European teaching hospital demonstrated a 4% rate of invasive aspergillosis in unselected autopsies (11). A culture-based population study in the San Francisco area (requiring cases to have, for example, two culture-positive bronchoscopy specimens or a sterile site positive by culture, which probably underrepresents invasive aspergillosis by perhaps 90%) showed increasing rates of disease over the last 25 years (12). Comparable figures from previous surveys had put the figure at 1.9 cases per million population in 1970, linearly rising to 12.4 per million in 1992/1993.

## II.  CLINICAL MANIFESTATIONS OF *ASPERGILLUS* INFECTION

### A.  Allergy

Wheezing in patients exposed to *Aspergillus* was recognized in the late 1800s, but was poorly understood. Allergic bronchopulmonary aspergillosis (ABPA), an extreme form of continuing local allergy to *Aspergillus*, was first reported in 1952 in 3 patients from the London Chest Hospital (13). Allergic bronchopulmonary aspergillosis complicates asthma and cystic fibrosis (CF), and patients develop exacerbations of the asthma, CF, or both. They are commonly "difficult-to-control" patients, in the pulmonary sense. Characteristic presentations include new pulmonary shadows, which resolve with steroids, and coughing up plugs of material. The diagnosis is made by a combination of criteria of which episodic wheezing (asthma), transient pulmonary shadows, elevated serum total immunoglobulin E (IgE) and *Aspergillus*-specific IgE, positive *Aspergillus* precipitins (IgG), and central bronchiectasis are the most important. Central bronchiectasis is not a useful diagnostic criterion in CF, as it is universally present. Some patients, especially those with long-standing disease, have barely detectable *Aspergillus* antibodies (14). Other criteria that have been used include peripheral eosinophilia and a positive immediate skin test to *Aspergillus*. Eventually, patients with untreated ABPA develop pulmonary fibrosis. Acute exacerbations are best treated with systemic corticosteroid therapy. Itraconazole (one of only two licensed oral antifungal agents available for aspergillosis) has shown benefit in several open studies (15, 16) and in a recent controlled trial (17).

Mold allergy is common among asthmatics, and this may represent a milder form of ABPA, although multiple other fungi are implicated. It is estimated that there are 275 million asthmatics worldwide (18). Of these, 5% have severe asthma, and there are around 100,000 deaths per year. Recent prevalence data in a population of asthmatics attending hospital regularly (19) compared the rates of fungal allergy in those admitted to the hospital at least twice annually with those not admitted. Skin testing to *Aspergillus*, *Alternaria*, *Cladosporium*, *Penicillium*, and *Candida* showed the frequency in the former to be 76%, but only 16% in the latter group. Rates of allergy to house-dust mite antigen were equivalent. Of interest, sensitization to *Aspergillus* is also common in patients with chronic granulomatous disease and hyper-IgE syndrome, both diseases usually being associated with invasive aspergillosis (20).

Allergic sinus disease caused by *Aspergillus* and other molds has been recognized in the last 15 or so years (21). The frequency of allergic fungal sinusitis or "eosinphilic fungal rhinosinusitis" is not known, and agreed-upon criteria for diagnosis are currently being evolved. Probably many cases of chronic sinusitis are related in part to fungal allergy, but this is hard to prove, as markers of allergy in blood are present in fewer than 30% of cases (21).

## B. Saprophytic or Chronic Invasive/Necrotizing *Aspergillus* Disease

"Aspergilloma" is the term given to the colonization of an intrathoracic cavity by *Aspergillus*. A fungus ball is formed when spores are deposited in the cavity and germinate on the wall, where mycelia and debris attach to create an amorphous mass. An aspergilloma may form in any pre-existing lung cavity. There are many causes of pulmonary cavities including tuberculosis, sarcoidosis, pneumoconiosis, histoplasmosis, and bullae. Some idea of the prevalence of aspergilloma can be gained from a review of 60,000 chest radiographs: aspergillomata were identified in 0.01% (22). During an 11-year period, 15 patients with aspergilloma were admitted to a Veteran's Administration Hospital, representing 0.02% of admissions (23). The frequency is much higher in patients with cavities of 2 cm or more in diameter. For example, in tuberculosis cavities of this size, 15% to 20% of British patients developed an aspergilloma (24). In a series of patients with pulmonary sarcoidosis, 10 of 19 (53%) patients with cystic parenchymal damage had aspergillomas, compared with none of 81 patients with noncystic pulmonary sarcoidosis (25). Many patients with aspergillomas have slowly progressive cavitary damage to the lung, which is termed "chronic necrotizing pulmonary aspergillosis (CPNA)." In AIDS, for example, progression of aspergillomas over time is seen with considerable morbidity and some mortality. This progression probably reflects invasion of cavity walls by *Aspergillus* and is, therefore, not strictly saprophytic *Aspergillus* disease.

The symptomatology of aspergilloma and CPNA is similar and variable in individual patients over time. Most patients are asymptomatic when an aspergilloma first forms. The most common presentation of aspergilloma is coughing up blood, which is because of new vessel formation in the vicinity of the aspergilloma. About 40% of patients are sensitized to *Aspergillus* and develop wheezing, weight loss, and malaise with or without fever. The patient is typically in the fourth to sixth decade of life and, as with all forms of aspergillosis, more men than women are affected. Chronic necrotizing pulmonary aspergillosis has similar clinical features, although weight loss is more profound.

Plain radiography of the chest in cases of aspergilloma shows a number of typical features, including a round solid mass that may be mobile within a cavity, separated from the wall by a rim of air. In classic aspergillomas, pleural thickening is also present and varies from several millimeters to 2 cm. All patients with CPNA have radiological evidence of a small or large cavitary lesion in the lung, usually in one or both upper lobes. Initially, infiltrates are ill-defined areas of consolidation or small cavities that progress to form well-defined multiple cavities with thickened walls—the cavities often contain an aspergilloma, debris, or fluid. Invasion of the pleura may follow.

*Aspergillus* precipitins are detectable in more than 95% of aspergilloma patients. The precipitin test may, however, be positive in some patients with cavities who do not have an overt aspergilloma, and these patients probably have CNPA. Two-thirds of patients have elevated levels of total IgE and *Aspergillus*-specific IgE. Most patients with aspergillomas have positive respiratory cultures of *Aspergillus*, usually *A. fumigatus*, and multiple variants and genotypes may be recovered. In a retrospective series, 25% of patients with pulmonary aspergillomas, 100% with a sinus aspergilloma, and 8% with disseminated aspergillosis had calcium oxalate crystals present in their tissue and, occasionally, renal oxalosis was observed (26). *Aspergillus niger* may be a more prolific producer of oxalate than *A. fumigatus* (27). Evidence of inflammation, such as the presence of C-reactive protein, is common in CPNA. We have recently observed mannose-binding protein deficiency as a probable association with CNPA (28). Pathologically, aspergillomas have multiple hyphae in cavities without invasion of cavity walls. In CPNA, hyphae may also be found in abnormal cavities without invading tissue. Distinguishing between aspergilloma and CNPA is clinically difficult and depends on progression over time.

Spontaneous resolution of aspergillomas occurs in 10% of cases within 3 years. However, the consequences of this type of *Aspergillus* infection can be dramatic. One report focused on the long-term outcome of 23 patients with aspergilloma and found that 5 died directly from complications of this infection (respiratory failure or hemoptysis) (29). Many patients with aspergillomas are elderly and have significant underlying disease, including severe respiratory compromise in addition to the aspergilloma. Thus, many patients die with an aspergilloma rather than of it. Treatment is problematic. Responses have been reported to intra-cavitary amphotericin B and older drugs, oral itraconazole, surgical resection, and corticosteroids. Bleeding is managed by resection or embolectomy. Likewise, CNPA runs a slowly progressive course over months or usually years, if not treated. Itraconazole or intravenous amphotericin B are appropriate for CNPA. Gamma-interferon has been used successfully in a few patients. Uncontrolled, cavities expand, reducing pulmonary capacity; then, local pulmonary fibrosis occurs and eventually the patient is left with little functional lung. Sometimes, the systemic features are more prominent and cachexia, mimicking carcinoma, is the eventual outcome.

## C. Invasive Aspergillosis

The incidence of invasive aspergillosis is variable in different patient groups (Table 1) and over time in the same institution. It is slightly more common in men, for reasons that are unexplained. In some studies, it has followed a seasonal distribution, but this has not been shown in most studies. Specific defects that put patients at risk include severe neutropenia, corticosteroid therapy (including transplantation), prior pulmonary damage or infection, neutrophil defects, such

**Table 1** Incidence of Invasive Aspergillosis in Different Host Groups

|  | Range (%) |
|---|---|
| Heart and lung or lung transplantation | 19–26[a] |
| Chronic granulomatous disease | 25–40[b] |
| Acute leukemia | 5–24 |
| Allogeneic BMT | 4–9 |
| Autologous BMT without growth factors | 0.5–6 |
| AIDS | 0–12 |
| Liver transplantation | 1.5–10 |
| Heart and renal transplantation | 0.5–10 |
| Severe combined immunodeficiency | 3.5 |
| Burns | 1–7 |
| Systemic lupus erythematosus | 1 |
| Autologous BMT with growth factors | <1 |

[a]Distinguishing colonization from disease is particularly difficult in these patients.
[b]Lifetime incidence.
BMT, Bone marrow transplantation.

as occur in diabetes mellitus, AIDS, chronic granulomatous disease, chronic liver disease, and because of excess alcohol consumption. Mitochrondrial defects (Pearson and MELAS syndromes) have recently been associated with invasive aspergillosis (30, 31). Some patients with no apparent risk factors develop invasive aspergillosis, often after respiratory or other infections.

The majority of patients (80%–90%) develop invasive pulmonary aspergillosis (IPA). A few have airways or sinus infection; other manifestations are rare. Interestingly, acute invasive sinusitis is extremely uncommon in solid organ transplant patients and occurs mostly in hematology patients or, in a more chronic form, in those with less extreme immunodeficiency. A classification of invasive aspergillosis is shown in Table 2. Local extension, as in sinus disease spreading to the orbit, is distinct from disseminated disease, which refers to distant spread from the portal of entry, presumably blood-borne.

Clinical manifestations of disease differ according to patient group. About two-thirds of patients have a fever. Of those with invasive pulmonary aspergillosis, a mild unproductive cough is the first manifestation in about 60% of patients, about 40% have chest pain, and 25% cough up blood. Most patients have no symptoms and 10% to 40% manifest no symptoms until a few days before death.

**Table 2**   Classification of Invasive *Aspergillus* Infection

1.  Infection associated with tissue damage, surgery or foreign body
    A.  Keratitis and/or endophthalmitis
    B.  Cutaneous infection, eg, burn wound aspergillosis
    C.  Operative site infection, eg, prosthetic valve endocarditis, wound infection after liver transplantation, and subdural empyema
    D.  Foreign body-associated, eg, Hickman or other IV line-associated and CAPD catheter
2.  Infection in the immunocompromised host
    A.  Primary cutaneous aspergillosis, esp. in leukemic children
    B.  Pulmonary aspergillosis
        i)   acute invasive
        ii)  chronic necrotizing
    C.  Airways aspergillosis:
        i)    obstructing bronchial aspergillosis
        ii)   invasive *Aspergillus* tracheobronchitis
        iii)  ulcerative *Aspergillus* tracheobronchitis
        iv)   pseudomembranous *Aspergillus* tracheobronchitis
    D.  Rhinosinusitis
    E.  Disseminated aspergillosis, esp. cerebral aspergillosis

CAPD, Continuous ambulatory peritoneal dialysis.

Early diagnosis depends on appropriate testing, and at present, 30% remain undiagnosed and untreated at death.

The approach to diagnosis depends on the type of patient. In neutropenic patients with leukemia, essentially all are hospitalized, and frequent observation and testing is possible. In these patients, a combination of galactomannan-antigen testing (32, 33) and early computed tomographic (CT) scanning of the lungs (34), followed up by bronchoscopy and surgery, if necessary, is the optimum approach. These patients also have elevated fibrinogen and often C-reactive protein or other acute phase response markers circulating in blood, but the sensitivity and specificity of each are unclear. The same approach is useful for bone marrow (hematopoietic stem cell) transplant patients, but as the risk period extends for months, regular testing is problematic. Polymerase chain reaction (PCR) diagnosis has been useful in some studies (35), although not all, probably reflecting both technical and patient differences. In other patients, culture of respiratory secretions followed by biopsy is the usual approach to diagnosis, with poor results overall. Generally, better diagnostic approaches are needed, especially in nonhematological patients.

The treatment of IA is problematic. There are only two established licensed drugs useful for its treatment—amphotericin B and itraconazole. The response rates to lipid-associated amphotericin B are not different from standard amphotericin B, but the therapeutic ratio has improved, certainly with respect to nephrotoxicity. Randomized studies between oral itraconazole and intravenous amphotericin B were unsuccessful, so comparative response rates can only be estimates. The response rate to amphotericin B varies from approximately 1% (cerebral aspergillosis in immunocompromised patients) to 60% to 80% in heart and renal transplant patients with pulmonary or cutaneous infection. Collected series suggest an overall response rate of approximately 35% (36–39). Response rates to itraconazole appear to be approximately 40% to 60%, again with a wide range in different patient groups and different settings (38, 40). Less data are available on the efficacy of itraconazole compared with amphotericin B in more immunocompromised settings (such as allogeneic bone marrow transplant patients) and persistently neutropenic patients, although the drug may be effective in these settings. It is well known that the response rates to amphotericin B do not exceed 15% in these difficult situations. Patients who fail amphotericin B may respond to itraconazole, and the converse is also true. Acute invasive *Aspergillus* sinusitis appears to respond better to amphotericin B than azoles. Azole resistance has been documented in *A. fumigatus* (41, 42). *Aspergillus terreus* is innately resistant to amphotericin B (41), and amphotericin B resistance has also been reported in some isolates of *A. flavus* and *A. fumigatus* (41).

The newer drugs being tested include voriconazole, posaconazole, ravuconazole, and three candins, caspofungin, micafungin and anidulafungin. All the azoles have the same mode of action as itraconazole, but may be slightly more active, especially posaconazole. Voriconazole is more active than amphotericin B in a recently completed comparative study (42a) and was licensed for the treatment of IA in 2002. The candins are noncompetitive inhibitors of glucan synthase, but do not appear to be fungicidal. They have been shown to be active in mice and patients.

## III.  VIRULENCE AND PATHOGENICITY

Certain species of the genus *Aspergillus* are remarkable opportunistic animal pathogens. As detailed above, the diversity of diseases produced by *A. fumigatus* alone is very broad and even broader if the other pathogenic Aspergilli are also considered. For instance, sinusitis is often caused by *A. flavus* and otitis externa is most often caused by *A. niger*. *Aspergillus terreus* causes disseminated aspergillosis in German Shepherd dogs. *Aspergillus clavatus* and *Aspergillus versicolor* are thought to be significant allergens and producers of unusual mycotoxins. Where these have been studied, differences between the pathogenic and nonpathogenic species of *Aspergillus* will be highlighted, especially between *A. fumigatus* and the model organism, *A. nidulans*. These differences will be important to bear in mind when investigating pathogenicity and in comparisons of the species' genomes.

Few data support the hypothesis that different strains of *A. fumigatus* are more or less pathogenic, partly because these depend on the definition of pathogenicity. Fiftyfold differences in inocula able to cause lethal infection have been documented between isolates from patients (43). Significant differences in virulence between environmental and clinical isolates also have been shown, although the differences between the clinical isolates were not tested and so the data probably just reflected the natural level of variation in the population (44). Molecular typing of large numbers of isolates does not reveal any grouping of the clinical isolates into types or clades as would be expected with an opportunistic pathogen (45, 46). These data could be the consequence of grouping together all the clinical isolates, despite the fact that patients with different forms of aspergillosis were included.

It is highly probable that different mechanisms of pathogenesis operate in the different clinical situations. However, in this section, only clinical and biological factors relevant to pathogenesis will be considered for IA.

### A.  Invasive Aspergillosis

The hallmark of IA is tissue invasion and damage by *Aspergillus*. The vast majority of cases occur in immunocompromised patients, but occasional cases in "normal" people also occur. Tissue damage, such as that caused by trauma or burns, may be followed by IA. Given this, *A. fumigatus* is best considered an opportunistic pathogen, but it can also be a rare primary pathogen. It may also be a primary pathogen in other vertebrates and occasionally in invertebrates as well. These observations lead to a key question:

> Are the pathogenicity factors, which must exist in *A. fumigatus* to cause disease in the normal host, also operative in the immunocompromised patient?

If *A. fumigatus* were a true primary pathogen, then it would have evolved along with its host(s) and would express factors specifically involved in the infection process that would enable it to avoid any host responses, invade tissue, adhere to certain cell types, etc. However, because *A. fumigatus* is such a rare pathogen of normal hosts and the epidemiological data suggest that all environmental isolates are pathogenic, then essentially it is an opportunistic pathogen. Therefore, the different factors that are required for growth in the host (pathogenicity factors) will presumably operate in any opportunistic situation regardless of the immune status of the patient. However, it might be possible, because of natural variation in the population, that more aggressive strains do exist and that only these strains are able to infect less immunocompromised people. It would be interesting to study the epidemiology and population genetics of such isolates and to test their virulence in animal models.

Invasive aspergillosis occurs in many different immunocompromised contexts, and the manifestations of infection may be different. For example, slowly progressive pneumonia with invasion of the chest wall is common in chronic granulomatous disease. This pattern of disease contrasts with focal pulmonary infection in neutropenic patients with leukemia, in which nodules or consolidation are common. Such lesions very rarely even cross the pleura and never invade the chest wall. This raises another key question:

> Is it the host response, is it *Aspergillus*-derived pathogenicity factors, or is it the interaction of the two that determines the manifestations of disease?

In the same patient group with essentially equivalent degrees of immunosuppression, the pace of progression and manifestation of IA is variable. For example, in allogeneic bone marrow transplant patients, about 40% get cerebral aspergillosis, and 60% do not. In heart transplant recipients, about one-third develop bilateral pulmonary aspergillosis, one-third disseminated disease, and one-third small pulmonary nodules. What factors determine the patterns of disease in different patients? Is it host genetic factors, isolate to isolate variation (with variability in pathogenicity factors), or both? The application of genome technologies derived from the human, *A. fumigatus*, and other fungal genome sequences, as well as epidemiological studies of patients and *A. fumigatus* isolates, should begin to address such complex issues. In fact, it could be argued that such studies of an opportunistic pathogen with so many manifestations of disease is only possible in this postgenomic era.

Another issue that requires consideration in any discussion of pathogenicity is whether an essential growth factor (such as para-aminobenzoic acid [47] or uridine and uracil [48]) can be legitimately termed a "pathogenicity factor." Deletion of any key enzyme necessary for growth in the mammalian host, which results in an avirulent mutant in a model system, is an important observation but does little to explain pathogenesis, as it indicates essentially the prototrophic requirement for survival in man. An extension of this argument is that the demonstration of the importance of any pathogenicity factor should include information on growth rate. Experiments conducted years ago concluded that reduced growth rate was associated with reduced virulence (49). However, no group has systematically studied growth rate in the host, and none in parallel with a demonstration of virulence. Clearly, the ability to grow at 37° C is critical to infection and only about 13 *Aspergillus* species are able to do so with any vigor (1). In addition, the small diameter (2–3 µm) of *A. fumigatus* conidia enables penetration, deep into the lung alveoli.

Infection by any micro-organism requires several steps. Pathogenicity should, therefore, be considered in the same way.

## B.  Adherence

The organism has to evade the surface defenses in the respiratory tract and the mucous ladder long enough to germinate. Normally, conidia would be removed from the lung by ciliated movement long before germination occurs. However, if this process is defective, fungal products, including gliotoxin, may slow ciliary beating and damage epithelia locally (50). Then attachment to epithethal cells, the basement membrane, or both will occur. Adhesion may be to carbohydrate or protein molecules. *Aspergillus fumigatus* conidia have been shown to bind to laminin (51), which is a major noncollagenous extracellular matrix protein. Conidial binding was mainly to the P1 fragment. After binding, production of elastase degraded laminin in situ. The external rodlet layer of conidia bound most strongly and the number of binding sites decreased as the conidia swelled with a concomitant loss of adherence (52). The laminin ligand was identified as a 72 kDa cell wall glycoprotein. The binding of spores to laminin and epithelial cells has been shown to be inhibited by surfactant protein D enriched-broncho-alveolar lavage and enhanced by diffusates from the spore surface (53).

Avid binding of fibrinogen to conidia, conidial heads, and hyphae has been shown in various pathogenic Aspergilli (54). Binding to individual conidia varied substantially and sometimes was absent. Conidia seemed to bind more fibrinogen than hyphae. *Aspergillus niger* spores bound the most fibrinogen molecules, followed by *A. fumigatus* and *A. flavus*, and then *A. terreus*. Nonpathogenic species, such as *A. nidulans* and *A. candidus*, bound relatively little fibrinogen. Conidia have been shown to bind to the D domain of fibrinogen (55). There are an average of 1,200 sites of a proteinaceous nature on a conidium, and the number of sites decreased during germination. Recent data have suggested that negatively charged carbohydrates, such as sialic acids, on the surface of conidia bind to the glycosaminoglycan-binding domain of fibronectin (56, 57). It was also demonstrated that *A. fumigatus* conidia bound significantly better to fibronectin than those of *A. flavus*, *A. wentii*, and *A. ornatus* (56), which reflected the relative sialic acid densities (57). Conidia have been shown to bind to pulmonary surfactant proteins through their carbohydrate recognition domains (58, 59). In addition, lactoferrin has been shown to bind both resting and swollen conidia (60). This may have implications for iron availability and, thus, for growth of the organism or for phagocyte killing of conidia.

Hydrophobins are small, secreted, moderately hydrophobic proteins that contain eight cysteine residues. They are structural proteins that make cell walls hydrophobic and are present in the outer spore wall. One hydrophobin, RodA, is the major component of the rodlet layer of *A. fumigatus* and *A. nidulans* conidia, and disruption of the gene results in conidia lacking this layer (61–63). RodA⁻ mutant colonies are more readily wettable with water, and conidia are not dispersed into the air on shaking. The mutant conidia still aggregate readily, and those of *A. fumigatus* do not have altered binding to pneumocytes, laminin, and fibrinogen (63). The spores do, however, adhere less to collagen (63). A detailed physicochemical comparison of *A. fumigatus* and *A. nidulans* rodA⁻ spores has shown differences between the two species (64). Although the rodA⁻ spores of both species were significantly less hydrophobic, removal of the rodlet layer reduced the hydrophobicity of *A. fumigatus* conidia less than those of *A. nidulans* and, for instance, no difference in adhesion to latex spheres was observed between wild-type and mutant *A. fumigatus* spores in contrast to a significant difference between those of *A. nidulans* (64). These data suggest that, although *A. fumigatus* and *A. nidulans* have an outer spore rodlet layer composed mainly of RodA, differences must exist in other components of this layer.

## C. Evasion of Phagocytosis by Conidia

The first immunological line of defense against *Aspergillus* is macrophage function—both resident and monocyte-derived macrophages. These cells are capable of ingesting and killing spores by an opsonin-independent nonoxidative method (65). Surfactant enhances phagocytosis of conidia. Macrophage killing is depressed by glucocorticoids—not through a T-cell-related mechanism, but possibly by causing failure of phagolysosomal fusion and affecting cytokine production. Granulocyte-macrophage colony stimulating factor reverses steroid depression in vitro (66), whereas gamma-interferon has been shown not to enhance macrophage defense. Other factors that negatively affect resident macrophage function include total body irradiation and some forms of chemotherapy.

Attachment of immune effector cells to *A. fumigatus* conidia has been studied and various receptors identified. A mannosyl-fucosyl-binding receptor is present on macrophages, which in mice is calcium dependent but is not in humans (67, 68). Another receptor, recognizing chitin components, is also likely given the pattern of inhibition seen (67). Beta-glucan inhibited conidial binding to human (but not murine) monocytes by 50% to 85% (68). Further work showed that oligosaccharides with either a β-1,3 or a β-1,4 linkage inhibited binding (68). *Aspergillus* conidia also bind to type II pneumocytes, probably through a lactosylceramide receptor (69). Finally, activated lymphocytes have been shown to reduce the adherence of germinating conidia to a surface (70).

Complement is also capable of binding *Aspergillus* conidia and hyphae. Binding to conidia by C3 is relatively rapid but weak in comparison to binding to swollen conidia or hyphae (71). More molecules of complement bound to "less pathogenic" species, such as *A. glaucus, A. niger, A. nidulans, A. ochraceus, A. terreus, A. versicolor,* and *A. wentii,* compared with *A. fumigatus* and *A. flavus* (72). Such differences between species could be part of the explanation for the relative frequency of infection. Isolate to isolate variation was also observed (72). A 54–58 kDa surface protein that binds C3 has been identified (73).

The blue-green pigment of the *A. fumigatus* conidium has been shown to be the product of a single biosynthetic pathway. At least seven genes have been identified that are involved in the conversion of acetate to a melanin. Six of these genes are clustered within a 19-kb region of the genome (74). Mutation in the first gene of the pathway (*pksP*), which codes for a polyketide synthase, results in white conidia (75, 76); mutation of the second gene in the pathway results in yellow conidia (77); mutation of the third and fourth genes in the pathway results in reddish-pink conidia (74, 78), and mutation in later genes results in brown conidia (74). Parts of the biosynthetic pathway have been elucidated and have been shown to be the same as parts of the 1,8 dihydroxynaphthalene (DHN)-melanin pigment pathway present in the black and brown-spored fungal species, *Magnaporthe grisea* and *Colletotrichum lagenarium* and so, like these two species, *A. fumigatus* produces a DHN-melanin. The first step, however, has been shown to be the same as that in *A. nidulans*. This step involves a polyketide synthase that converts acetyl CoA and malonyl CoA to the yellow naphthopyrone YWA1 (77). The two pathways diverge after this, and it has been shown that the *A. nidulans* pathway does not involve a DHN intermediate (79). As with *A. fumigatus,* mutation of the *A. nidulans* polyketide synthase gene (*wA*) results in white conidia. These white conidia, however, differ from those of *A. fumigatus* *pksP⁻* mutants, as they exhibit the ornamentation characteristic of wild-type conidia, whereas *A. fumigatus* white conidia are smooth (80). The pigments of both *A. fumigatus* and *A. nidulans* conidia have been shown to quench reactive oxygen species produced by phagocytes, to inhibit the stimulation of phagocytes (80) and, in *A. fumigatus*, to inhibit C3 binding and subsequent phagocytosis (76, 78). *Aspergillus fumigatus* mutants with white conidia have significantly reduced virulence in immunocompetent murine models (76, 81), and mutants with yellow and reddish-pink conidia are also significantly less virulent (44). Interestingly, though, a white-spored clinical isolate of *A. fumigatus* has been reported (82).

These data suggest that pigmented conidia are important for survival in the host, but that the exact nature of the pigment is not so crucial. For instance, as with *A. nidulans*, the production of the green pigments of *A. flavus* and *A. parasiticus* is not prevented by inhibitors of DHN-melanin biosynthesis (83).

## D. Germination

Once adherent, conidia will germinate and enter logarithmic phase growth. Two studies have shown that tissue extracts can decrease the germination time of *A. fumigatus* conidia (84, 85). Data from our unpublished experiments have shown that a complex medium containing lecithin can decrease the time when germ tubes are first observed by 30 minutes. Hydrocortisone, however, has no effect on time to germination (86). A few studies of *A. fumigatus* have investigated the changes in morphology that occur on germination. During the first few hours after addition to a suitable medium, the conidia undergo a phase of isotropic growth (swelling). Major changes have been observed in the cell wall during this phase with the disintegration of an outer convoluted layer and associated carbohydrates, and the unmasking or synthesis of other carbohydrates (87). These changes result in an increased hydrophobicity and cause the conidia to stick together (agglutinate). Inhibition of protein synthesis was shown to reduce agglutination, but not swelling. After this phase of isotropic growth, polar growth occurs with the production of a germ tube. Factor(s), possibly of a proteinaceous nature, released by leukocytes during the inflammmatory process have been shown to limit germ tube formation (88).

In a comparative study, definite differences were observed in morphological events during germination between a strain of *A. fumigatus* and a strain of *A. nidulans* (89). For instance, after the first mitosis, more than 90% of *A. fumigatus* cells had established polar growth, whereas more than 90% of *A. nidulans* cells were polar only after the second mitosis. By the fifth mitosis, just 19% of *A. fumigatus* cells possessed a second germ tube, whereas 98% of *A. nidulans* cells possessed a second germ tube. Septum formation was never observed in *A. fumigatus* germlings with fewer than 16 nuclei, and of these germlings, 21% had a septum. In *A. nidulans*, septum formation was observed in germlings with 8 nuclei, and 55% of those germlings with 16 nuclei had a septum. The first septum formed in *A. fumigatus* is not laid down at the base of the germ tube, as is the case with *A. nidulans*, but can be found anywhere within about 20 μm of the base. Finally, this study showed that approximately 4% to 5% of the cells of both species are in mitosis at any time.

Germination in *A. fumigatus* has also been investigated at the molecular level. Glucan synthase activity is important for germ tube formation and hyphal growth, because treatment with an inhibitor of glucan synthase resulted in profound morphological changes (90). The glucan synthase complex, along with newly synthesized β-1,3 glucans, is located at the apex of the germ tube (91). Using antibodies to mammalian myosin, an 180 kDa immuno-analogue has been detected in resting and swollen conidia, and in germ tubes (92). A role for myosin in conidial swelling and germination is indicated by the fact that BDM, a potent general inhibitor of myosin ATPase, significantly affected swelling and abolished germination. The involvement of actin was also investigated in the same study and a 43 kDa immuno-analogue was similarly detected in resting and swollen conidia, and in germ tubes. An actin inhibitor, cytochalasin B, had no effect on swelling, but partially inhibited germination. These data suggest that actin might be required for apical growth, as has been previously described for *A. nidulans* (93).

## E.   Growth Rate and Morphology

Factors affecting growth rate and patterns of growth could determine pathogenicity, but are more likely to influence the manifestation of disease. Such factors would include the intrinsic growth rate of any given isolate, physical barriers to hyphal extension, immune attack, hormonal effects, nutrient supply including oxygen, and antifungal therapy. The three-dimensional pattern of growth in the human host may differ between isolates, and the extent of tissue invaded will be determined by this. One aspect of this is branching frequency, which has not been extensively studied to date in pathogenic Aspergilli. One of the chitin synthases of *A. fumigatus* (ChsG) has been shown to be important in determining branching frequency (94). Disruption of the *chsG* gene results in a mutant that, although its growth rate is unaltered, produces denser colonies with almost twice as many hyphal tips as normal. This mutant behaves differently in a murine model with a delay in the development of respiratory illness and a statistically significant reduction in mortality. The mutant is capable of invasive growth, but the colonies within the lung tissue are smaller and more highly branched.

The specific growth rates of *Aspergillus* species are approximately the same and, for instance, have been measured in the range of 0.43 to 0.48 h$^{-1}$ on modified Vogel's medium at 32° C (86). This equates to a hyphal extension rate of 1–2 mm/hr. Addition of hydrocortisone in physiological concentrations increased the specific growth rate by about 40% for *A. fumigatus*, about 30% for *A. flavus* (86), and not at all for *A. niger, A. oryzae*, and *A. nidulans* (86, Anderson, unpublished data). Other human (sex) hormones had no effect.

Immune attack on *Aspergillus* hyphae appears to be mediated by complement (ineffectually) (71), phagocytic cells (especially neutrophils) (65, 95, 96) and platelets (97). Given that those patients whose primary immune defect is phagocytic cell dysfunction (as opposed to profound neutropenia), have more slowly progressive IA, it is likely that the in vivo growth rate of *Aspergillus* is substantially reduced by phagocytic attack. Such effects have not been measured.

Aspergilli are capable of growth at oxygen concentrations as low as 0.1% (98). Rare instances of isolates being grown in anaerobic conditions have been described. The invasion of devitalized tissue after trauma and the persistence of the organism in infarcted lung, as a result of IA, requires such micro-aerophilic growth. In addition, the action of amphotericin B on *Aspergillus* may be different (ie, protective) at very low oxygen tensions (41).

## F.  Tissue Invasion

It is not clear that *Aspergillus* respects physical barriers to growth. However, demarcation of infection by the pleura is common in neutropenic patients, and this may constitute a physical barrier for the organism in the absence of phagocyte attack. This hypothesis remains to be proven. Once germinated, hyphae penetrate epithelial tissue. It is not known how this occurs. Does *Aspergillus* cause separation of epithelial cells by disrupting tight intercellular junctions? Do conidia get ingested by epithelial cells, germinate within the vacuole, and then penetrate the lung interstitium from there? Do germinating conidia damage and kill epithelial cells, leaving a gap for hyphal penetration? Elongating hyphae have been demonstrated in tissue culture to penetrate pneumocytes (69).

In many, but not all, instances of clinical disease, hyphae penetrate blood vessels, particularly small arteries, leading to distal (hemorrhagic) infarction. Vessel wall penetration is not the result of elastic degradation, but is rapid and effected simply by turgor pressure (99). Hyphae grow within the lumen of the vessel, leading to further thrombosis. These pathological features are similar to icthyma gangrenosum and pneumonia caused by *Pseudomonas aeruginosa*, to clostridial infection and to mucormycosis. Common pathogenetic factors are likely, as the pathology is so similar. Phospholipases may be responsible, at least in part, for this pathology (100), by inference from established pathogenetic mechanisms in clostridial infection.

Some patients with IA do not have evidence of vessel invasion. Vessel invasion appears to be more common in neutropenic patients, but is not universal. This variation in pathology attests to both organism and host related factors being responsible for pathological appearances. Those patients without vessel invasion tend to have nodules with surrounding acute or chronic inflammation, sometimes with Langerhans' giant cells (multinucleated macrophages). This pathology is much more frequent in patients with milder forms of immunosuppression, often with phagocyte dysfunction. Even in those patients with the most severe immuno-suppression, the prognosis seems to be better, all things considered, when there is nodular disease without vessel invasion. Thus certain pathologies appear to be associated with different radiological appearances and outcomes, for unexplained reasons. One possible candidate for inducing granuloma formation is phthioic acid, which is also produced by *Mycobacterium tuberculosis*. Phthioic acid is produced by a minority of *A. fumigatus isolates* (101). A knock-out murine model of chronic granulomatous disease demonstrated exuberant granulomatous re-sponses in the lungs in response to killed *Aspergillus* hyphae—possibly an exag-gerated "foreign body" reaction (102).

## G.  Dissemination

In a minority of cases, IA is a disseminated infection, with involvement of brain, skin, kidneys, heart valve, and other organs. Dissemination occurs through the bloodstream. The viable hyphal fragments (no conidia are produced in tissue) presumably travel in the bloodstream, and two clinical observations are pertinent to evaluating this. First, positive blood cultures of *Aspergillus* do occur, but are uncommon (103, 104). Secondly, detection of *Aspergillus* DNA in blood occurs almost universally in allogeneic bone marrow transplant recipients with IA (35), and such patients have a high dissemination rate. Hence, "DNA-emia" is com-mon, but fungemia is not. Patients who have had splenectomies may be more likely to disseminate and have multiple sites of infection, and so the spleen may filter out some hyphal fragments without becoming infected. Intermittent fungemia is, therefore, likely.

## IV.  TOOLS FOR GENETIC ANALYSIS

## A.  Transformation

To manipulate the genomes of *Aspergillus* species in a targeted way, transforma-tion systems suitable for use in these organisms were required. Historically, the model fungus *A. nidulans* has been the subject of many genetic studies. The genes and approaches for mutant selection and transformation procedures in *A. nidulans* have since been used in other Aspergilli, such as *A. niger* and *A. fumigatus* (105). Enzymatic removal of the fungal cell wall from recently germinated spores is used to produce protoplasts before DNA can be taken up into the fungal cells using polyethylene glycol (PEG) fusion/heat shock-type methodologies. The first such use of transformation to disrupt a gene in *A. fumigatus* was reported in 1992 (106). Another approach, which is technically simpler, is the use of electroporation; this has been carried out successfully in *A. fumigatus* with swollen spores (107, 108).

In general, either a dominant marker for antibiotic resistance, such as hygromycin resistance, or complementation of an auxotrophic mutation, such as *pyrGI* (108), is used for selection of the transformants. We have encountered problems with the use of antibiotic markers in clinical isolates of *A. fumigatus*. Sensitivity to individual antibiotics can vary greatly with different strains, and we have commonly observed false positives during culture. These positives are resistant to the antibiotic in comparison with the parental strain, but do not contain an antibiotic resistance marker gene. Furthermore, unlike true transformants, the resistance is lost almost immediately when selection is removed, suggesting a possible up-regulation of efflux pumps in these *A. fumigatus* strains. This phenomenon is likely to be a strain variation effect as other workers find the hygromycin marker to be very useful. Indeed, nearly all reports of gene replacements or disruptions in *A. fumigatus* have mainly used hygromycin resistance and, occasionally, phleomycin resistance as the selectable marker. These studies have principally been carried out to investigate whether potential pathogenicity factors actually have an effect on pathogenicity, by using the mutants in animal models of infection. Only three studies have been reported that have sequentially disrupted two genes, in all cases using both hygromycin and phleomycin resistance as the selectable markers (94, 109, 110).

A common target for an auxotrophic marker in filamentous fungi and yeasts is the *pyrG* gene; this encodes the enzyme orotidine 5′-monophosphate decarboxylase. Mutations in this gene result in strains auxotrophic for uridine or uracil. This system is particularly attractive for use in molecular genetics, as the URA$^+$ phenotype can also be selected against with the toxic analogue, 5-fluoro-orotic acid (111). As a consequence, *URA3* or *pyrG* blaster cassettes have been developed for several fungal species, which permits the sequential disruption of genes. A *pyrG* blaster consisting of the *A. niger pyrG* gene flanked by a direct repeat that encodes the neomycin phosphotransferase of transposon Tn5 has been constructed for use in *A. fumigatus* and other *Aspergillus* species (112). Promoters from *A. niger* and *A. nidulans* function in *A. fumigatus*, presumably because of the close evolutionary relationship of the species, and so heterologous genes from other Aspergilli will complement auxotrophies in *A. fumigatus* and vice-versa, although the complementation is not always thought to be as efficient as with the homologous form of the gene.

Weidner et al. (108) have developed a homologous transformation system for *A. fumigatus* that permits complementation of *pyrG* mutations using its own *pyrG* gene. This system has been adapted by introducing a mutation into the cloned *pyrG* gene (108). An integrative vector using this mutant allele in combination with promoter-reporter gene fusions can be constructed and used to select prototrophic transformants from a strain containing a different mutant allele of *pyrG*. Some of the transformants will contain an integration at the *pyrG* locus, and these transformants can be used to study gene expression in *A. fumigatus*.

An example of a gene reporter study in *A. fumigatus* involved the fusion of the *alp* gene promoter to the *lacZ* reporter gene and integration at the *alp* locus (110). Expression of the gene, which encodes an extracellular alkaline protease, was demonstrated during colonization of the lung in a murine model.

Transformation in *Aspergillus* species is generally achieved through integration of the transforming DNA into the genome. Autonomously replicating plasmids have been described in *A. nidulans*, although they are not yet widely used. Further development of these systems would provide extremely useful tools for the Aspergilli (e.g., see 113). Recently, May and coworkers described the use of an autonomously replicating plasmid-based genomic library in the model organism *A. nidulans*. They have used this library to correlate gene dosage with drug resistance (114) and to characterize cell-signaling pathways (115). They also showed that these plasmids may have some utility in *A. fumigatus* (114).

## B.  Complementation of Mutations

Banks of *A. fumigatus* mutants have been generated through the traditional methods of UV (44, 81) or chemical mutagenesis (48, 78, 116) and screened using easy-to-score phenotypes such as spore color (44, 78, 81, 116), colony morphology (116), or auxotrophy (48). Two studies have performed screens for complementation of spore color mutations by transforming the *A. fumigatus* mutants with a cosmid library using the selectable marker hygromycin resistance. Blue-green wild-type spores were obtained when an integrated cosmid clone complemented the mutations that resulted in either white conidia (75) or reddish-pink conidia (78) and enabled the investigators to identify the mutated gene.

Complementation of mutations can also be used to establish if genes are orthologues. A few studies have used either *A. fumigatus* or *A. nidulans* genes to complement a mutation in the other species (62, 108, 117). For instance, a hydrophobin gene from *A. fumigatus*, whose product is 86% identical to the RodA protein of *A. nidulans*, was shown to complement the rodletless phenotype in *A. nidulans* (62, 117). A homologous gene present in *A. fumigatus*, which was identified using an *A. nidulans* gene-specific probe, was able to complement a *wA* mutation in *A. nidulans* and, therefore, presumably encodes the orthologous polyketide synthase (117). The reverse complementation was also demonstrated to work where the *wA* gene of *A. nidulans* complemented a white conidial mutation in *A. fumigatus* (117).

## C.  Parasexual Analysis

*Aspergillus fumigatus*, *A. flavus*, and *A. niger* are haploid organisms that do not show a natural sexual cycle. *Aspergillus nidulans* differs in this respect and undergoes a sexual cycle producing diploids under natural conditions at a low but measurable rate. A sexual cross is required to generate linkage data between genetic loci. Genetic analyses such as these were made easier in *A. nidulans* and in other Aspergilli by the discovery of the parasexual cycle in *A. niger* by Pontecorvo and in *A. nidulans* by Roper, both in 1952 (117a,b).

In the parasexual cycle, two fungal strains, regardless of compatability, can be induced to form heterokaryons when grown in close proximity. The haploid nuclei will then fuse to form a diploid nucleus that can be maintained in a stable form. This diploid can be returned to the haploid state by treatment with destabilizing agents, such as fluorophenylalanine or benomyl. Selection of the haploid and diploid forms usually relies on the use of spore color mutants or mutants in metabolic processes, such as nitrate assimilation.

The long history and relative ease, compared with other *Aspergillus* species, of genetic study in *A. nidulans* means that many mutants are available and that mutant alleles have been assigned to chromosomes by linkage studies. The situation in *A. fumigatus* is nowhere as advanced in the numbers of mutants available and no linkage data are known. Spore color mutants have been reported (44, 74, 76, 78, 81, 116) and data have been published on the generation of diploids using the parasexual cycle (116, 118, 119, Brookman, unpublished data). The use of such diploids will enable heterozygotes to be produced with one copy of a gene disrupted. If the gene is essential, then the haploid progeny produced after haploidization will not be viable, whereas nonessential disruptions will produce viable progeny. Haploid progeny generated from the nondisrupted allele will be viable in all cases, but can be distinguished from the disrupted progeny by the absence of the selectable marker (e.g., *pyrG* or hygromycin resistance) (119).

## D. Molecular Mutagenesis

Genome-wide molecular mutagenesis approaches are required to assist in the understanding of important biological questions in *A. fumigatus*, such as the identification of pathogenicity factors required in the infection process. A number of approaches have been suggested for use with pathogens, and these are discussed below.

Brown et al. (107) have developed a restriction enzyme-mediated integration (REMI) approach for use in *A. fumigatus*. Transformation by PEG-fusion in the presence of *Xho*I or *Kpn*I resulted in a high level (>75%) of transformants with a single-copy integration into the *A. fumigatus* genome. Electroporation of spores without the addition of restriction enzyme resulted in up to 10 times higher transformation frequencies than were obtained with protoplasts and between 53% and 73% of the transformants had single copy integrations (107). Production of libraries of such mutant strains could then be used in the investigation of gene function as has been described in plant fungal pathogens (120, 121). The gene responsible for the altered phenotype observed can be identified using plasmid rescue or inverse polymerase chain reaction (PCR). In addition, the use of insertional mutagenesis in combination with a diploid strain will enable screens to be made for essential genes (119). Recently, de Souza et al. (122) have used a REMI system to investigate gene systems responsible for multidrug resistance in *A. nidulans*. They were able to characterize these mutants further than is currently possible with *A. fumigatus* by performing genetic crosses with other drug-resistant strains. They were able to identify 4 new REMI-generated mutant loci from a total of 12 different complementation groups for the multidrug resistance/sensitivity phenotype in *A. nidulans*.

Signature-tagged mutagenesis (STM) has been used for genome-wide searches for virulence factors in pathogenic bacteria (123). This methodology produces insertional mutants with tagged sequences to allow monitoring of entire populations. This can be useful for studying changes in a population during infection of an animal model and can identify organisms with altered pathogenic phenotypes. The genetic inactivation giving rise to the changed phenotype can be determined through the use of the signature tag on the insertion.

An STM screen has been carried out in *A. fumigatus* (47). Two signature-tagged mutant libraries, one made using REMI protoplast transformation and the other using non-REMI electroporation, were screened in a murine model of IPA. Pools of up to 96 mutants were screened at a time. After two rounds of STM, 35 out of 4648 (0.8%) mutant strains failed to give strong hybridization signals from the fungi recovered from murine lungs. These strains were tested against a reference strain by infection in which the inoculum consisted of approximately 50% of each strain. The 2 strains with the lowest recovery rates from the lung were investigated further. Plasmid rescue was used to identify where the insertional mutagenesis had occurred. There were no DNA or protein database hits to one of the sequences, but the other sequence was identified as the promoter of the para-aminobenzoate synthase gene. A mutant defective in the synthesis of an essential growth factor might be expected to have an attenuated virulence, although not all such auxotrophic mutants present in the STM libraries were identified by the screen as having a reduced virulence (0.25% of the strains were found to be auxotrophs or have abnormal morphology). Another important observation the authors made about these screens was that, although several strains were repeatedly identified as attenuated in pools of 96 strains, they were not effectively outcompeted when present as 50% of the inoculum. This contrasts with the virulence determinants of bacterial pathogens that have specifically evolved to play an important role for survival in the host. The degree of attenuation of the corresponding mutant strains of bacteria is, therefore, more likely to be independent of the inoculum. In contrast, because *A. fumigatus* is an opportunistic pathogen where many factors play a more nutritional or physiological role for survival in the host, a slight reduction in the fitness of a mutant will probably result in it being outcompeted in a mutant pool, but not when there is just one other strain present.

## E. Additional Genetic Tools Required

Development of additional tools will assist in the genetic analysis of function and pathogenicity in *A. fumigatus*. Steps have already been taken in setting up some of these. For instance, insertional mutagenesis using a transposon might result in a bank of mutants containing only single insertions rather than the various possibilities that can result when using transformation-mediated insertional mutagenesis (tandem insertions, multiple integrations, and DNA rearrangements). Nicosia et al. (124) have recently described the transposition of a *Fusarium oxysporum* transposon, *impala*, in *A. nidulans*. These authors report the failure of a transposon trap in this organism to isolate active homologous transposons. We have had a similar lack of success in *A. fumigatus* despite exhaustive efforts. The extension of the approach of heterologous transposition into *A. fumigatus* and, if successful, development of an active transposon-tagging system would be most beneficial for the elucidation of genes responsible for important phenotypes, such as resistance to host defenses and antifungal drug resistance.

It would be desirable, also, to have a systematic procedure for knocking out genes. It will not be possible to use a system similar to that developed for *Saccharomyces cerevisiae,* in which short-flanking (<50 bp) or long-flanking (a few 100 bp) homologous regions are added to either end of a selection cassette using PCR (125, 126), as recombination in filamentous fungi requires longer regions of homology. Generally, the longer the extent of homologous overlap, the more efficient is the recombination. A system based on in vitro transposition has been developed in *Magnaporthe grisea* for mutating genes that should be applicable to other filamentous fungi (127). Individual cosmid bacterial artificial chromosome (BAC) clones or entire libraries can be mutagenized in vitro using a transposon carrying a selectable marker. Sequence analysis of the transposon insertion sites will permit selection of the required insertional mutation, and the entire cosmid/BAC insert can be used in a transformation to efficiently generate homologous recombinants.

It will also be necessary to develop regulatable titratable promoters for use in *A. fumigatus* to assist in the study of essential genes. Promoters, such as *alcA*, which is repressed by glucose, have been used in other *Aspergillus* species. These are generally very strong promoters, and the overexpression of a gene product in a pathogenic species could cause problems with genetically modified organism regulations. Such problems might be circumvented by using a disabled strain of *A. fumigatus*, which has, for instance, been made auxotrophic for essential factors required for growth in the host.

## V.  TOOLS FOR PHENOTYPIC ANALYSIS

### A.  Growth Rate

Measurement of growth rate in *Aspergillus* is problematic, especially in the animal host. In vitro, the traditional approach has been to determine dry weight, after broth cultures. This method is useful for a large mass of *Aspergillus*, but inaccurate for small quantities. Measurement of growth rate directly in liquid cultures is hindered by the clumping of the hydrophobic spores, which results in pellet formation. Pellet formation can be reduced sufficiently to use optical density readings to generate growth curves by adding the polymer Junlon (128, A. Mousavi, personal communication). It is possible to obtain filamentous growth by growing *Aspergillus* in a thin liquid layer (129), and this behavior has been exploited in minimum inhibitory concentration (MIC) testing in microtiter plates. Reproducible growth curves have been obtained for *A. fumigatus* by using microtiter plates in a plate reader and measuring optical density automatically (130). Colony diameter measurement on agar is not useful, as it does not reflect branching frequency. However, hyphal extension rates on agar can be measured microscopically, using video capture, and this provides a highly accurate means of determining specific growth rate (86).

Measurement of cell wall mass can be done by assaying chitin directly (as has been performed in host tissues) (131–134) or by measuring incorporation of radiolabeled glucosamine, a precursor of chitin (86,94). In many animal model systems, colony forming units (cfu) are measured, usually as an indicator of antifungal drug activity (43 and others). There is some uncertainty about how accurate such cfu measurements are, as ungerminated conidia and hyphal fragments may yield individual colonies. Small differences (e.g., ~ one log) are probably not significant. It should be possible to measure indirectly the amount of actively respiring fungus in infected tissue using molecular techniques, such as quantitative reverse-transcriptase PCR. The relative amounts of a fungal and a mammalian housekeeping gene should provide a measure of the progress of infection in the host.

### B.  Animal Models

Multiple different animal models have been used in the investigation of the pathogenesis of IA and other forms of aspergillosis. In many instances, animal models have been a critically important component of the evaluation of new antifungal drugs.

The pulmonary route of infection is the most common in humans, but most murine models have used intravenous inoculation. Such infections lead to a disseminated infection primarily involving the kidneys, liver, brain, and to a lesser extent, the lung (135, 136). Pulmonary models usually result in very severe disease and no dissemination to other organs (137). Mice, like most animals, including humans, are intrinsically resistant to *Aspergillus*. Large inocula are required to establish infection by the intravenous route and no inoculum is high enough to infect normal mice by the pulmonary route. For this reason, mice (and most other animals) need to be immunosuppressed to establish infection. This is typically done with chemotherapy agents that induce neutropenia, such as cyclophosphamide or corticosteroids (138). From an immunological perspective, these modes of immunosuppression are very different and need to be considered separately (136). The models that have been established in mice are usually rapidly fatal (3–7 days). Approaches to reducing suffering in mice include timed sacrifice, sacrifice of mice that are obviously very ill, and determination of a temperature threshold that predicts death (Denning, unpublished data).

From the perspective of assessing pathogenicity determinants, the "whole mouse" mortality assay has been used most commonly. Substantial differences in the virulence of different isolates of *A. fumigatus* can be discerned (43, 44), but mutant strains containing individual or even double gene disruptions have not yielded any meaningful differences in mortality (139). The exceptions to this have been the demonstration of reduced or nonvirulent auxotrophic mutants, defective in their ability to synthesize para-aminobenzoic acid, pyrimidine, and lysine, or to regulate nitrogen metabolism (47, 48, 140, 141), and that pigmentless mutants are more readily killed and therefore less virulent (76, 81). However, these observations, although important, do not explain the differences in virulence between isolates.

The study of immune mechanisms of both allergic and invasive disease has been done in different strains of mice and in mutant (knock-out) mice. For example, cytokine concentrations have been measured in the pulmonary fluids of mice after challenge (142). The administration of cytokine inhibitors has also been revealing (143, 144). The role of interleukin-10 (IL-10) in IA has been studied in IL-10 gene knock-out mice (145). Allergic responses have been studied in several gene knock-out mice, including C–C chemokine receptor 1, chemoattractant protein-1 receptor, and interleukin-4 (146–148). A recent development has used knock-out mice deficient in one or more genes critical for the oxygen-dependent respiratory burst, which mimic some phenotypes of chronic granutomatous disease (102, 149). Remarkably, the inoculum used in these models is several logs lower than in neutropenic or corticosteroid-treated mice and can be modulated to produce acute or chronic infection (102). These mutant murine models will be valuable for studying many aspects of the interaction of host with *Aspergillus*, such as surfactant, cytokine, and mannose-binding protein polymorphisms.

Species other than mice used in models include the rat, rabbit, guinea pig, chick, turkey poult, and pigeon. Reasons for these choices are summarized as follows. Rats have been used because direct tracheal inoculation is possible, thus standardizing the inoculum to the lung. Rats have, in addition, been used for delivering aerosolized antifungal drug therapy (150, 151). Rabbits, as larger species, allow multiple sequential blood samples to be taken. This has facilitated pharmacokinetic and pharmacodynamic studies (152) and temporal studies of antigenemia (38). Scanning of the lungs with CT or magnetic resonance has also been pioneered to follow the course of the infection (153). Guinea pig models have been used in antifungal drug studies to facilitate better drug exposure, especially with the azoles that are metabolized very quickly in mice and rats. This has been particularly useful for the comparative evaluation of voriconazole (154). Turkey poults have been used to study pathogenesis and immunological protection (155).

## VI.  *ASPERGILLUS* GENOMICS

### A.  The *Aspergillus fumigatus* Genome Sequencing Project

An international project has been initiated to sequence the genome of *A. fumigatus*. A pilot project was started at the Wellcome Trust Sanger Institute and the University of Manchester in July 1999, with funding from the Wellcome Trust, to make a BAC library, map the clones into contigs, and sequence at least one megabase (Mb) of DNA.

A clinical isolate (AF293) from the collection of Denning was selected as the strain to be sequenced (156). As the primary purpose of the sequencing effort is to assist in the understanding of pathogenicity, it was essential to select an isolate of proven and known pathogenicity. Unfortunately, the details for almost all isolates in culture collections are paltry and insufficient to be confident about the pathological features of a given infection. It was also essential to select an isolate grown from a sterile source rather than respiratory fluids to ensure that it was not a colonizing isolate. Certain features of pathogenicity were important to "capture," including vascular invasion and lung infarction. This isolate was grown from the lung of a 57-year-old woman from Shrewsbury, UK, with aplastic anemia who died of undiagnosed invasive pulmonary aspergillosis. The aplastic anemia may have been caused by gold therapy for rheumatoid arthritis. Her lung revealed hemorrhagic nodules at autopsy, which indicates that there was vascular invasion and infarction. The isolate is susceptible to itraconazole and amphotericin B, and has been shown to be virulent in two murine models. The isolate was confirmed as *A. fumigatus* Fresenius by both traditional taxonomy and molecular phylogeny. The isolate has been deposited in the National Collection of Pathogenic Fungi (UK), the Centraal Bureau voor Schimmelcultures (Netherlands), and the Fungal Genetics Stock Center (USA). Resolution has not been optimal, but electrophoretic karyotyping has determined that AF293 possibly contains eight chromosomes (Anderson, unpublished data and J.-A. van Burik, personal communication). There are two small chromosomes of 1.7 and 1.8 Mb (chromosomes 8 and 7). Chromosomes 6, 5, and 4 are all around 3.5 Mb and form a "blob" on pulsed-field gels. Chromosome 3 is 4.0 to 4.3 Mb in size and chromosomes 2 and 1 are both 5.0 to 5.3 Mb.

Three BAC libraries have been used for the physical mapping stage of the project. Two were constructed at the Sanger Institute (Quail and Woodman) as part of the pilot project, using chromosomal DNA provided by the University of Manchester (Anderson). The vector used was pBACe3.6; one library (B28) was made using *Sau*3AI partially digested restriction enzyme fragments and the other (B46) using *Eco*RI. The B28 library has an average insert size of 80 kb with a range from 15 to 160 kb. Random clones (36 × 96) have been picked into replicate microtiter plates and gridded onto membranes from a total of about 24,000, and paired end-sequences have been generated. The B46 library has an average insert size of 75 to 80 kb and a titer of 100,000. The third library was constructed at The Centre d'Etude du Polymorphisme Humain (Billaut, France) using DNA provided by the Institut Pasteur (Paris). The vector used was pBeloBac11 and the enzyme *Hind*III. The mean insert size is 77 kb, with a range from 25 to 190 kb. Paired end-sequences have been generated from 90 × 96 clones at the Institut Pasteur (Glaser and Kunst).

As library B28 was likely to have a more random clone distribution as the consequence of using a 4-base-pair-specific restriction enzyme, this library was used as the primary library for the physical mapping. The clones were digested with the restriction enzyme *Pst*I, which gave the optimal number of fragments to enable fingerprint contigs to be built (20–50 per clone). In addition, 36 × 96 clones have also been fingerprinted from the two other BAC libraries. Fingerprint contigs were built up using the fingerprinted contigs (FPC) software developed at the Sanger Institute. A contig centered around the *niaD* locus has been sequenced as part of the pilot project. This contig was extended using the end-sequencing and fingerprinting data until it was more than 900 kb in size.

Five sequencing centers are involved in sequencing the genome of *A. fumigatus*. The Institute for Genomic Research (TIGR) (USA) (Nierman and Feldblyum) has been awarded a grant by the National Institutes of Health (NIH) (in collaboration with the University of Manchester) under the National Institute of Allergy and Infectious Diseases (NIAID)-funding mechanism for sequencing small genomes. Continuation of the sequencing at the Sanger Institute (Hall and Barrell) has been funded by an additional grant from the Wellcome Trust. The Spanish government, through its research agency Fondo de Investigaciones Sanitorias (FIS), has awarded a grant to three collaborating centers: Salamanca University (Sánchez-Pérez, Vazquez, and del Rey), Complutense University (Arroyo and Nombela), and the Centro de Investigaciones Biológicas (Peñalva and Rodriguez).

The sequencing phase of the project will involve a whole genome shotgun approach followed by gap closure and annotation. Shotgun sequencing will be done at the Sanger Institute and TIGR to a genome coverage of 10 times, and was initiated in April 2001. The sequence will be completed by closing sequencing and physical gaps. The intention is to generate as complete a sequence as possible (excluding the ribosomal DNA repeat region, centromeres, and telomeres), and both the Sanger Institute and TIGR will annotate this complete sequence.

The genome sequencing project is being coordinated by a committee, which includes Dr. Denning (chair), Dr. Nieman (TIGR), Dr. Hall (Sanger), Professor Bennett (Tulane), Dr. Latgé (Pasteur), Professor Turner (Sheffield), Dr. Dixon (NIH), and Dr. Anderson (secretary). Progress about the project is available at the Sanger, TIGR, and *Aspergillus* websites.

## B.  Comparative Genomics

When the complete genomic sequence of *A. fumigatus* has been generated, and even when whole genome shotgun assemblies are released, it will be possible to compare this genome with the sequences available for other *Aspergillus* and fungal species. Researchers will be able to carry out many different types of analyses from metabolic, regulatory, evolutionary viewpoints, among others, and so only some of the possibilities within the Aspergilli are discussed here.

Because the species within the genus *Aspergillus* have recently evolved from each other, it is expected that there will be much conservation in the organization of their genomes. Already, for instance, the evidence from genetics and electrophoretic karyotyping suggests that most species have eight chromosomes. Therefore, one of the first questions that will be resolved from the sequencing of *A. fumigatus* is how many chromosomes it also has. There should, in addition, be a high level of conservation in gene order between species. It will be possible to make a direct comparison with the sequenced cosmids from chromosome VIII of *A. nidulans*, but the sequence of chromosome IV and the whole genome shotgun of *A. nidulans* will be of only limited use in comparing gene order unless the contigs are bigger than the average gene. However, extensive genetic and physical maps are available for *A. nidulans*, and so it should be possible to carry out comparisons at this level. The only preliminary evidence that there will be conservation in chromosome structure is that the nitrate-assimilation gene cluster has been mapped to the largest chromosomes in both species (157, 158).

A direct comparison of the gene complements in various *Aspergillus* species will be of greatest interest for many researchers, and here, even information from partial whole genome shotgun and EST sequencing will be valuable. The total number of genes is not expected to vary much between species, but the differences will be most revealing. It will be interesting to see which additional genes are present in the more pathogenic species, such as *A. fumigatus*. For example, it has been shown that the gene, which encodes the cytotoxic ribonuclease mitogillin, is not present in *A. flavus*, *A. niger*, or *A. nidulans* (159).

Many differences between the species might be expected to occur in the so-called "dispensible" metabolic pathways (160), which includes biosynthetic pathways involved in the production of secondary metabolites (e.g., 74, 161–163). These dispensable metabolic pathways are often arranged into gene clusters, where the genes are coregulated and contain definable cis-acting elements in their promoters. It is possible, perhaps, when these genes are present in clusters, that they are more readily gained or lost. It will initially be easy to see if *A. fumigatus* contains pathways or portions of pathways that have been studied in other Aspergilli: an obvious example would be the aflatoxin biosynthetic pathway, as *A. fumigatus* produces an intermediate in this pathway (sterigmatocystin). An example of a difference between *A. fumigatus* and *A. nidulans* in a disposable metabolic pathway is the conidial pigment biosynthetic pathway. The first enzyme in the pathway is the same in both species; however, after that the genes and enzymes are different. The two genes in *A. nidulans*, *wA* and *yA*, are located on different chromosomes, whereas most of the genes for pigment synthesis in *A. fumigatus* are located in a 19-kb cluster. Interestingly, the *wA* orthologue is located at one end of this cluster and the largest intergenic distance (1.1 kb) in the cluster lies between it and the next gene (74).

Another area where differences have been reported between *Aspergillus* species is in the number of copies of a protein family. For instance, five chitin synthases have been detected in *A. nidulans*, whereas seven genes have been detected in *A. fumigatus* (164). Two extracellular metalloproteinases have been reported in *A. fumigatus*, but only one of these is present in *A. flavus* (165, 166).

One obvious difference between *A. fumigatus* and *A. nidulans* is the lack of a detectable sexual cycle in *A. fumigatus*, and it will be interesting to see if this is due to missing or nonfunctional genes. It had been assumed that the yeast *Candida albicans* was an asexual organism, but a mating-type-like locus was identified in the whole genome shotgun sequence, and subsequent experiments have demonstrated that *C. albicans* is able to undergo mating (167, 168). Additional analysis of the *C. albicans* sequence, comparing it with *S. cerevisiae*, indicated that *C. albicans* should be able to undergo meiosis as well (169).

It will, of course, be possible to use the gene micro-arrays that will be produced as a consequence of the *A. fumigatus* genome sequencing project, to carry out cross-species comparisons and screen for those genes that are unique to *A. fumigatus*. Electrophoretic karyotyping has shown that there is considerable variation in chromosome size between isolates of *A. fumigatus* (Anderson, unpublished data), and so these microarrays could also be used to check the gene complements of isolates as well.

Even though relatively little research has been carried out on the genes and proteins of *A. fumigatus*, there has been much research published on other *Aspergillus* species, especially *A. nidulans*. Therefore, it will be possible to identify orthologues in other species and start with the premise that the protein performs the same role in *A. fumigatus*. This has generally been shown to be the case with the orthologous genes that have been studied, such as the hydrophobin RodA (62), the polyketide synthase involved in pigment biosynthesis (77), the *pyrG* gene (108, 112), the *pdmA* gene (114), and the *brlA fluG, flbA,* and *trpC* genes (117). One notable exception is the *A. fumigatus* chitin synthase, ChsG and its orthologue in *A. nidulans*, ChsB. Both chitin synthases are class III enzymes, and they are 90% identical at the protein level and 76% identical at the DNA level over the coding region. At the level of gene structure, there is complete conservation in the positions of the start and stop codons and the splice junctions (except for the 5′ splice site of intron 1), and the exon sizes vary by multiples of 3 (*A. fumigatus/A. nidulans* exon 1: 84/93; exon 2: 475/481; exon 3: 362/362; exon 4: 847/847; exon 5: 830/830; exon 6: 138/138). Both enzymes have been shown to have important roles in hyphal tip growth, but the severity of the phenotype in the disrupted mutants differed. The *A. fumigatus chsG⁻* mutant was more highly branched than normal, and its conidia failed to mature properly (94), whereas the phenotype of the *A. nidulans chsB⁻* mutant was more severe, with the highly branched hyphae also displaying enlarged tips with bulges along their lengths; in addition, no conidiophores or conidia were formed (170, 171). The differences in phenotype is not the consequence of *A. fumigatus* containing an additional class III enzyme, ChsC, because a mutant disrupted in both genes displays the same phenotype as a *chsG⁻* mutant (94). It is still possible, however, that functional redundancy exists, and that one of the other chitin synthases in *A. fumigatus* is important for some of the roles carried out only by ChsB in *A. nidulans*. It will, therefore, require complementation experiments to prove formally that these two enzymes can perform the same functions in both organisms.

## VII.  FUTURE APPROACHES TO UNDERSTANDING PATHOGENICITY

It is highly likely that there are multiple pathogenicity determinants, some operative in certain host settings and others in other settings. So the ability of *Aspergillus* to invade the cornea of the eye will likely depend on different organism-related factors than those required for the invasion of a burn wound or a lung in a neutropenic patient. Given this, highly specialized model systems will need to be developed to dissect out the detailed mechanisms involved. A transcriptome analysis of *Aspergillus* will be of little use if very broad questions are asked, such as which genes are involved in pulmonary infection and RNA is derived from a rapidly lethal murine model, two or more days after infection. If, on the other hand, RNA is derived from each and every step of the infection process, then the data derived will be much more useful. Knock-outs of genes thought to be important can then be produced and studied in each stage of the infection process. The use of different models, with different wild-type *A. fumigatus* isolates may also be revealing, in that alternative pathogenetic mechanisms may be revealed. Multiple knock-outs, with phenotypic correlates, could start to unravel the complexities of why *Aspergillus* is such a common allergic and opportunistic pathogen, and yet can also cause disease in normal hosts.

## REFERENCES

1. JI Pitt, RA Samson, JC Frisvad. List of accepted species and their synonyms in the family Trichocomaceae. In: RA Samson, JI Pitt, eds. Integration of Modern Taxonomic Methods for *Penicillium* and *Aspergillus* Classification. Amsterdam: Harwood, 2000, pp 9–49.
2. N Nolard, M Detandt, H Beguin. Ecology of *Aspergillus* species in the human environment. In: H Vanden Bossche, DWR Mackenzie, G Cauwenbergh, eds. *Aspergillus* and Aspergillosis. New York: Plenum Press, 1988, pp 35–42.
3. WR Solomon. HP Burge, JR Boise. Airborne *Aspergillus fumigatus* levels outside and within a large clinical center. J Allergy Clin Immunol 62:56–60, 1978.
4. PM Arnow, M Sadigh, C Costas, D Weil, R Chudy. Endemic and epidemic aspergillosis associated with in-hospital replication of *Aspergillus* organisms. J Infect Dis 164:998–1002, 1991.
5. DR Hospenthal, KJ Kwon-Cbung, JE Bennett. Concentrations of airborne *Aspergillus* compared to the incidence of invasive aspergillosis: lack of correlation. Med Mycol 36:165–168, 1998.
6. RC Summerbell, S Krajden, J Kane. Potted plants in hospitals as reservoirs of pathogenic fungi. Mycopathologia 106:13–22, 1989.
7. NE Rankin. Disseminated aspergillosis and moniliasis associated with agranulocytosis and antibiotic therapy. Br Mod J 25:918–919, 1953.
8. DW Denning. Invasive aspergillosis. Clin Infect Dis 26:781–803, 1998.
9. AH Groll, PM Shah, C Mentzel, M Schneider, G Just-Nuebling, K Huebner. Trends in the postmortem epidemiology of invasive fungal infections at a university hospital. J Infect 33:23–32, 1996.
10. T Yamazaki, H Kume, S Murase, E Yamashita, M Arisawa. Epidemiology of visceral mycoses: analysis of data in annual of the pathological autopsy cases in Japan. J Clin Microbiol 37:1732–1738, 1999.
11. M Vogeser, A Wanders, A Haas, G Ruckdeschel. A four-year review of fatal aspergillosis. Eur J Clin Microbiol Infect Dis 18:42–45, 1999.
12. JR Rees, RW Penner, RA Hajjeh, ME Brandt, AL Reingold. The epidemiological features of invasive mycotic infections in the San Francisco bay area, 1992–1993: results of population based laboratory active surveillance. Clin Infect Dis 27:1138–1147, 1998.
13. KFW Hinson, AJ Moon, NS Plummer. Broncho-pulmonary aspergillosis. A review and a report of eight new cases. Thorax 7:317–333, 1952.
14. AP Kurup, A Kumar. Immunodiagnosis of aspergillosis. Clin Microbiol Rev 4:439–456, 1991.

15.  DW Denning, J Van Wye, NJ Lewiston, DA Stevens. Adjunctive therapy of allergic bronchopulmonary aspergillosis with itraconazole. Chest 100:813–819,1991.
16.  F Salez, A Brichet, S Desurmont, JM Grosbois, B Wallaert, AB Tonnel. Effects of itraconazole therapy in allergic bronchopulmonary aspergillosis. Chest 116:1665–1668, 1999.
17.  DA Stevens, HJ Schwartz, JY Lee, BL Moskovitz, DC Jerome, A Catanzaro, DM Bamberger, AJ Weinmann, CU Tuazon, MA Judson, TA Platts-Mills, AC DeGraff Jr. A randomized trial of itraconazole in allergic bronchopulmonary aspergillosis. N Engl J Med 342:756–762, 2000.
18.  WHO report, 1995.
19.  BR O'Driscoll, LC Hopkinson. Mould allergy is common in patients with severe asthma. Amer Rev Respir Dis 157:A623, 1998.
20.  TM Eppinger, PA Greenberger, DA White, AE Brown, C Cunningham-Rundles. Sensitization to *Aspergillus* species in the congenital neutrophil disorders chronic granulomatous disease and hyper-IgE syndrome. J Allergy Clin Immunol 104: 1265–1272, 1999.
21.  JU Ponikau, DA Sherris, EB Kern, HA Homburger, E Frigas, TA Gaffey, GD Roberts. The diagnosis and incidence of allergic fungal sinusitis. Mayo Clin Proc 74:877–884, 1999.
22.  P Macpherson. Pulmonary aspergillosis in Argyll. Br J Chest 59:148–157, 1965.
23.  B Varkey, HD Rose. Pulmonary aspergilloma. A rational approach to treatment. Am J Med 61:626–631, 1976.
24.  Research committee of the British Thoracic and Tuberculosis Assocation. Aspergilloma and residual tuberculous cavities—the results of a re-survey. Tubercle 51: 227, 1970.
25.  C Wollschlager, F Kan. Aspergilloma complicating sarcoidosis. A prospective study in 100 patients. Chest 86:585–588, 1984.
26.  FA Nime, GM Hutchins. Oxalosis caused by *Aspergillus* infection. Johns Hopkins Med J 133:183–194, 1973.
27.  LC Severo, GR Geyer, N da Silva Porto, MB Wagner, AT Londero. Pulmonary *Aspergillus niger* colonisation. Report of 23 cases and a review of the literature. Rev Iberoam Micol 14:104–110, 1997.
28.  DJ Crosdale, KV Poulton, WE Ollier, W Thomson, DW Denning. Mannose-binding lectin gene polymorphisms as a susceptibility factor for chronic necrotising pulmonary aspergillosis. J Infect Dis 184:653–656, 2001.
29.  P Rafferty, BA Biggs, GK Crompton, IWB Grant. What happens to patients with pulmonary aspergilloma? Analysis of 23 cases. Thorax 38:579–583, 1983.
30.  A Warris, PE Verweij, R Barton, DC Crabbe, EL Evans, J Meis. Invasive aspergillosis in two patients with Pearson syndrome. Pediatr Infect Dis J 18:739–741, 1999.
31.  DH McKee, PN Cooper, DW Denning. Invasive aspergillosis in a patient with MELAS syndrome. Brit J Neurol Neurosurg Psychiatry 68:765–767, 2000.
32.  P Rohrlich, J Sarfati, P Mariani, M Duval, A Carol, C Saint-Martin, E Bingen, J-P Latgé, E Vilmer. Prospective sandwich enzyme-linked immunosorbent assay for serum galactomannan: early predictive value and clinical use in invasive aspergillosis. Pediatr Infect Dis J 15:232–237, 1996.

33. J Maertens, J Verhaegen, H Demuynck, P Brock, G Verhoef, P Vandenberghe, JV Eldere, L Verbist, M Booghaerts. Autopsy-controlled prospective evaluation of serial screening for circulating galactomannan by a sandwich enzyme-linked immunosorbent assay for hematological patients at risk for invasive aspergillosis. J Clin Microbiol 37:3223–3228, 1999.

34. D Caillot, O Casasnovas, A Bernard, JF Couaillier, C Durand, B Cuisenier, E Solary, F Piard, T Petrella, A Bonnin, G Couillault, M Dumas, H Guy. Improved management of invasive pulmonary aspergillosis in neutropenic patients using early thoracic computed tomographic scan and surgery. J Clin Oncol 15:139–147, 1997.

35. H Hebart, J Löffler, C Meisner, F Serey, D Schmidt, A Böhme, H Martin, A Engel, D Bunjes, WV Kern, U Schumacher, L Kanz, H Einsele. Early detection of *Aspergillus* infection after allogeneic stem cell transplantation by polymerase chain reaction screening. J Infect Dis 181:1713–1719, 2000.

36. DW Denning, DA Stevens. Antifungal and surgical treatment of invasive aspergillosis: review of 2121 published cases. Rev Infect Dis 12:1147–1201, 1990.

37. R Bowden, P Chandrasekar, MH White, X Li, L Pietrelli, M Gurwith, J-A van Burik, M Laverdiere, S Safrin, JR Wingard. A double-blind, randomized, controlled trial of amphotericin B colloidal dispersion versus amphotericin B for treatment of invasive aspergillosis in immunocompromised patients. Clin Infect Dis 35:359–366, 2002.

38. TE Patterson, WR Kirkpatrick, M White, J Hiemenz, JR Wingard, B Dupont, MG Rinaldi, DA Stevens, JR Graybill. Invasive aspergillosis—disease spectrum, treatment practices and outcomes. Medicine 79:250–260, 2000.

39. S-J Lin, J Schranz, M Teutsch. Aspergillosis case-fatality rate: systematic review of literature. Clin Infect Dis 32:358–366, 2001.

40. DA Stevens, JY Lee. Analysis of compassionate use itraconazole therapy for invasive aspergillosis by the NIAID Mycoses Study Group criteria. Arch Intern Med 157:1857–1862, 1997.

41. CB Moore, N Sayers, J Mosqua, J Slaven, DW Denning. Antifungal drug resistance in *Aspergillus*. J Infection 41:203–220, 2000.

42. E Dannaoui, E Borel, MF Monier, MA Piens, S Picot, F Persat. Acquired itraconazole resistance in *Aspergillus fumigatus*. J Antimicrob Chemother 47:333–340, 2001.

42a. R Herbrecht, DW Denning, TF Patterson, et al. Voriconazole versus amphotericin B for primary therapy of invasive aspergillosis. N Engl J Med 347:408–415, 2002.

43. DW Denning, SA Radford, K Oakley, L Hall, EM Johnson, DW Warnock. Correlation between in-vitro susceptibility testing to itraconazole and in-vivo outcome for *Aspergillus fumigatus* infection. J Antimicrob Chemother 40:401–414, 1997.

44. A Aufauvre-Brown, JS Brown, DW Holden. Comparison of virulence between clinical and environmental isolates of *Aspergillus fumigatus*. Eur J Clin Microbiol Infect Dis 17:778–780, 1998.

45. JP Debeaupuis, J Sarfati, V Chazalet, J-P Latgé. Genetic diversity among clinical and environmental isolates of *Aspergillus fumigatus*. Infect Immun 65:3080–3085, 1997.

46. E Bart-Delabesse, IF Humbert, E Delabesse, S Bretagne. Microsatellite markers for typing *Aspergillus fumigatus* isolates. J Clin Microbiol 36.2413–2418, 1998.

47. JS Brown, A Aufauvre-Brown, J Brown, JM Jennings, H Arst Jr, DW Holden. Signature-tagged and directed mutagenesis identify PABA synthetase as essential for *Aspergillus fumigatus* pathogenicity. Mol Microbiol 36:1371–1380, 2000.

48. C d'Enfert, M Diaquin, A Delit, N Wuscher, JP Debeaupuis, M Huerre, J-P Latgé. Attenuated virulence of uridine-uracil auxotrophs of *Aspergillus fumigatus*. Infect Immun 64:4401–4405, 1996.

49. GR Smith. Experimental aspergillosis in mice: aspects of resistance. J Hyg (Camb) 70:741–753, 1972.

50. T Murayama, R Amitani, Y Ikegami, R Nawada, WJ Lee, F Kuze. Suppressive effects of *Aspergillus fumigatus* culture filtrates on human alveolar macrophages and polymorphonuclear leucocytes. Eur Respir J 9:293–300, 1996.

51. G Tronchin, J-P Bouchara, G Larcher, J-C Lissitzky, D Chabasse. Interaction between *Aspergillus fumigatus* and basement membrane laminin: binding and substrate degradation. Biol Cell 77:201–208, 1993.

52. G Tronchin, K Esnault, G Renier, R Filmon, D Chabasse, J-P Bouchara. Expression and identification of a laminin-binding protein in *Aspergillus fumigatus* conidia. Infect Immun 65:9–15, 1997.

53. Z Yang, SM Jaeckisch, CG Mitchell. Enhanced binding of *Aspergillus fumigatus* spores to A549 epithelial cells and extracellular matrix proteins by a component from the spore surface and inhibition by rat lung lavage fluid. Thorax 55:579–584, 2000.

54. J-P Bouchara, A Bouali, G Tronchin, R Robert, D Chabasse, JM Senet. Binding of fibrinogen to the pathogenic *Aspergillus* species. J Med Vet Mycol 26:327–334, 1988.

55. V Annaix, JP Bouchara, G Larcher, D Chabasse, G Tronchin. Specific binding of human fibrinogen fragment D to *Aspergillus fumigatus* conidia. Infect Immun 60: 1747–1755, 1992.

56. JA Wasylnka, MM Moore. Adhesion of *Aspergillus* species to extracellular matrix proteins: evidence for involvement of negatively charged carbohydrates on the condidial surface. Infect Immun 68:3377–3384, 2000.

57. JA Wasylnka, Ml Simmer, MM Moore. Differences in sialic acid density in pathogenic and non-pathogenic *Aspergillus* species. Microbiology 147:869–877, 2001.

58. T Madan, P Eggleton, U Kishore, P Strong, SS Aggrawal, P Usha Sarma, KBM Reid. Binding of pulmonary surfactant proteins A and D to *Aspergillus fumigatus* conidia enhances phagocytosis and killing by human neutrophils and alveolar macrophages. Infect Immun 65:3171–3179, 1997.

59. MJ Allen, R Harbreck, B Smith, DR Voelker, RJ Mason. Binding of rat and human surfactant proteins A and D to *Aspergillus fumigatus* conidia. Infect Immun 67: 4563–4569, 1999.

60. SM Levitz, TP Farrell. Human neutrophil degranulation stimulated by *Aspergillus fumigatus*. J Leukoc Biol 47:170–175, 1990.

61. MA Stringer, RA Dean, TC Sewall, WE Timberlake. Rodletless, a new *Aspergillus* developmental mutant induced by directed gene inactivation. Genes Dev 5:1161–1171, 1991.

62. M Parta, Y Chang, S Rulong, P Pinto-DaSilva, KH Kwon-Chung. *HYP1* a hydrophobin gene from *Aspergillus fumigatus*, complements the rodletless phenotype in *Aspergillus nidulans*. Infect Immun 62:4389–4395, 1994.

63. N Thau, M Monod, B Crestani, C Rolland, G Tronchin, J-P Latgé, S Paris. Rodlet-less mutants of *Aspergillus fumigatus*. Infect Immun 62:4380–4388, 1994.
64. H Girardin, S Paris, J Rault, M-N Bellon-Fontaine, J-P Latgé. The role of the rodlet structure on the physicochemical properties of *Aspergillus* conidia. Letts Appl Microbiol 29:364–369, 1999.
65. A Schaffner, H Douglas, A Braude. Selective protection against conidia by mononuclear and against mycelia by polymorphonuclear phagocytes in resistance to *Aspergillus*: observations on these two lines of defense in vivo and in vitro with human and mouse phagocytes. J Clin Invest 69:617–631, 1982.
66. E Roilides, C Blake, A Holmes, PA Pizzo, TJ Walsh. Granulocyte-macrophage colony-stimulating factor and interferon-gamma prevent dexamethasone-induced immunosuppression of antifungal monocyte activity against *Aspergillus fumigatus* hyphae. J Med Vet Mycol 34:63–69, 1996.
67. VL Kan, JE Bennett. Lectin-like attachment sites on murine pulmonary alveolar macrophages bind *Aspergillus fumigatus* conidia. J Infect Dis 158:407–414, 1988.
68. VL Kan, JE Bennett. Beta 1,4-oligoglucosides inhibit the binding of *Aspergillus fumigatus* conidia to human monocytes. J Infect Dis 163:1154–1156, 1991.
69. DJ DeHart, DE Agwu, NC Julian, RG Washburn. Binding and germination of *Aspergillus fumigatus* conidia on cultured A549 pneumocytes. J Infect Dis 175:146–150, 1997.
70. MD Martins, LJ Rodriguez, CA Savary, ML Grazziutti, D Deshpande, DM Cohen, RE Cowart, DG Woodside, BW McIntyre, EJ Anaissie, JH Rex. Activated lymphocytes reduce adherence of *Aspergillus fumigatus*. Med Mycol 5:281–289, 1998.
71. TR Kozel, MA Wilson, TP Farrell, SM Levitz. Activation of C3 and binding to *Aspergillus fumigatus* conidia and hyphae. Infect Immun 57:3412–3417, 1989.
72. S Henwick, SV Hetherington, CC Patrick. Complement binding to *Aspergillus* conidia correlates with pathogenicity. J Lab Clin Med 122:27–35, 1993.
73. JE Sturtevant, J-P Latgé. Interactions between conidia of *Aspergillus fumigatus* and human complement component C3. Infect Immun 60:1913–1918, 1992.
74. H-F Tsai, MH Wheeler, YC Chang, KJ Kwon-Chung. A developmentally regulated gene cluster involved in conidial pigment biosynthesis in *Aspergillus fumigatus*. J Bacteriol 181:6469–6477, 1999.
75. K Langfelder, B Jahn, H Gehringer, A Schmidt, G Wanner, AA Brakhage. Identification of a polyketide synthase gene (pksP) of *Aspergillus fumigatus* involved in conidial pigment biosynthesis and virulence. Med Microbiol Immunol 187:79–89, 1998.
76. H-F Tsai, YC Chang, RG Washburn, MH Wheeler, KJ Kwon-Chung. The developmentally regulated *alb1* gene of *Aspergillus fumigatus*: its role in modulation of conidial morphology and virulence. J Bacteriol 180:3031–3038, 1998.
77. H-F Tsai, I Fujii, A Watanabe, MH Wheeler, YC Chang, Y Yasuoka, Y Ebizuka, KJ Kwon-Chung. Pentaketide-melanin biosynthesis in *Aspergillus fumigatus* requires chain-length shortening of a heptaketide precursor. J Biol Chem 276:29292–29298, 2001.
78. H-F Tsai, RG Washburn, YC Chang, KJ Kwon-Chung. *Aspergillus fumigatus arp1* modulates conidial pigmentation and complement deposition. Mol Microbiol 26:175–183, 1997.

79. MH Wheeler, AA Bell. Melanins and their importance in pathogenic fungi. In: RH McGinnis, ed. Current Topics in Medical Mycology. New York: Springer, 1998, pp 338–387.

80. B Jahn, F Boukhallouk, J Lotz, K Langfelder, G Wanner, AA Brakhage. Interaction of human phagocytes with pigmentless Aspergillus conidia. Infect Immun 68: 3736–3739, 2000.

81. B Jahn, A Koch, A Schmidt, G Wanner, H Gehringer, S Bhakdi, AA Brakhage. Isolation and characterization of a pigmentless-conidium mutant of *Aspergillus fumigatus* with altered conidial surface and reduced virulence. Infect Immun 65: 5110–5117, 1997.

82. A Schmidt, MH Wolff. Morphological characteristics of *Aspergillus fumigatus* strains isolated from patient samples. Mycoses 40:347–351, 1997.

83. MH Wheeler, MA Klich. The effects of tricyclazole, pyroquilon, phthalide and related fungicides on the production of conidial wall pigments by *Penicillium* and *Aspergillus* species. Pesticide Biochem Physiol 52:125–136, 1995.

84. LO White, H Smith. Placental locations of *Aspergillus fumigatus* in bovine mycotic abortion: enhancement of spore germination in vitro by foetal tissue extracts. J Med Microbiol 7:27–33, 1974.

85. PF Lehmann, LO White. Rapid germination of *Aspergillus fumigatus* conidia in mouse kidneys and a kidney extract. Sabouraudia 16:203–209, 1978.

86. TTC Ng, GD Robson, DW Denning. Hydrocortisone-enhanced growth of *Aspergillus* spp.: implications for pathogenesis. Microbiology 140:2475–2480, 1994.

87. G Tronchin, JP Bouchara, M Ferron, G Larcher. D Chabasse. Cell surface properties of *Aspergillus fumigatus* conidia: correlation between adherence, agglutination and rearrangements of the cell wall. Can J Microbiol 41:714–721, 1995.

88. MT Larocco, HR Buckley, RJ Mandle. Inhibition of germ tube development of *Aspergillus fumigatus* in cell-free transudate produced in subcutaneous chambers in rabbits. J Med Vet Mycol 25:153–164, 1987.

89. M Momany, I Taylor. Landmarks in the early duplication cycles of *Aspergillus fumigatus* and *Aspergillus nidulans*: polarity, germ tube emergence and septation. Microbiology 146:3279–3284, 2000.

90. MB Kurtz, IB Heath, J Marrinan, S Dreikorn, J Onishi, C Douglas. Morphological effects of lipopeptides against *Aspergillus fumigatus* correlate with activities against $(1,3)$-β-D-glucan synthase. Antimicrob Agents Chemother 38:1480–1489, 1994.

91. A Beauvais, JM Bruneau, PC Mol, MJ Buitrago, R Legrand, J-P Latgé. Glucan synthase complex of *Aspergillus fumigatus*. J Bacteriol 183:2273–2279, 2001.

92. K Esnault, B El Moudni, JP Bouchara, D Chabasse, G Tronchin. Association of a myosin immunoanalogue with cell envelopes of *Aspergillus fumigatus* conidia and its participation in swelling and germination. Infect Immun 67:1238–1244, 1999.

93. SD Harris, JL Morrell, JE Hamer. Identification and characterization of *Aspergillus nidulans* mutants defective in cytokinesis. Genetics 136:517–532, 1994.

94. E Mellado, A Aufauvre-Brown, NA Gow, DW Holden. The *Aspergillus fumigatus* chsC and chsG genes encode class III chitin synthases with different functions. Mol Microbiol 20:667–679, 1996.

95. RD Diamond, R Krzesicki, B Epstein, W Jao. Damage to hyphal forms of fungi by human leukocytes in vitro: a possible host defense mechanism in aspergillosis and murcormycosis. Am J Pathol 91:313–328, 1978.

96. E Roilides, A Holmes, C Blake, D Venzon, PA Pizzo, TJ Walsh. Antifungal activity of elutriated human monocytes against *Aspergillus fumigatus* hyphae: enhancement by granulocyte-macrophage colony-stimulating factor and interferon-gamma. J Infect Dis 170:894–899, 1994.

97. L Christin, DR Wysong, T Meshulam, R Hastey, ER Simons, RD Diamond. Human platelets damage *Aspergillus fumigatus* hyphae and may supplement killing by neutrophils. Infect Immun 66:1181–1189, 1998.

98. LA Hall, DW Denning. Oxygen requirements of *Aspergillus* species. J Med Microbiol 41:311–315, 1994.

99. DW Denning, PN Ward, LE Fenelon, EW Benbow. Lack of vessel wall elastolysis in human invasive pulmonary aspergillosis. Infect Immun 60:5153–5156, 1992.

100. M Birch, G Robson, D Law, DW Denning. Evidence of multiple phospholipase activities of *Aspergillus fumigatus*. Infect Immun 64:751–755, 1996.

101. M Birch, DB Drucker, V Boote, DW Denning. Prevalence of phthioic acid in *Aspergillus* spp. J Med Vet Mycol 35:143–145, 1997.

102. DE Morgenstern, MAC Gifford, LL Li, CM Doerschuk, MC Dinauer. Absence of respiratory burst in X-linked chronic granulomatous disease mice leads to abnormalities in both host defense and inflammatory response to *Aspergillus fumigatus*. J Exp Med 185:207–218, 1997.

103. R Duthie, DW Denning. *Aspergillus* fungemia. Two cases and review. Clin Infect Dis 20:598–605, 1995.

104. DP Kontoyiannis, D Sumoza, J Tarrand, GP Bodey, R Storey, II Raad. Significance of aspergillemia in patients with cancer: a 10-year study. Clin Infect Dis 31:189–191, 2000.

105. C d'Enfert, G Weidner PC Mol, AA Brakhage. Transformation systems of *Aspergillus fumigatus*. Contrib Microbiol 2:149–166, 1999.

106. CM Tang, J Cohen, DW Holden. An *Aspergillus fumigatus* alkaline protease mutant constructed by gene disruption is deficient in extracellular elastase activity. Mol Microbiol 6:1663–1671, 1992.

107. JS Brown, A Aufauvre-Brown, DW Holden Insertional mutagenesis of *Aspergillus fumigatus*. Mol Gen Genet 259:327–335, 1998.

108. G Weidner, C d'Enfert, A Koch, PC Mol, AA Brakhage. Development of a homologous transformation system for the human pathogenic fungus *Aspergillus fumigatus* based on the *pyrG* gene encoding orotidine 5'-monophosphate decarboxylase. Curr Genet 33:378–385, 1998.

109. K Jaton-Ogay, S Paris, M Huerre, M Quadroni, R Falchetto, G Togni, J-P Latgé, M Monod. Cloning and disruption of the gene encoding an extracellular metalloprotease of *Aspergillus fumigatus*. Mol Microbiol 14:917–928, 1994.

110. JM Smith, CM Tang, S Van Noorden, DW Holden. Virulence of *Aspergillus fumigatus* double mutants lacking restriction and an alkaline protease in a low-dose model of invasive pulmonary aspergillosis. Infect Immun 62:5247–5254, 1994.

111. BJ Boeke, J Trueheart, G Natsoulis, GR Fink. 5-Fluoro-orotic acid as a selective agent in yeast molecular genetics. Methods Enzymol 154:164–175, 1987.

112. C d'Enfert. Selection of multiple disruption events in *Aspergillus fumigatus* using the orotidine-5′-decarboxylase gene, *pyrG*, as a unique transformation marker. Curr Genet 30:76–82, 1996.

113. A Aleksenko, JA Clutterbuck. Autonomous plasmid replication in *Aspergillus nidulans*: AMA1 and MATE elements. Fungal Genet Biol 21:373–387, 1997.

114. N Osherov, DP Kontoyiannis, A Romans, GS May. Resistance to itraconazole in *Aspergillus nidulans* and *Aspergillus fumigatus* is conferred by extra copies of the *A. nidulans* P-450 14alpha-demethylase gene, *pdmA*. J Antimicrob Chemother 48: 75–81, 2001.

115. N Osherov, G May. Conidial germination in *Aspergillus nidulans* requires RAS signaling and protein synthesis. Genetics 155:647–656, 2000.

116. V Chalupová. Morphological and colour mutants of *Aspergillus fumigatus*. Acta Univ Palacki Olomuc Fac Med 137:11–14, 1994.

117. PT Borgia, CL Dodge, LE Eagleton, TH Adams. Bidirectional gene transfer between *Aspergillus fumigatus* and *Aspergillus nidulans*. FEMS Microbiol Letts 122: 227–232, 1994.

117a. A Pontecorvo. Non-random distribution of multiple mitotic crossing-over among nuclei of heterozygous diploid *Aspergillus*. Nature 170:204, 1952.

117b. JA Roper. Production of heterozygous diploids in filamentous fungi. Experentia 8:14–15, 1952.

118. JR de Lucas, AI Dominguez, A Mendoza, F Laborda. Use of flow-cytometry to distinguish between haploid and diploid strains of *Aspergillus fumigatus*. Fung Genet Newslett 45:7–9, 1998.

119. A Firon, A Beauvais, J-P Latgé, E Couvé, M-C Grosjean-Cournoyer, C d'Enfert. Characterization of essential genes by parasexual genetics in the human fungal pathogen *Aspergillus fumigatus*: impact of genomic rearrangements associated with electroporation of DNA. Genetics 161:1077–1087, 2002.

120. M Bolker, HU Bohnert, KH Braun, J Gorl, R Kahmann. Tagging pathogenicity genes in *Ustilago maydis* by restriction enzyme-mediated integration (REMI). Mol Gen Genet 248:547–552, 1995.

121. SH Yun, BG Turgeon, OC Yoder. REMI-induced mutants of *Mycosphaerella zea-maydis* lacking the polyketide PM-toxin are deficient in pathogenesis to corn. Physiol Mol Plant Pathol 52:53–66, 1998.

122. CC de Sousa, MH de S Goldman, GH Goldman Tagging of genes involved in multidrug resistance in *Aspergillus nidulans*. Mol Gen Genet 263:702–711, 2000.

123. M Hensel, JE Shea, C Gleeson, MD Jones, E Dalton, DW Holder. Simultaneous identification of bacterial virulence genes by negative selection. Science 269:400–403, 1995.

124. MGL Nicosia, C Brocard-Masson, S Dumais, AH Van, M-J Daboussi, C Scazzocchio. Heterologous transposition in *Aspergillus nidulans*. Mol Microbiol 39:1330–1344, 2001.

125. A Wach, A Brachat, R Pohlmann, P Philippsen. New heterologous modules for classical or PCR-based gene disruptions in *Saccharomyces cerevisiae*. Yeast 10: 1793–1808, 1994.

126. A Wach. PCR-synthesis of marker cassettes with long flanking homology regions for gene disruptions in *S. cerevisiae*. Yeast 12:259–265, 1996.

127. L Hamer, K Adachi, MV Montenegro-Chamorro, MM Tanzer, SK Mahanty, C Lo, RW Tarpey, AR Skalchunes, RW Heiniger, SA Frank, BA Darveaux, DJ Lampe, TM Slater, L Ramamurthy, TM DeZwaan, GH Nelson, JR Shuster, J Woessner, JE Hamer. Gene discovery and gene function assignment in filamentous fungi. Proc Natl Acad Sci USA 98:5110–5115, 2001.

128. MP Wainwright, APJ Trinci, D Moore. Aggregation of spores and biomass of *Phanerochaete chrysosporium* in liquid culture and the effect of anionic polymers on this process. Mycol Res 97:801–806, 1993.

129. BL Cohen. Growth of *Aspergillus nidulans* in a thin liquid layer. J Gen Microbiol 76:277–282, 1973.

130. J Meletiadis, JF Meis, JW Mouton, PE Verweij. Analysis of growth characteristics of filamentous fungi in different nutrient media. J Clin Microbiol 39:478–484, 2001.

131. PF Lehmann, LO White. Chitin assay used to demonstrate renal localization and cortisone-enhanced growth of *Aspergillus fumigatus* mycelium in mice. Infect Immun 12:987–992, 1975.

132. CL Spreadbury, T Krausz, S Pervez, J Cohen. Invasive aspergillosis: clinical and pathological features of a new animal model. J Med Vet Mycol 27:5–15, 1989.

133. L de Repentigny, S Petitbois, M Boushira, E Michaliszyn, S Senechal, N Gendron, S Montplaisir. Acquired immunity in experimental murine aspergillosis is mediated by macrophages. Infect Immun 61:3791–3802, 1993.

134. B Mehrad, RM Strieter, TJ Standiford. Role of TNF-alpha in pulmonary host defense in murine invasive aspergillosis. J Immunol 162:1633–1640, 1999.

135. DW Denning, DA Stevens. Efficacy of cilofungin in a murine model of disseminated aspergillosis. Antimicrob Agents Chemother 35:1329–1333, 1991.

136. JR Graybill, R Bocanegra, LK Najvar, D Loebenberg, MF Luther. Granulocyte colony-stimulating factor and azole antifungal therapy in murine aspergillosis: role of immune suppression. Antimicrob Agents Chemother 10:2467–2473, 1998.

137. DW Denning, L Hall, M Jackson, S Hollis. Efficacy of D0870 compared with those of itraconazole and amphotericin B in two murine models of invasive aspergillosis. Antimicrob Agents Chemother 39:1809–1814, 1995.

138. DM Dixon, A Polak, TJ Walsh. Fungus dose-dependent primary pulmonary aspergillosis in immunosuppressed mice. Infect Immun 57:1452–1456, 1989.

139. J-P Latgé. *Aspergillus fumigatus* and aspergillosis. Clin Microbiol Rev 12:310–350, 1999.

140. CM Tang, JM Smith, HN Arst Jr, DW Holden. Virulence studies of *Aspergillus nidulans* mutants requiring lysine or p-aminobenzoic acid in invasive pulmonary aspergillosis. Infect Immun 62:5255–5260, 1994.

141. M Hensel, HN Arst Jr, A Aufauvre-Brown, DW Holden. The role of the *Aspergillus fumigatus areA* gene in invasive pulmonary aspergillosis. Mol Gen Genet 258:553–557, 1998.

142. E Cenci, A Mencacci, CF d'Ostiani, G Del Sero, P Mosci, C Montagnoli, A Bacci, L Romani. Cytokine- and T helper-dependent lung mucosal immunity in mice with invasive pulmonary aspergillosis. J Infect Dis 178:1750–1760, 1998.

143. E Cenci, P Stefano, K-H Enssle, P Mosci, J-P Latge, L Romani, F Bistoni. Th1 and Th2 cytokines in mice with invasive aspergillosis. Infect Immun 65:564–570, 1997.

144. E Cenci, A Mencacci, G Del Sero, A Bacci, C Montagnoli, CF d'Ostiani, P Mosci, M Bachmann, F Bistoni, M Kopf, L Romani. Interleukin-4 causes susceptibility to invasive pulmonary aspergillosis through suppression of protective type I responses. J Infect Dis 180:1957–1968, 1999.

145. KV Clemons, G Grunig, RA Sobel, LF Mirels, DM Rennick, DA Stevens. Role of IL-10 in invasive aspergillosis: increased resistance of IL-10 gene knockout mice to lethal systemic aspergillosis. Clin Exp Immunol 122:186–191, 2000.

146. VP Kurup, HY Choi, PS Murali, JQ Xia, RL Coffman, JN Fink. Immune responses to *Aspergillus* antigen in IL-4-/- mice and the effect of eosinophil ablation. Allergy 54:420–427, 1999.

147. K Blease, B Mehrad, TJ Standiford, NW Lukacs, J Gosling, L Boring, IF Charo, SL Kunkel, CM Hogaboam. Enhanced pulmonary allergic responses to *Aspergillus* in CCR2-/- mice. J Immunol 165:2603–2611, 2000.

148. K Blease, B Mehrad, TJ Standiford, NW Lukacs, SL Kunkel, SW Chensue, B Lu, CJ Gerard, CM Hogaboam. Airway remodelling is absent in CCR1-/- mice during chronic fungal allergic airway disease. J Immunol 165:1564–1572, 2000.

149. YC Chang, BH Segal, SM Holland, GF Miller, KJ Kwon-Chang. Virulence of catalase-deficient *Aspergillus nidulans* in p47 (phox)-/- mice. Implications for fungal pathogenicity and host defense in chronic granulomatous disease. J Clin Invest 101:1843–1850, 1998.

150. Y Niki, EM Bernard, HJ Schmitt, WP Tong, FF Edwards, D Armstrong. Pharmacokinetics of aerosol amphotericin B in rats. Antimicrob Agents Chemother 34:29–32, 1990.

151. FJ Ruijgrok, AG Vulto, EW Van Etten. Aerosol delivery of amphotericin B desoxycholate (Fungizone) and liposomal amphotericin B (AmBisome): aerosol characteristics and in-vivo amphotericin B deposition in rats. J Pharm Pharmacol 52:619–627, 2000.

152. R Petraitiene, V Petraitis, AH Groll, T Sein, S Piscitelli, M Candelario, A Field-Ridley, N Avila, J Bacher, TJ Walsh. Antifungal activity and pharmacokinetics of posaconazole (SCH 56592) in treatment and prevention of experimental invasive pulmonary aspergillosis: correlation with galactomannan antigenemia. Antimicrob Agents Chemother 45:857–869, 2001.

153. TJ Walsh, K Garrett, E Feurerstein, M Girton, M Allende, J Bacher, A Francesconi, R Schaufele, PA Pizzo. Therapeutic monitoring of experimental invasive pulmonary aspergillosis by ultrafast computerized tomography, a novel, noninvasive method for measuring responses to antifungal therapy. Antimicrob Agents Chemother 39:1065–1069, 1995.

154. FC Odds, M Oris, P Van Dorsselaer, F Van Gerven. Activities of an intravenous formulation of itraconazole in experimental disseminated *Aspergillus, Candida* and *Cryptococcus* infections. Antimicrob Agents Chemother 44:3180–3183, 2000.

155. JL Richard, WM Peden, JM Sacks. Effects of adjuvant-augmented germling vaccines in turkey poults challenged with *Aspergillus fumigatus*. Avian Dis 35:93–99, 1991.

156. MJ Anderson, DW Denning. *Aspergillus fumigates* isolate AF293 (NCPF 7367). www.aspergillus.man.ac.uk, 2001.

157. RA Prade, J Griffith, K Kochut, J Arnold, WE Timberlake. In vitro reconstruction of the *Aspergillus (=Emericella) nidulans* genome. Proc Natl Acad Sci USA 94: 14564–14569, 1997.

158. YG Amaar, MM Moore. Mapping of the nitrate-assimilation gene cluster (*crnA-niiA-niaD*) and characterization of the nitrite reductase gene (*niiA*) in the opportunistic fungal pathogen *Aspergillus fumigatus*. Curr Genet 33:206–215, 1998.

159. LK Arruda, BJ Mann, MD Chapman. Selective expression of a major allergen and cytotoxin, Asp f1, in *Aspergillus fumigatus*. Implications for the immunopathogenesis of *Aspergillus*-related diseases. J Immunol 149:3354–3359, 1992.

160. NP Keller, TM Hohn. Metabolic pathway gene clusters in filamentous fungi. Fungal Genet Biol 21:17–29, 1997.

161. F Trail, N Mahanti, M Rarick, R Mehigh, SH Liang, R Zhou, JE Linz. Physical and transcriptional map of an aflatoxin gene cluster in *Aspergillus parasiticus* and functional disruption of a gene involved early in the aflatoxin pathway. Appl Environ Microbiol 61:2665–2673, 1995.

162. DW Brown, J-H Yu, HS Kelkar, M Fernandes, TC Nesbitt, NP Keller, TH Adams, TJ Leonard. Twenty-five coregulated transcripts define a sterigmatocystin gene cluster in *Aspergillus nidulans*. Proc Natl Acad Sci USA 93:1419–1422, 1996.

163. J Kennedy, K Auclair, SG Kendrew, C Park, JC Vederas, CR Hutchinson. Modulation of polyketide synthase activity by accessory proteins during lovastatin biosynthesis. Science 284:1368–1372, 1999.

164. H Horiuchi, M Takagi. Chitin synthase genes of *Aspergillus* species. Contrib Microbiol 2:193–204, 1999.

165. MV Ramesh, T Sirakova, PE Kolattukudy. Isolation, characterization, and cloning of cDNA and the gene for an elastinolytic serine proteinase from *Aspergillus flavus*. Infect Immun 62:79–85, 1994.

166. TD Sirakova, A Markaryan, PE Kolattukudy. Molecular cloning and sequencing of the cDNA and gene for a novel elastinolytic metalloproteinase from *Aspergillus fumigatus* and its expression in *Escherichia coli*. Infect Immun 62:4208–4218, 1994.

167. CM Hull, RM Raisner, AD Johnson. Evidence for mating of the "asexual" yeast *Candida albicans* in a mammalian host. Science 289:307–310, 2000.

168. BB Magee, PT Magee. Induction of mating in *Candida albicans* by construction of *MTLa* and *MTLα* strains. Science 289:310–313, 2000.

169. KW Tzung, RM Williams, S Scherer, N Federspiel, T Jones, N Hansen, V Bivolarevic, L Huizar, C Komp, R Surzycki, R Tamse, RW Davis, N Agabian. Genomic evidence for a complete sexual cycle in *Candida albicans*. Proc Natl Acad Sci USA 98:3249–3253, 2001.

170. K Yanai, N Kojima, N Takaya, H Horiuchi, A Ohta, M Takagi. Isolation and characterization of two chitin synthase genes from *Aspergillus nidulans*. Biosci Biotechnol Biochem 58:1828–1835, 1994.

171. PT Borgia, N Iartchouk, PJ Riggle, KR Winter, Y Koltin, CE Bulawa. The *chsB* gene of *Aspergillus nidulans* is necessary for normal hyphal growth and development. Fungal Genet Biol 20:193–196, 1996.

# 2

# Developmental Processes in Filamentous Fungi

**Reinhard Fischer**
*Max-Planck-Institute for Terrestrial Microbiology and Philipps-University of Marburg, Marburg, Germany*

**Ursula Kües**
*ETH Zurich, Zurich, Switzerland, and Georg-August-University Göttingen, Göttingen, Germany*

## I. INTRODUCTION

Based on structural features specific to the sexual cycle, four different phyla are distinguished within the fungal kingdom: the Chytridiomycota, Zygomycota, Ascomycota, and Basidiomycota. Fungi without a known sexual cycle are collected as mitosporic fungi (Deuteromycota, "Fungi Imperfecti"). According to hyphal fine structures, DNA analysis, or both, many of the latter organisms relate to ascomycetes and less often to basidiomycetes (Hawksworth et al. 1995). Species proliferating solely or partly in the form of independent single cells are known in all four phyla (Esser 2000), but the majority of fungal isolates are filamentous. Filamentous fungi are coenocytes characterized by hyphae that increase in length by tip growth (Harold 1999; Wessels 1999). Hyphae of chytridiomycetes and zygomycetes enlarge typically without septation, whereas in ascomycetes and basidiomycetes, cellular divisions occur at regular distances under formation of hyphal septa (Webster 1980; Esser 2000). Septa in ascomycetes have a simple porus (Kaminskyj and Hamer 1998; Kaminskyj 2000), and septa in basidiomycetes a barrel-shaped doliporus covered by a membranous body called the parenthosome (Moore 1985). The pores within the septa enable nutrient flow and organelle exchange between neighboring cellular segments within a hypha (Shepherd et al. 1993). Moreover, hyphae of different mycelia might fuse, resulting in a transfer of cellular material between colonies and in an exchange of organelles in particular nuclei (Raudaskowski 1998; Glass et al. 2000), but normally not mitochondria (Röhr et al. 1998).

Hyphal interactions between different individuals together with the coenocytic structure of filamentous fungi bring about cytological and genetic dynamics not found in other eukaryotic kingdoms (Rayner 1991; Ainsworth et al. 1992; Glass et al. 2000; Kües 2002; Kües et al. 2002b). Haploid nuclei in filamentous fungi rarely undergo karyogamy directly upon hyphal fusion (anastomosis) but exist in vegetative proliferating mycelium as individual units next to each other. Mycelia containing only genetic identical haploid nuclei are distinguished from those with mixed populations of haploid nuclei by the terms homokaryons and heterokaryons (Hawksworth et al. 1995). Heterokaryons might disintegrate into homokaryons by nuclear sorting (Kumata et al. 1996) or escape of nuclei through asexual (mitotic) spore formation (Polak 1999; Kitamoto et al. 2000). In addition, heterokaryons might fuse with unrelated homokaryons or heterokaryons. This can lead to the replacement of nuclei through others of a different genetic identity, resulting in the formation of new heterokaryons (Buller 1930; Kües 2000).

Sooner or later, hyphal fusion between different individuals might lead to a fertilization event when the fusing cells belong to a different mating-type. Sexual cycles eventually initiate with karyogamy, the fusion of two haploid nuclei to one diploid nucleus. In filamentous fungi, karyogamy often occurs after a prolonged vegetative growth phase within specialized cells. In many species, karyogamy is directly followed by meiosis serving genomic recombination and the formation of sexual (meiotic) spores. In contrast to meiotic spores, mitotic spore formation is not restricted to a fertile heterokaryotic stage but typically occurs abundantly on the usually sterile homokaryotic mycelia. Mitotic spores might have fertilization function. Moreover, the abundant production of such spores on mycelium of mitosporic fungi indicates that mitotic spores act as major distribution units and in duration to overcome periods of adverse environmental conditions. However, meiotic spores can also exert these functions (Coppin et al. 1997; Kües 2000).

To understand the lifestyle of filamentous fungi, a handful of model species are currently studied by molecular and classic genetic means. Some of these were specifically chosen because of the ease of following sexual and asexual development within the laboratory (Fig. 1, Fig. 2). Others undergo interesting symbiotic or pathogenic interactions. Here we discuss aspects of fungal differentiation, giving insight into common mechanisms in development. However, in details fungi differ considerably in the genetic and cellular routines of differentiation, showing how variable these processes can be in the different phyla and even among organisms in the same phylum. Most of our current genetic and molecular knowledge on differentiation comes from asco- and basidiomycetes; little is available from zygomycetes (see Wöstemeyer et al. 1995; Franken and Requena 2001), and chytridiomycetes still await analysis. Consequently, we will focus on the first two phyla. Nowadays, complete or nearly complete genome sequences

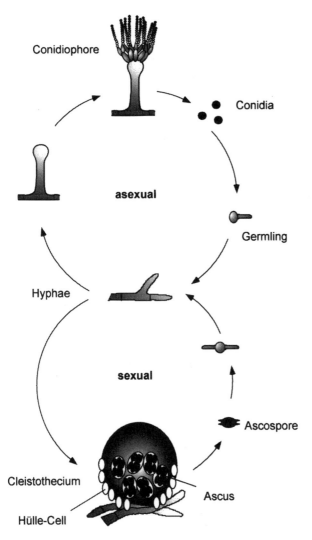

**Figure 1**   Lifecycle of *Aspergillus nidulans*. A conidium germinates and produces the vegetative mycelium. Competent mycelium can undergo the asexual developmental cycle and generate conidiophores or enter the sexual cycle in which cleistothecia are formed. (From Krüger et al., 1997.)

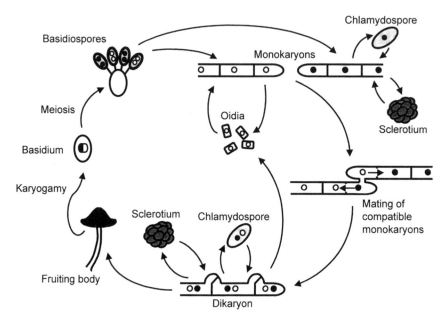

**Figure 2**  Lifecycle of *Coprinus cinereus*. Meiotic basidiospores germinate and produce the primary vegetative mycelia, monokaryons. Upon fusion of two monokaryons compatible in their mating types, a secondary mycelium, the dikaryon, arises with the typical clamp cells at the hyphal septa and two haploid nuclei per cellular segment, one from each monokaryotic parent. On the dikaryon, fruiting bodies develop with basidia in which karyogamy and meiosis occur, leading to the sexual basidiospores. Uninucleate haploid asexual spores, oidia, are constitutively formed on the monokaryon and upon light induction also on the dikaryon. Both types of mycelia may produce the thickwalled mitotic chlamydospores and the multicellular sclerotia. (Modified from Kües, 2000.)

are already at hand from a number of fungal species. The first sequence to be completed was, of course, from the ascomycetous yeast *Saccharomyces cerevisiae* (Goffeau et al. 1996). Other genome sequences are following, and the projects will be far more advanced by the time of appearance of this book. Examples are the sequences from the filamentous ascomycetous molds *Aspergillus nidulans* (http://www.cereon.com), *Neurospora crassa* (http:// www.genome.wi.mit.edu/ annotation/fungi/neurospora/) and *Podospora anserina* (http://cgdc3.igmors.u-psud.fr/Podospora/index_english.htm), the rice blast fungus *Magnaporthe grisea* (http://www.dbs.ucdavis.edu/seminars/index.cfm3519), the human pathogen *Candida albicans* (http://www.sequence.stanford.edu/group/candida/), the fission yeast *Schizosaccharomyces pombe* (http://www.sanger.ac.uk/Projects/ S_pombe/EUseqgrp.shtml), the cotton pathogen *Ashbya gossypii* (Dietrich et al. 1999) and the basidiomycetous human-pathogenic yeast *Cryptococcus neo-*

*formans* (http://www. sequence.stanford.edu/group/C.neoformans/), the maize pathogen *Ustilago maydis* (Kämper et al. 2001), and the filamentous wood-rotting species *Phanerochaete chrysosporium* (http://www.biosino.org/pages/hot5.htm). Several more fungal genome sequences will be available in the near future (Birren et al. 2002; Casselton and Zolan 2002) and will allow genome-wide analyses of transcriptional regulatory circuits. Transcriptional profiling will be extremely valuable to accelerate the understanding of developmental processes, because differential gene expression is one main principle of how morphogenetic changes can be brought about. However, the new methodology should not be overestimated, as more and more examples are known that posttranscriptional regulation and modifications of protein activities are crucial for the regulation and fine-tuning of the various processes. This demonstrates that genome-wide approaches have to be implemented by thorough phenotypic analyses and protein-based approaches.

## II. HYPHAL DEVELOPMENT

Filamentous fungi are typical microorganisms that grow on solid surfaces, such as organic matter in soil. In contrast, only some filamentous species populate aquatic environments, where generally single cell, yeast-like fungi dominate. Hyphae are fairly fragile elongated tubes with a conical shape at the tip, which defines the growth zone. Cell wall material is continuously delivered to the apex and allows the tip to extend with high speeds of several micrometers per minute. A characteristic structure, an accumulation of vesicles close to the hyphal tip of filamentous fungi, is regarded as a vesicle supply center and named Spitzenkörper (Reynaga-Peña et al. 1997). The position of this organelle determines the growth direction of the fungus and is dependent on the microtubule cytoskeleton (Riquelme et al. 2000; Riquelme et al. 1998). The delivery of vesicles to this organelle might be achieved by the action of a conventional kinesin motor protein (Seiler et al. 1997). Most likely, other motors are involved and serve overlapping functions, as deletion mutants are still able to extend their hyphae. There is also good evidence that the actin-myosin system is crucial for hyphal extension. Myosin mutants are not viable and hyphae swell at their tips upon down-regulation of the motor concentration (McGoldrick et al. 1995). This transportation system could be involved in the transportation of the vesicles from the Spitzenkörper to the cell surface. However, the exact arrangement and the contribution of each cytoskeleton are not well understood at the moment.

The fungal cell wall is an essential structure that not only protects cells from environmental stress, including osmotic stress, but also provides the specific cellular morphology reflecting the lifestyle of a fungus. Moreover, cell walls are places of enzymatic activities for nutritional uptake, hyphal-hyphal interactions, and adherence to other living organisms and inert surfaces. The protein and glycan composition of the *S. cerevisiae* cell wall is well known (Kapteyn et al. 1999), and proteomic approaches are now undertaken to study cell wall assembly and remodeling (Pardo et al. 2000). Filamentous fungi in their cell walls use more chitin as the microfibrillar component, and the matrix component consists of 1,3-/1,6-β-glucan, but little is known about the linkage between these polymers (Sietsma et al. 1995). New cell wall deposition requires continuously a weakening of the cell wall mesh and a restoration with new material embedded. This is a risky process, as hyphae would burst because of the high osmotic pressure inside the cells, if the integrity of the cell wall would not be maintained. Recent findings of proteins involved in Worronin-body composition nicely demonstrate this. Worronin bodies are typical organelles of filamentous fungi that are located close to the septa and, in case of damage of a hyphal compartment, in the front seal of the septa to prevent further damage of the organisms through loss of the cytoplasm (Calvo and Agut 2002; Momany et al. 2002). One major component of this structure, Hex1, apparently is also required for normal hyphal growth. If the protein is missing, hyphae swell at their tips and eventually lyse (Jedd and Chua 2000; Tenney et al. 2000).

Hyphal growth is characteristic for filamentous fungi, and differential gene expression is not likely to be crucial for polarized growth once the process is initiated. The questions, what determines the polarity and how do the proteins required for tip extension find the tip, among others, are fundamental (Bartnicki-Garcia 2002). Answers are more likely to be found on the protein level. However, knowledge of the genomic DNA sequence allows specific questions about the roles of genes with established functions in other organisms. This approach was recently undertaken in the group of Philippsen (University of Basel). They studied the role of some *S. cerevisiae* genes—with defined roles in polarity establishment—in the filamentous fungus *A. gossypii*, and found that the homologues are also required for polarized growth (Ayad-Durieux et al. 2000; Wendland and Philippsen 2000; 2001; Alberti-Segui et al. 2001).

## III.  HYPHAL INTERACTIONS

### A.  Mating and Fertilization

Fungal species are classified heterothallic when two different haploid cells or cells of two different sterile, that is, self-incompatible, homokaryons need to fuse to form a fertile zygote or a fertile heterokaryotic mycelium, respectively (Hawksworth et al. 1995). Only a few fungi exert sexual dimorphism with specialized male and female cells found separately on different organisms (Webster 1980; Esser 2000).

In the filamentous ascomycetes, female organs and male cells, either mitotic spores or hyphal tip cells, commonly occur on the same individual. However, in heterothallic species, male cells cannot fuse with female structures (trichogynes) formed on the same strain. Instead, gametes must come from two mycelia having physiological differences manifested by the mating-type genes (Coppin et al. 1997; Hiscock and Kües 1999; Pöggeler 2001). Such sexual fusion (fertilization) initiates development of fruiting bodies in which the hook-shaped ascogenous hyphae with dikaryotic cells are formed, specific heterokaryotic cells with two distinct haploid nuclei, one from each parent. The uppermost hook cell of an ascogenous hypha, separated from the rest of the organism by true cell walls, will differentiate into an ascus mother cell in which karyogamy eventually occurs. Nuclear fusion is directly followed by the two meiotic reduction divisions and, depending on the species, by one or more postmeiotic mitoses before ascospores are endogenously formed (Coppin et al. 1997; Debuchy 1999).

In the basidiomycetes, distinct female and male cells usually do not occur, and sexually undifferentiated cells of identical shape fuse to form a dikaryon. After periods of vegetative proliferation, this secondary mycelium gives rise to specific cells (basidia) in which karyogamy and meiosis takes place (Banuett 1995; Kües 2000) (Fig. 2; Sec. IV.B.2). Also, in heterothallic basidiomycetes, mating-type-determined physiological conditions decide whether dikaryon formation and fertilization happen (Casselton and Olesnicky 1998; Hiscock and Kües 1999).

Mating-type genes are the master regulators of sexual development and, therefore, have been cloned and analyzed already from a number of fungal species. They represent a perfect example how variable organisms can act to achieve a same task. A handful of different regulatory proteins have been found to be encoded by fungal mating-type genes, which in various combinations are used by the diverse species to control sexual development. In ascomycetes, these regulatory proteins (appear to) control mating-type-specific pheromone expression, whereas in basidiomycetes, pheromone and pheromone receptor genes themselves adopted mating-type function (Coppin et al. 1997; Casselton and Olesnicky 1998; Hiscock and Kües 1999; Pöggeler 2001). In heterothallic filamentous ascomycetes, where cloned, pheromone genes are differentially expressed (Zhang et al. 1998; Shen et al. 1999) in contrast to the pheromone receptor genes (Pöggeler and Kück 2001).

## 1. The Mating-Type Loci of Ascomycetes Have Two Idiomorphic Forms

With the one exception of *Glomerella cingulata* that appears to have three (Cisar and TeBeest 1999), heterothallic filamentous ascomycetes do have two distinct mating types (Coppin et al. 1997; Turgeon and Yoder 2000). Mating-type specificity is controlled by a single locus with two different forms that have now been cloned from a number of species belonging to three different classes of ascomycetes (Hiscock and Kües 1999; Turgeon and Yoder 2000). Mating-type loci of filamentous ascomycetes are idiomorphic (Metzenberg and Glass 1990), denoting that the two forms within a species (*Mat1-1* and *Mat1-2*) carry genes of evolutionary unrelated sequence. However, a closer look shows that short stretches of sequence similarities exist between idiomorphs (Turgeon et al. 1993; Pöggeler et al. 1997; Pöggeler and Kück 2000). Such islands of identity within the dissimilar mating-type sequences might have been places of ancient recombination events in heterothallic species from which homothallic species arose carrying sequences homologous to both mating-type idiomorphs within a single genome (Turgeon 1998; Yun et al. 1999). Homothallic species are self compatible and can enter the sexual cycle without the need to mate to another strain (Hawksworth et al. 1995). Phylogenetic analysis supports that within the genus *Cochliobolus*, heterothallism is ancient to homothallism and that the switch from heterothallism to homothallism within ascomycetes occurred repeatedly (Yun et al. 1999; Pöggeler 1999).

The genes of the two mating-type idiomorphs in filamentous ascomycetes encode different sets of transcription factors (reviewed in detail by Coppin et al. 1997; Kronstad and Staben 1997; Hiscock and Kües 1999; Shiu and Glass 2000; Pöggeler 2001). In the simplest case, *Cochliobolus heterostrophus*, the two mating-type idiomorphs carry just one gene each. Gene *MAT1-1-1* encodes a protein with an αl-box defined by homology to the *S. cerevisiae* α1 transcription factor, gene MAT1-2-1, an HMG-box transcription factor related to the Mc protein of *S. pombe*. *MAT1-1-1* and *Mat1-2-1* genes (respectively, *mat A-1* and *mat a-1* in *N. crassa* and *FMR1* and *FPR1* in *P. anserina*) are also present in the idiomorphs of other species. In addition, several species carry a gene for a protein with an amphipathic α helix at the *Mat1-1* locus (*Mat1-1-2* in *Gibberella fujikuroi* and *M. grisea*, *mat A-2* in *N. crassa* and *SMR1* in *P. anserina*) and a gene for another HMG box transcription factor (*MAT1-1-3* in *G. fujikuroi* and *Pyrenopezziza brassica*, *mat A-3* in *N. crassa* and *SMR2* in *P. anserina*). In place of *MAT1-1-3*, *M. grisea* has a gene for an unknown product and *P. brassica* in place of *Mat1-1-2* a gene *MAT1-1-4* for a metallothionein whose relation to mating-type function is not known (Turgeon and Yoder 2000). Not found in mating-type loci of filamentous ascomycetes are genes for homeodomain transcription factors, unlike in ascomycetous yeasts and in the basidiomycetous fungi (Hiscock and Kües 1999; Section III.A.2). In the *MATa* locus, *S. cerevisiae* has one functional homeodomain transcription factor gene (a1), and in the *MATα* locus, another (α2), together with the gene 1 for the α1 box transcription factor (Johnson 1995). *S. pombe* contains in its *mat1-P* idiomorph the genes *mat1-Pc* for an αl box protein and *mat1-Pm* for a homeodomain transcription factor. The *mat1-M* idiomorph has the HMG box transcription factor encoding gene *mat1-Mc* and *mat1-Mm* for an as yet unclassified putative transcription factor (Kelly et al. 1988; Coppin et al. 1997).

As might be expected by such variations within the loci, there are differences between organisms in the functions exerted by the mating-type products in gamete fusion and sexual development. In the haploid *MATα* cells of *S. cerevisiae*, homodimers of protein α2 act as repressors of **a** cell-specific genes and α1 as an activator of α cell-specific genes (together with the general transcription factors MCM1 and STE12). In contrast, **a** cell-specific genes are constitutively expressed in the haploid *MATa* cells. Among the differentially regulated α- and **a**-specific genes are those for mating-type pheromones and pheromone receptors, needed for recognition of respective mates. Once a diploid cell is formed through mating, their expression is turned off by the actions of the α2 homodimers (**a** specific genes) and of the newly formed α2/**a1** heterodimers that repress gene *α1* and all mating-type-independent haploid specific functions (Johnson 1995). In *S. pombe*, the products of *mat1-Pc* and *mat1-Mc* are essential for nutritionally regulated conjugation and control of mating-type-specific pheromone and pheromone receptor expression. In contrast, genes *mat1-Pm* and *mat1-Mm* are activated as a response on pheromone signaling and their protein products are needed together with Pc and Mc for post-fusion events, including meiosis (Kelly et al. 1988; Willer et al. 1995). It is interesting to note that the homeodomain transcription factor Pm of *S. pombe* acts as an activator unlike the homologous protein α2 that always is a repressor in *S. cerevisiae*. Pm and Mc both appear to bind to the promoter of the *mei3* gene needed for entry into meiosis (van Heeckeren et al. 1998). For filamentous ascomycetes, DNA binding so far was only demonstrated for the MAT a-1 protein of *N. crassa* (Philley and Staben 1994), that is, the main regulator of sexual development from Mat a cells. The major regulator of sexual development in *N. crassa* A strains is encoded by the *mat A-1* gene (Ferreira et al. 1998). In *C. heterostrophus*, the two genes *MAT1-1-1* and *MAT1-2-1* from the different idiomorphs are enough to initiate and complete the whole sexual cycle (Turgeon et al. 1993); however, the extra genes *mat A-3* and *mat A-2* in *N. crassa* and *SMR2* and *SMR1* in *P. anserina* have important roles in ascosporogenesis (Zickler et al. 1995; Ferreira et al. 1998; Coppin and Debuchy 2000; Arnaise et al. 2001). FMR1 and SMR2 coming from the *mat-* idiomorph have been shown to interact in the yeast two-hybrid system but not proteins from different mating-type idiomorphs (Coppin et al. 1997). This contrasts a recent finding in which the *S. macrospora* SMTA-1 and SMTa-1 mating-type proteins interacted in the yeast two-hybrid system (Jacobson et al. 2002), those proteins that correspond to Mat A-1 and Mat a-1 of *N. crassa* and FMR1 and FPR1 of *P. anserina*, respectively (Pöggeler et al. 1997). In *S. macrospora*, the genes for these proteins have been brought together onto one chromosome while the species is homothallic (Pöggeler et al. 1997). Notwithstanding the genetic identity, individual nuclei in *S. macrospora* appear to be able to adopt different mating-type preferences corresponding to the two mating-type statuses of the genetically different nuclei in *N. crassa* and *P. anserina* (Pöggeler 2001). In *S. macrospora*, the genes for the two types of mating-type pheromones are both expressed (Pöggeler 2000); however, do the transcripts come from the same or different nuclei? Nuclear identity in *P. anserina* is controlled by the mating-type proteins FPR1 in case of the *mat+* identity and FMR1 and SMR2 together in case of the *mat-* nuclei (Zickler et al. 1995; Arnaise et al. 1997; Coppin and Debuchy 2000; Arnaise et al. 2001). Localized gradients of these proteins around their parental nuclei and, as a direct consequence of these gradients, selective nuclear entrance have been proposed to be the basis of determining nuclear identity, the prerequisite for formation of the dikaryotic ascogeneous hyphae (Debuchy 1999).

## 2. The Multiple Mating Type Loci of Basidiomycetes

Special attention has been drawn to the mating-type loci of basidiomycetes by the fact that there are often more than two mating types within a given species (Whitehouse 1949; Hiscock and Kües 1999). Estimates of different mating types in nature reach, for example, 12,000 for the mushroom *Coprinus cinereus* and 18,000 for the wood-rotting fungus *Schizophyllum commune* (Casselton and Olesnicky 1998; Kothe 1999). In comparison, the number of 50 different mating types in the smut fungus *U. maydis* is rather low (Banuett 1995), but still amazing when thinking about the two in ascomycetes. Classic genetics established that expression of the multiple mating types in basidiomycetes is governed by either one or two different mating-type loci. In meiosis, distribution of their alleles gives rise to spores of two or four distinct mating types, respectively. Species with one mating-type locus are called bipolar, and those with two different mating-type loci, tetrapolar (Raper 1966).

Two mating-type loci control different steps in the developmental process of dikaryon formation. For instance, in *U. maydis*, different alleles at the *a* mating-type locus are essential for fusion of the haploid yeast-like cells, whereas heterozygocity at the *b* locus is required for dikaryotic filament formation and pathogenicity (Banuett 1995). In higher basidiomycetes, including *C. cinereus* and *S. commune*, mating-type genes are not needed for hyphal fusion. This has been observed to occur also between hyphae of identical mating-type specificity (for extensive discussion, see Raudaskoski 1998, Hiscock and Kües 1999). Mating-type genes, however, are essential in post-fusion events. After hyphal fusion, the genes of the *B* mating type locus enable invading a homokaryotic mycelium by nuclei of different mating type. When migrating nuclei reach hyphal tip cells within the foreign mycelium, the genes of the *A* mating-type locus control pairing between two nuclei of different origin, the formation of the dikaryon-typical clamp cells at future places of septum formation (compare Fig. 2), and the synchronized division of the nuclei (Casselton and Olesnicky 1998; Kües 2000). Clamp cell formation serves equal distribution of the two types of nuclei in cellular divisions, as one daughter nucleus of one pair of dividing nuclei migrates into this special cell. Upon septum formation within the hyphal cell and between hyphal and clamp cell, a pair of nuclei of different mating type remains in the apical hyphal cell and another nucleus is trapped in the clamp cell, whereas the second nucleus of the opposite mating type places in the new subapical cell (Iwasa et al. 1998). Because of action of the *B* genes, the nucleus within the clamp cell is released into the subapical cell upon cellular fusion, combining also two nuclei of different mating type in this cell (Casselton and Olesnicky 1998; Kües 2000).

In tetrapolar higher basidiomycetes, the *A* and *B* mating loci may be further divided into two closely linked, functionally redundant subloci that will recombine at low frequency during meiosis (Raper 1966; Kües et al. 2001). Molecular analysis in *C. cinereus* (Kües et al. 1992, 1994a; Pardo et al. 1995) and *S. commune* (Stankis et al. 1992; Specht et al. 1994; Shen et al. 1998; Wendland et al. 1995; Vaillaincourt et al. 1997) revealed that these subloci contain sets of paralogous genes with limited sequence similarity, but are linked by stretches of conserved DNA. When directly linked, as in the case of the *B* genes in *C. cinereus* (O'Shea et al. 1998; Halsall et al. 2000), there will be no meiotic recombination because of the low sequence homologies. Thus, the sets of paralogous *C. cinereus* *B* genes are nonseparable in meiosis (Casselton and Olesnicky 1998). Currently, it is not known whether a close linkage between *A* and *B* mating-type genes buries presence of these very different genetic control functions in dikaryon formation in bipolar higher basidiomycetes. However, a precedent for this is found in the heterobasidiomycete *U. hordei*, where equivalents of the *U. maydis a* and *b* genes are linked by DNA sequences of 430–500 kb. Recombination in this linker-region is repressed and, as a consequence, the two sets of different mating-type genes behave in crosses as one genetic trait (Lee et al. 1999).

Acquiring paralogous mating-type genes that freely recombine(d) is seen in the higher basidiomycetes as one secret in the evolution of multiple mating-type specificities. Each set of paralogous genes can have several allelic forms, characterized by a high proportion of sequence divergence, which is the second secret to increase the numbers of mating-type specificities. Multiplying the numbers of alleles of all paralogous sets of genes for both mating-type loci and the results for the two with each other gives the absolute figure of possible mating types within a given species. In species such as *U. maydis* lacking paralogous mating type genes, the number of absolute mating types just calculates from each one set of allelic forms and is thus comparably low (Kües and Casselton 1993; Casselton and Olesnicky 1998; Hiscock and Kües 1999).

*One Mating-Type Locus Codes for Homeodomain Transcription Factors.* The *A* locus of the studied higher basidiomycetes contains two or more paralogous pairs of divergently transcribed genes, the *b* locus of *Ustilago* species being one such pair. These gene pairs encode two distinct classes of homeodomain transcription factors (Gillissen et al. 1992; Bakkeren and Kronstad 1993; Kües et al. 1992; Stankis et al. 1992; Shen et al. 1996; Kües et al. 2001) related to α2 and **a1** of *S. cerevisiae*, respectively, and referred to as HD1 and HD2 (Kües and Casselton 1992; Hiscock and Kües 1999). As there is nomenclature confusion between different basidiomycetes (eg, the *A* locus of higher basidiomycetes corresponds to the *b* locus of *Ustilago* species), it has been proposed to call all mating-type loci encoding HD1 and HD2 proteins *matT* (Hiscock and Kües 1999).

Transcription factors of the HD1 class have an atypical homeodomain with a less-conserved DNA recognition sequence and variable spacing between the three helices formulating the tertiary structure of this DNA binding domain. In contrast, homeodomains of the HD2 class of proteins are in all aspects classic (Casselton and Kües 1994). HD1 and HD2 products of different mating-type specificity interact to form a transcription factor complex (Banham et al. 1995; Kämper et al. 1995; Magae et al. 1995) that control dikaryon formation and, in smut fungi, pathogenicity (Gillissen et al. 1992; Specht et al. 1992; Pardo et al. 1995). Mutations and deletions within the HD1 motif are tolerated in *C. cinereus* and *S. commune* (but not in *U. maydis*!) unlike within the HD2 motif, suggesting the HD2 class of proteins to mediate the DNA contact (Kües et al. 1994b; Asante-Owusu et al. 1996; Luo et al. 1995; Schlesinger et al. 1997). Binding to specific sequence motifs within mating-type-regulated promoters has so far been demonstrated in *U. maydis* (Romeis et al. 2000; Brachmann et al. 2001). Good candidates for binding a HD1-HD2 transcription factor complex in *C. cinereus* are the promoters of gene *clp1* and of galectin genes shown to be controlled by the *A* mating-type proteins (Boulianne et al. 2000; Inada et al. 2001; Kües et al. 2002a; see Secs. IV.B.2.1, IV.B.3, and IV B.6). A nuclear localization signal present in the *C. cinereus* and *S. commune* HD1 proteins obviously mediates transfer of the protein complex into the nucleus. HD2 proteins have no nuclear localization signal and are, therefore, unable to enter a nucleus in absence of a compatible HD1 protein (Spit et al. 1998; Robertson et al. 2002). Most important for mating-type specificity are the N-terminal regions of both protein classes, as these mediate heterodimer formation between compatible HD1 and HD2 proteins coming from allelic gene pairs (Banham et al. 1995; Kämper et al. 1995; Magae et al. 1995). Heterodimer formation between incompatible protein combinations coming either from the same or from paralogous gene pairs, if occurring, is unspecific at best. In *S. commune*, some unspecific protein interactions have been observed mediated by C-terminal protein regions (Asada et al. 1997). Shown in *C. cinereus*, the N-terminal discrimination function becomes surplus in HD1-HD2 fusion proteins generated either in vivo by mutation or by in vitro techniques. Such HD1-HD2 fusion proteins induce *A*-regulated development without heterodimerization to an unrelated *A* mating-type protein (Kües et al. 1994a; Asante-Owusu et al. 1996; for further details on these interesting regulator proteins, see the reviews by Casselton and Olesnicky 1998; Hiscock and Kües 1999; Kües 2000).

*The Second Mating-Type Locus Encodes Pheromones and Pheromone Receptors.*    The second mating-type locus in basidiomycetes contains sets of genes either encoding for a pheromone receptor or for short peptide pheromones (Bölker et al. 1992; Bakkeren and Kronstad 1994; Wendland et al. 1995; O'Shea et al. 1998; Vaillancourt et al. 1997; Halsall et al. 2000) that bind to receptors of the seven-transmembrane type coming from allelic sets of genes of different mating type (Spellig et al. 1994; Fowler et al. 1999, 2001; Hegner et al. 1999; Olesnicky et al. 1999, 2000). In *Ustilago* species, there are only two alleles of the *a* locus with each one active pheromone gene and one active pheromone receptor gene (Bölker et al. 1992; Spellig et al. 1994; Urban et al. 1996). In contrast, in the single *B* locus of *C. cinereus*, there are three closely linked sets (also called sub families) of each one receptor and two pheromone genes (O'Shea et al. 1998; Halsall et al. 2000), whereas the two *B* subloci in *S. commune* each contain one receptor gene with variable numbers (2 to 8) of pheromone genes (Wendland et al. 1995; Vaillancourt et al. 1997; Fowler et al. 2001; Kothe 2001). Because of the encoded pheromone and receptor functions, these loci are also referred to as *matP* (Hiscock and Kües 1999).

Specificity of the systems is determined by both pheromones and receptors. Analogous to the case of the HD1 and HD2 *A* mating-type proteins, specificity is mediated by discrimination between gene products coming from the same set of genes. Recent mutant gene constructions transformed specificities of receptors as well as specificites of pheromones (Fowler et al. 1999, 2001; Gola et al. 2000; Olesnicky et al. 2000). Pheromone sequences are highly variable between those recognizing the same given heterologous receptor and also between those that come from the same set of genes and do not recognize the respective receptor from their own set of genes, making it very difficult to recognize what is important for receptor interaction (Fowler et al. 1999, 2001; Olsenicky et al. 2000). Within a receptor, different domains seem to contribute to pheromone recognition. However, construction of chimeric receptors so far has not given a clear-cut answer to what makes the difference from one specificity to another (Gola et al. 2000). Elegant heterologous expression studies in yeast suggest pheromones are possibly posttranslationally modified by addition of a lipophilic farnesyl tail at the cystein moiety in the terminal CAAX amino acid motif (A = aliphatic, X = any amino acid) (Fowler et al. 1999; Olesnicky et al. 1999). Heterologous expression further revealed that the basidiomycete pheromone receptors couple to a heterotrimeric G protein in *S. cerevisiae* and, when activated either by a respective pheromone or constitutively through mutation, induce the yeast pheromone response pathway (Fowler et al. 1999; Olesnicky et al. 1999, 2000; further reading in Casselton and Olesnicky 1998; Hiscock and Kües 1999; Kothe 1999, 2001; Brown and Casselton 2001). Knowing this, the hunt is on the downstream modules of the pheromone and related signaling chains in the higher basidiomycetes, whereas in *U. maydis*, several of these elements have already been identified and characterized (Kahmann et al. 1999; Lengeler et al. 2000; Loubradou et al. 2001; Banuett 2002).

## B.  Vegetative Incompatibility

Among strains of the same fungal species, two fundamentally different incompatibility reactions have been observed upon hyphal fusion: sexual incompatibility sometimes called homogenic incompatibility, and vegetative incompatibility or heterogenic incompatibility (Leslie 1993; Esser and Blaich 1994). Sexual incompatibility serves out breeding, because it hinders successful mating between genetically like strains (Hiscock and Kües 1999). Examples for sexual incompatibility are hyphal reactions between colonies of basidiomycetes (Hennebert et al. 1994)—the responsible mating-type genes are discussed in Section III.A.2. In vegetative incompatibility systems, a genetic difference of at least one single gene inhibits the coexistence of nuclei in a common cytoplasm. In some incompatibilty reactions, a "barrage" zone consisting of dead hyphal segments is observed in the contact zone of two strains. As a result of such events, outbreeding might be restricted and evolution of isolated groups within a single species favored (Esser and Blaich 1994; Hiscock and Kües 1999). Vegetative incompatibility has rarely been observed in basidiomycetes (Aimi et al. 2002); in *C. cinereus* it has been connected to mitochondrial differences between dikaryons (May 1988). In ascomycetes, however, vegetative incompatibility reactions are common, and usually, it is determined by several different chromosomal loci called *het*. *A. nidulans*, for example, has at least 8, *N. crassas* 10, and *P. anserina* 9 *het* loci. Vegetative incompatibility systems might be allelic or nonallelic where alternative alleles of a single *het* locus and specific alleles at two different loci, respectively, trigger the incompatibility reaction. Strong intracellular incompatibility reactions on anastomosis include septal plugging, vacuolization of the cytoplasm, organelle degradation, shrinkage of the plasma membrane from the cell wall, and DNA fragmentation—features associated with programmed cell death (recently reviewed by Glass et al. 2000; Saupe 2000).

Several *het* loci have now been cloned from *N. crassa* and *P. anserina* and have been seen to encode different proteins with unrelated cellular roles. It appears that possibly any kind of cellular protein might be captivated into a vegetative incompatibility reaction, and it is an intriguing task to understand how this can happen in so many different ways. In *N. crassa*, one vegetative incompatibility reaction is linked to the mating-type proteins mat A-1 and mat a-1, but to protein domains that are dispensable for mating function. Therefore, nuclei of different mating type cannot coexist in vegetative hyphal cells, although fusion between gametes to enter into the sexual cycle is possible. *het-C* of *N. crassa* is a polymorphic locus with multiple alleles of divergent sequences. The gene product has a signal peptide and glycine-rich repeats and is possibly secreted. The corresponding locus in *P. anserina (hch)* is not polymorphic and obviously does not act in vegetative incompatibility. However, when transformed with an *N. crassa het-C* allele, *P. anserina* shows typical intracellular incompatibility reactions. Unlike the former case, genes *het-6* of *N. crassa* and *het-e* of *P. anserina* appear somehow to be related. Their products have sequence similarities over a 140 amino acid region, whereas other parts are totally different. The HET-6 product possibly interacts with the large subunit of type I ribonucleotide reductase encoded by the essential gene *un-24*, also localized at the *het-6* locus. In *P. anser-*

*ina*, the product of the multiallelic *het-c* gene (only coincidently named like the unrelated *het-C* locus of *N. crassa*) displays similarity to a mammalian glycolipid transfer protein. The gene is nonessential in vegetative proliferation but affects ascospore maturation. Alleles of *het-c* undergo incompatibility reactions with specific *het-e* alleles. Together with Het-D proteins of *P. anserina*, the *het-e* products form a new familiy of WD40 proteins. Most fascinating are the products of the *het-s* locus of *P. anserina*, as these have been shown to have prion-like activity (more details in the reviews of Glass et al. 2000; Saupe 2000; Saupe et al. 2000 and in Coustou-Linares et al. 2001; Espagne et al. 2002; Maddelein et al. 2002; Muirhead et al. 2002).

## IV. SPORE FORMATION

Hyphae are very useful for the exploration for nutrients of the intimate environment, and if nutrients are available extension is indefinite, but most filamentous fungi undergo developmental processes to cope with the changing environmental conditions and nutrient availability. Therefore, many fungi develop reproductive structures arising from the substrate and reaching into the air. The formation of spores allows the distribution of the organism over large distances. Two model organisms will be discussed in detail, the ascomycetous fungus *A. nidulans* and the basidiomycete *C. cinereus*.

## A. Spore Formation in the Ascomycetous Fungus *A. nidulans*

*Aspergillus nidulans* was first described by Eidam in 1883, and the strain, which is used today in most laboratories, was isolated in Glasgow by Pontecorvo (Eidam 1883; Pontecorvo et al. 1953). From the Glasgow laboratory this mold started its successful travels into many laboratories worldwide. *Aspergillus nidulans* grows well on artificial media at 37° C. Developmental processes are easily visible on an agar plate and can be analyzed with the light microscope and, for more details, with the electron microscope (Mims et al. 1988; Oliver 1972). From the genetic point of view, *A. nidulans* is amenable to mutagenesis experiments because the conidiospores contain only one haploid nucleus. However, diploid strains can be generated and propagated through conidiospores (Käfer 1977). This allows the study of essential genes and dominance assays. In contrast to many other molds, *A. nidulans* is able to reproduce in a sexual manner (Fig. 1). It is a homothallic ascomycetous fungus, but the sexual cycle also allows crosses between different strains (Fig. 3). This is especially convenient because *A. nidulans* does not possess any true mating type. Hundreds of different mutants have been generated over the time and subsequently mapped to one of the eight linkage

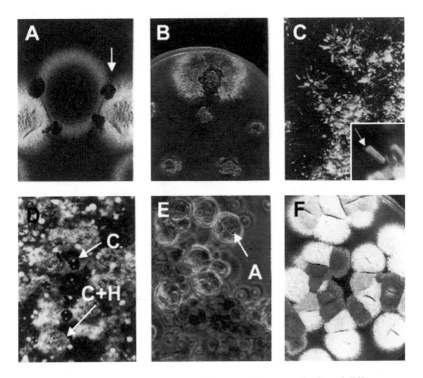

**Figure 3**   Illustration of a cross of *A. nidulans*. (A) Two colonies of different genotypes (different nutritional requirements and one strain green and the other one white [dark and light gray]) were grown on an agar plate until the hyphae made contact. Small pieces of the contact zone were transferred to a minimal agar plate where none of the parent strains can grow individually. (B) One of transferred pieces gave rise to the growth of a hetero-karyon. (C) Conidiophores of the heterokaryon with different colors (gray or white). Inset: Mixed conidiophores with green spore chains (dark gray) and some white chains (*arrow*). (D) Cleistothecia (C) and cleistothecia covered with Hülle cells (C + H). (E) Broken cleistothecia with intact asci (A) and free ascospores. (F) Ascospores were plated on com-plete media. Progeny colonies display different colors (green [dark grey], white and yellow [light grey]). The yellow phenotype is the result of a new combination of the chromosomes during meiosis. The white parent strain harbors a double mutation, white and yellow, and white is epistatic over yellow. (From Sievers et al., 1997.)

groups (http://aspergillus-genomics.org) (Prade 2000). One great advantage of *A. nidulans* as a model for developmental processes is that conidiation can be synchronized (Fig. 4). If conidia are germinated and grown in liquid culture, they do not produce conidiophores until they are exposed to an air interphase. Experimentally one can filter vegetative hyphae and place the filter on top of an

**Filter hyphae**

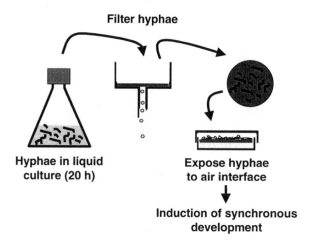

| Hyphae in liquid | Expose hyphae |
|---|---|
| culture (20 h) | to air interface |

**Induction of synchronous
development**

**Figure 4**   Synchronization of development in *A. nidulans*.

agar surface. This will lead to an immediate, synchronous initiation of development and, thus, mycelium of identical developmental stage can be harvested and
analyzed, for example, for gene expression. This synchronization will be very
helpful for genome-wide analyses for differentially expressed genes, because it
allows the generation of stage-specific RNA. For detailed analyses of cellular
processes in *A. nidulans*, it is important that for almost 20 years, it has been
possible to introduce DNA into the mold to construct genetically engineered
strains (Timberlake and Marshall 1989; Yelton et al. 1984). The combination of
different classic and modern molecular biological methods allows an analysis of
the differentiation process at all levels.

## 1.   Asexual Sporulation

Conidiation in *A. nidulans* is initiated from a thick-walled hyphal cell, the foot.
This cell extends into the air and produces a stalk 50 to 70 µm in length before
it swells terminally to form the vesicle (Fig. 5). In a budding-like process, up to
70 metulae are formed on top of the vesicle. These cells in turn generate two to
three phialides, the spore-forming cells. The phialides repeatedly undergo mitosis
and provide the emerging conidia with one nucleus each. Therefore, the youngest
conidium is located close to the sterigmatum and the oldest at the tip of a conidia
chain. Gradually conidia become dark green. The conidiophore, as a relatively
simple, reproductive structure, consists only of four different cell types, the foot
with the stalk, the metulae, the phialides, and the conidia. They are all easily
distinguishable in the light microscope and thus, abnormal morphologies are easy

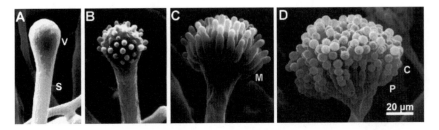

**Figure 5** Conidiophore development as observed in the scanning electron microscope. An aerial hypha (S = stalk) swells terminally to a vesicle (= V) (A), which nearly synchronously form metulae (= M) in a budding-like process (B, C). Metulae produce two to three phialides (= P), which continuously generate conidia (= C) (D). Pictures were taken from Fischer and Timberlake, 1995; Karos and Fischer, 1996.

to detect. Despite the simple organization, many interesting aspects can be considered. How is development intiated and how are environmental signals processed and transduced into cellular actions? What determines the height of the conidiophore stalk or the diameter of the vesicle? How is the switch between polarized hyphal growth and budding-like growth in the conidiophore regulated? What determines the number of metulae, the cell volume of metulae and phialides, or the number of spores? How is the cell cycle coordinated to developmental requirements? All these questions touch basic cell biological problems, and the study of conidiation in *A. nidulans* might help understanding of some of the questions, which are of general importance.

## 2.  Sexual Spore Formation

The process of sexual spore formation occurs in specialized morphological structures, fruiting bodies called cleistothecia (Fig. 6). The developmental process is initiated by the production of a rather profuse clump of hyphae, which originate often at the base of a conidiophore. Some of these hyphae swell terminally to form specialized cells, named Hülle cells. The precise function of these cells is still unknown, but they hide the further development of the cleistothecium. The mature fruiting bodies consist of an outer shell of fused and highly melanized cells and a lumen with asci with eight ascospores each (Benjamin 1955; Zonneveld 1977). The formation of such fruiting bodies can be initiated in homokaryons upon prolonged incubation. To initiate sexual development successfully and complete the differentiation by the production of mature ascospores, it is necessary that sufficient carbon was available during vegetative growth. Glucose is converted into α-1,3 glucan, a polymer, which is deposited onto the cell wall and, on induction of development, metabolized by an α-1,3 glucanase (Zonneveld 1972; Zonneveld 1974). This enzyme is specifically induced during early stages

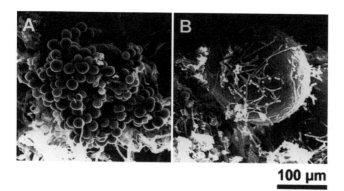

**100 µm**

**Figure 6** Sexual development in *A. nidulans*. (A) Young cleistothecium embedded in a layer of Hülle cells. (B) Mature cleistothecium.

of sexual development and is mainly expressed in Hülle cells (Wei et al. 2001). A second prerequisite for a successful completion of the sexual cycle is the availability of amino acids. In mutants deficient for the synthesis of a certain amino acid, development is not completed even if the corresponding amino acid is supplied in the media (Eckert et al. 1999). Recent findings demonstrate that the regulatory pathways triggering amino acid metabolism are also involved in fruiting body formation (Hoffmann et al. 2000).

## 3. Signals, Signal Transduction, and Decisions

The two developmental cycles, asexual and sexual development, are linked in a way that under some conditions, one or the other pathway is favored and the other is inhibited. This implies that the organism reacts on environmental or internal signals in different ways with regard to the different processes. The first step for asexual development is the transformance of an undifferentiated vegetative cell into a cell committed to enter the differentiation program. Several signals were described that trigger the developmental decisions. One essential condition is the exposure to a water-air interface. One can grow *A. nidulans* in liquid culture, which only allows vegetative growth. Display of the mycelium to the air-exposed surface of a medium induces synchronous induction of asexual development. However, *A. nidulans* requires about 18 hours of vegetative growth until induction of development is possible. This period was defined as the time to acquire developmental competence (Axelrod et al. 1973). Not much is known of what happens to the mycelium during this process, but mutants were isolated with different time length of competence. One gene, the developmental modifier *stuA* is transcriptionally induced when developmental competence is achieved (Miller et al. 1991), indicating a possible role of StuA in this process (see below).

Filamentous fungi are sessile organisms many of which live in the soil. Because organic nutrients are not evenly distributed, fungi need to explore the terrestrial environment to find and colonize new substrates. One important strategy to achieve this is the production of airborne spores. Therefore, it is not surprising that differentiation processes are triggered by the nutritional status of the mycelium. It was long believed that conidiation of *A. nidulans* is programmed into the lifecycle rather than strictly dependent on unfavorable conditions. Skromne et al. (1995) showed that carbon and nitrogen starvation leads to conidiophore development under noninducing conditions, namely in liquid culture. The complexity of the conidiophores was bigger in strains starved for nitrogen than starved for glucose, although the number of conidia produced was higher in glucose-starved mycelium. How starvation induces development is not known yet. However, in other eukaryotes, carbon starvation involves the actions of the ras-pathway and adenylate cyclase (Mösch et al. 1999; Alspaugh et al. 2000). In *A. nidulans* a *ras*-homologue has been isolated, and dominant active alleles were created. It appears that different threshold concentrations of active RAS must exist, which allow development to proceed to certain points (Som and Kolaparthi 1994). In addition, it was found recently that RAS signaling is required during conidia germination (Osherov and May 2000). In addition to RAS, the level of cyclic adenosine monophosphate (cAMP), synthesized by adenylate cyclase, has been shown to trigger developmental processes in *S. pombe* and *N. crassa* (Kays et al. 2000; Kunitomo et al. 2000; Mochizuki and Yamamoto 1992). Whether the cAMP level changes during different developmental stages in *A. nidulans* is not yet clearly shown. Early experiments related the cAMP level to the nutritional status of the mycelium and induction of sexual development (Zonneveld 1976). Recent molecular analyses of adenylate cyclase function suggest several important roles in the lifecycle of *A. nidulans* (Fillinger et al. 2002). Whether nutritional limitation is important for induction of conidiation on agar plates or under natural conditions has not yet been demonstrated.

Besides the exposure to an air-interface and nitrogen and carbon starvation, conidiation is only effectively initiated in cultures that are exposed to light. When *A. nidulans* mycelium is induced for development, it takes 6 hours until the first vesicles appear and another 6 hours for the production of mature conidia. When wild type *A. nidulans* mycelium is induced for development in the dark, aerial hyphae are produced, but they do not differentiate mature conidiophores. However, after exposure to light, development occurs. Mooney and Yager (1990) defined a period of 6 hours after induction in which *A. nidulans* is susceptible to light. Within this critical period, a light pulse of 15 to 30 minutes is sufficient to elicit vesicle formation and, subsequently, conidiation. Determination of the light quality revealed red light of 650 to 700 nm to be very effective. Interestingly, the effect of red light can be reversed by far red light (720 nm), a phenomenon that resembles the phytochrome system of higher plants. Besides the action of red light, in certain *A. nidulans* mutant strains blue light (436 nm) appears to be effective, suggesting that both red and blue light are important triggers for developmental processes in *A. nidulans* (Yager et al. 1998). After asking how developmental competence is achieved, the question arises how the physical parameter of light quality is being detected and transduced to initiate the developmental program. In *A. nidulans*, one gene has been known for many years and is believed to play a central role in light sensing or transduction, the *velvet* (*veA*) gene, located on chromosome VIII (Käfer 1965). Mutation of this gene causes light-independent asexual development. Because asexually derived conidiospores are very useful for many experiments and because *velvet* mutant strains conidiate well in the dark, most of the common laboratory strains harbor this mutation. In contrast to the wide use of the mutation, little is known about the molecular function of the gene. It has been cloned recently, independently in two different laboratories, and. the results were somewhat contradictory. One group claimed that deletion of the gene was lethal (Yager, personal communication); however, it was not in the other group (Chae, personal communication). *veA* encodes a protein with a nuclear localization signal and might act as a repressor of asexual and an inducer of sexual development. Induced expression of *veA* in liquid culture led to the formation of sexual structures (Kim et al. 2002). Some evidence for blue light regulation of conidiation in *A. nidulans* came from comparisons with the related fungus *N. crassa*, in which many different cellular processes depend on light (Linden 2002). The light receptor has likely now been identified (Froehlich et al. 2002; He et al. 2002; Linden 2002) as part of a very interesting regulatory system (Ballario and Macino 1997; Linden et al. 1997). Two transcription factors have been identified, WC-1 and WC-2 (white collar 1 and 2), which contain a PAS dimerization and a LOV domain (Ballario et al. 1998). The latter was shown to be crucial for blue light responses (He et al. 2002). There is evidence that a flavin is the photoactive component, and that this is associated with one of the proteins, WC-1 (Froehlich et al. 2002). This means that the photoreceptor would be part of one of the regulatory proteins (Talora et al. 1999; Linden 2002; Macino, personal communication). The activity of WC-1 might be modified through phosphorylation by protein kinase C (Arpaia et al. 1999). Interestingly, WC-1 and WC-2 are also part of the light-coordinated circadian clock of *N. crassa* (Crosthwaite et al. 1997). In agreement with a general role of WC-1 and WC-2 in filamentous ascomycetes is the recent identification of homologous genes in *A. nidulans* (H. Haas, Innsbruck, personal communication). Vice versa, a gene with very high homology to *A. nidulans velvet* appeared in the *N. crasssa* sequencing project (http://www.mips.biochem.mpg.de/proj/neurspora/). Apart from the white collar system of *N. crassa*, the sensors for red and blue light in fungi are still unknown, as is the signaling cascade transmitting the light to cellular actions. It will be very interesting to analyze the relation between the *velvet* light-sensing system and the blue light response mediated through WC-1 and WC-2. Taken together, sensing of an air-interface and light guarantees that the soil-borne fungus *A. nidulans* undergoes asexual sporulation when it reaches the soil surface. This enables the mold to most efficiently disperse the conidiospores into the air or into free water.

In addition to developmental competence, the nutritional status, and light, asexual development appears to be triggered also by a pheromone system. Although mating in *A. nidulans* is not dependent on a pheromone-based partner selection, a pheromone system was described several years ago (Champe et al. 1987). Ethylacetate extraction of the mycelium of a certain aconidial *A. nidulans* strain allowed the characterization of a compound that had severe effects on development. Applying it to a lawn of growing *A. nidulans* caused a block of asexual development and precocious sexual differentiation. It thus behaved much like a pheromone and was named PSI (precocious sexual induction). The molecule can exist mainly in four different but related forms. Each of them had slightly different agonistic or antagonistic signaling properties (Champe and El-Zayat 1989). The structures were solved as unsaturated C-18 fatty acid derivatives (Mazur et al. 1990; Mazur and Nakanishi 1992). The questions of how PSI factor is recognized by the fungus, whether a membrane-bound receptor is required, how the signaling occurs, and whether a G-protein and a MAP-kinase cascade is involved in signal transduction, as it is in *S. cerevisiae* pheromone signaling (Lengeler et al. 2000), remain to be determined. In *N. crassa*, a homologue of the *S. cerevisiae* Ste11 MAPKK kinase was recognized recently to be involved in the repression of the onset of conidiation (Kothe and Free 1998). Studies of the function of a homologue in *A. nidulans* are also under way. Deletion of the Ste11 homologue, SteC, caused sexual sterility (Wei et al. 2002). The defect was caused by at least two distinct developmental processes, hyphal fusion and cleistothecium development. The requirement of hyphal fusion for sexual development could be circumvented through protoplast fusion, but the heterokaryons still failed to produce cleistothecia. These findings are very interesting, because *A. nidulans* is a homothallic fungus and it raises the question which molecules would be recognized and lead to hyphal fusion and what the role of the MAP kinase cascade in a later stage of cleistothecium development would be. One possibility is that it is involved in internuclear recognition during ascus formation (Debuchy 1999).

A protein, which might be directly or indirectly involved in early signal integration and processing, could be PhoA, a cyclin-dependent kinase, which is a homologue of *S. cerevisiae* Pho85. In yeast, this kinase is involved in the regulation of the cell cycle but serves many additional functions, such as in glycogen metabolism, phosphate acquisition, or morphogenesis (Andrews and Measday 1998; Measday et al. 2000; Measday et al. 1997). In *A. nidulans*, deletion of the gene had an effect on early decisions between asexual and sexual development (Bussink and Osmani 1998). The reaction was dependent on pH, phosphate concentration, and cell density; the latter condition resembles the effects with the PSI factor (see above). Other insights into the early events of conidiation came from the analysis of a class of mutants, named "*fluffy*" (Fig. 7). The current knowledge has been excellently reviewed recently (Adams et al. 1998) and will only briefly be summarized here.

## 4. The Fluffy Genes

One candidate for a signaling molecule came from the analysis of the *fluG* mutant, which produces masses of aerial mycelium but no mature conidiophores. The developmental defect was overcome by growing the strain close to wild-type, suggesting the lack of a signaling molecule to be responsible for the differentiation block. This cross-feeding also worked when the two strains were separated through a dialysis membrane with a 6000 to 8000 Dalton pore size, suggesting that the molecule is a low-molecular-weight, soluble, and diffusible compound whose biochemical nature is still unknown. Gene *fluG* was cloned through com-

**Figure 7**   Isolation of mutants as a tool to study asexual development at the molecular level. Colonies of the wild type (green colony) and several pigmentation or morphological mutants on an agar plate (central panel). The scanning electron microscope pictures show the wild type and different mutant conidiophores. For details see the text. (From Krüger et al., 1997.)

plementation of the recessive mutation (Adams et al. 1992; Yager et al. 1982). Sequence comparisons of the deduced polypeptide identified homology to bacterial glutamine synthetase I (Lee and Adams 1994a). Interestingly, there is a genetic link between the *fluG* and the *veA* gene, as the extracellular rescue of *fluG* mutant strains was shown to be dependent on the *velvet* gene. Furthermore, *fluG* was isolated as an extragenic suppressor of the *velvet* mutation (Yager et al. 1998). Although FluG appears to be involved in the regulation of a specific step in development, the expression is only slightly up-regulated upon induction of development. The protein is abundant in the cytoplasm of hyphae grown in liquid culture as well as in developing conidiophores. However, the protein itself negatively regulates its expression (Lee and Adams 1994a).

Like the *fluG* mutant, all *fluffy* mutants are characterized by the proliferation of undifferentiated aerial hyphae, which give the colony a white, cottonlike or *fluffy* appearance. However, different types can be distinguished. Whereas some mutants overcome the defect upon longer incubation, other strains undergo lysis of the hyphae (Fig. 8). Mutants were generated over the years in numerous intelligent genetic screening approaches. Early screening methods based on the mutagenesis of a wild-type strain and visual inspection of recovered strains for a *fluffy*-like phenotype (Wieser et al. 1994; Yager et al. 1982). The mutations were recessive and allowed epistasis analyses among them and determination of the induction of the central regulatory cascade through the activation of the *bristle* transcription factor (Br1A) (see below). In another approach, genes were classified according to their potential to induce *brlA* (measured as a *brlA(p)::lacZ* gene fusion) and named *flb* (= *fluffy* low *brlA* expression), *fmb* (= *fluffy* moderate *brlA* expression) and *fhb* (= *fluffy* high *brlA* expression). One interesting mutant obtained in this screen was *flbA*, whose hyphae autolyse as the colonies mature (Lee and Adams 1994b). The gene *flbA* encodes a protein with similarity to *S. cerevisiae* SST2 (Lee and Adams 1994b), a protein involved in the pheromone response pathway (Dietzel and Kurjan 1987). The proteins share a 120 amino acids long domain, RGS (regulator of G-protein signaling), defining a large protein family of G-protein interacting regulators (Druey et al. 1996). Forced expression of F1bA induced expression of BrlA and the formation of simple conidiophores in liquid culture, similar to the structures obtained through overexpression of BrlA in submersed culture (see Sec. IV.A.5). These results suggest an important regulatory role during early stages of induction. Novel insights into the function of F1bA came from a mutant screen for dominant mutations. A diploid *A. nidulans* wild-type strain was mutagenized, and fluffy autolytic strains have been isolated. The mutants were named *fad* (= *fluffy* autolytic dominant), and *fadA* has been characterized in detail. The FadA protein shares high homology to α-subunits of heterotrimeric G proteins. The dominant active mutant allele caused proliferation and inhibited conidiation. Because the gain-of-function mutation of *fadA* led to a similar phenotype than a loss-of-function mutation of *flbA*, Adams

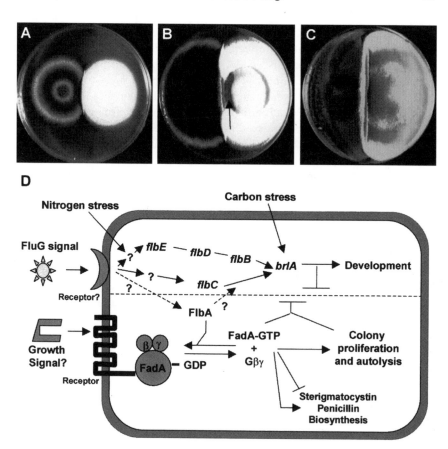

**Figure 8** *Fluffy* mutants are characterized by a cotton-like appearance on agar plates. (A) A wild type (left colony) and a *fluffy* mutant (right colony) were inoculated on an agar plate and incubated for 3 days at 37° C. (B) One class of *fluffy* mutants undergoes lysis of the hyphae after prolonged incubation. After 5 to 6 days, lysis is obvious in the middle of the colony (*arrow*) and after (C) eight days, almost all mycelium disappeared. (D) Scheme showing the regulatory interactions of some of the *fluffy* genes. Two antagonistic signaling pathways appear to regulate *A. nidulans* growth and development. Growth signaling is mediated by the FadA G protein α-subunit. Activation of FadA by exchange of GDP for GTP results in a proliferative phenotype and blocks sporulation. To induce conidiation, FlbA has to be activated, which in turn deactivates the growth signaling and favors the differentiation pathway. Scheme from Adams et al., 1998.

and coworkers suggested that the role of F1bA is in controlling growth and activating sporulation by negatively affecting FadA signaling (Yu et al. 1996). Furthermore, a suppressor analysis of *flbA* identified *sfaB*, which encodes the β-subunit of a heterotrimeric G-protein (Rosén et al. 1999). Taken together, these experiments show that G-protein signaling is of crucial importance for the switch between vegetative growth and initiation of development. This raises the yet-unanswered question to the signal feeding into this cascade. Because other heterotrimeric G-proteins interact with seven-transmembrane receptor molecules, it is likely that such a receptor type for signal perception exists also in *A. nidulans* (Bölker 1998). Interestingly, the same G-protein signaling pathway involved in morphogenetic regulations acts also in the coupling of the production of secondary metabolites, such as sterigmatocystin and penicillin, to the developmental processes (Hicks et al. 1997; Tag et al. 2000; Calvo et al. 2002).

Another approach for isolation of genes responsible for activating the central regulatory pathway took advantage of the vegetative growth inhibition after activation of the pathway. Hence, genes upstream of *brlA*, which lead to its activation, should result in a block of hyphal growth. Therefore, Marhoul and Adams (1995) constructed a DNA library under the control of the carbon source-regulated *alcA* promoter. Transformants could be isolated that do not form colonies under inducing conditions. They were named *fig* (= forced expression inhibition of growth) or *fab* (forced expression activation of *brlA*). One of them, *figA*, was analyzed and the corresponding protein displayed homology to Bni1 form *S. cerevisiae*, a protein required for polarized growth (Evangelista et al. 1997). One member of the *fab* class of mutants, *fabM*, encodes a poly(A)-binding protein that is essential for viability. Because it activates development when overexpressed, it defines a class of genes that are required for vegetative functions but are, in addition, necessary for certain development-specific steps (Marhoul and Adams 1996). Other *flb* genes, such as *flbD*, encode nucleic acid-binding proteins, whose function could be directly in activating downstream developmental genes such as *brlA* (Wieser and Adams 1995). However, target sequences have not yet been determined and the exact roles of F1bD and the other potential regulators (F1bB and F1bC) in conidiation remains to be determined.

## 5.  Bristle and Abacus are Development-Specific Transcription Factors

Among the early mutants of *A. nidulans*, isolated by J. Clutterbuck, were strains with defects in the genes *brl*, *aba* and *wet*, which were later shown to be central regulators of conidiophore development (Clutterbuck 1969; Martinelli and Clutterbuck 1971) (Fig. 7). *brlA* mutants initiate conidiophore formation, but after elongation of the stalk, swelling of the stalk does not occur and the mutants fail to elaborate the next cell generation, the metulae (Fig. 9). The resulting elongated bristle-like structures gave the mutants their name, *brlA*. The severeness of the

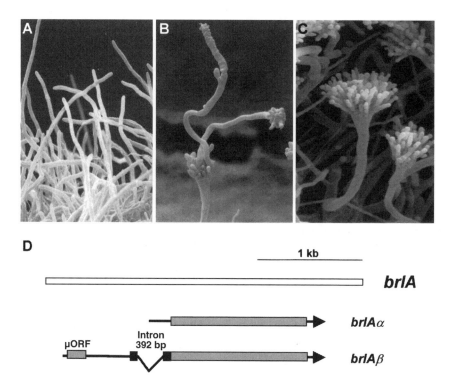

**Figure 9**  The *brlA* mutation. The phenotype of *brlA* mutants depends on the molecular defect. In complete deletion mutants, only elongated stalks are formed (A), whereas *brlAβ* mutants secondary conidiophores may arise from the vesicles of aberrant primary conidiophores (B). In *brlAα* mutants, development proceeds further but conidia are not produced (C). (D) Scheme of the *brlA* locus. The transcripts are indicated with an *arrow* and the open reading frames are shown by *gray boxes*. The N-terminal extension of the BrlAβ protein, disrupted by the intron, is drawn with a black box. Modified from Prade and Timberlake, 1993.

phenotype is dependent on the corresponding mutant allele, suggesting a rather complex interaction of this gene with the developmental program (Griffith et al. 1999). Cloning of the corresponding gene, by complementation of the recessive mutation, proved that *brlA* encodes a regulatory protein, a transcriptional activator with a typical TFIII Zn-finger DNA binding domain (Adams et al. 1988; Adams et al. 1990; Boylan et al. 1987; Johnstone et al. 1985). The Br1A protein has not yet been purified and tested for specific DNA binding in vitro; however, an in vivo assay in *S. cerevisiae* has been established (Chang and Timberlake 1992). This assay demonstrated that BrlA is sufficient for specific gene activation. Furthermore, the system has been used to define consensus target sequences 5′-

(C/A)(G/A)AGGG(G/A)-3′ in the promoter of *brlA*-dependent genes. Detailed analysis of the *brlA* locus revealed that the locus consists of two overlapping transcription units, α and β. The initiation of the α-transcript is located in a region, which is spliced out in the β-transcript. The transcripts are 2.1 and 2.5 kb in size, respectively. The β-transcript initiates about 850 bp upstream of the initiation site of the α-transcript and is characterized by an unusually large intron of 392 bp. The two corresponding proteins are mostly identical but the one derived from the β-transcript has a 23 amino acid long extension at the N-terminus. The two proteins are almost identical, but each appears to fulfill specific functions during development. Deletion of either causes abnormal conidiophore morphology, which was different from the phenotype of *brlA* null mutants. Both individual mutants developed further than the null mutant and formed metulae and abnormal sterigmata or secondary stalks and vesicles (Fig. 9). Interestingly, the α- and β-transcripts could substitute for each other when overexpressed. This means the two transcripts serve specific but overlapping functions. Besides the complex organization of the gene locus, the regulation of expression appears to be complex and fine-tuned. The promoter region is, with 2.9 kb, very long. In addition to other development-specific genes, BrlA activates its own expression, namely the α-transcript. In contrast, the β-transcript is present at any time, in hyphae and during development. However, translation of the β-transcript is suppressed in hyphae and induced upon induction of conidiophore formation. The regulation of translation occurs through a 41 amino acid small open reading frame in the 5′-leader of the β-transcript (Han et al. 1993). Removal of the initiation codon of the μORF results in inappropriate induction of development, suggesting again very distinct roles of the proteins in cellular processes. The importance of BrlA for asexual development is not only demonstrated by mutagenesis experiments. When BrlA is overexpressed in submerged culture where conidiogenesis normally is suppressed (see above), vegetative growth ceases and conidia differentiate directly from hyphal tips (Adams et al. 1988).

One of the target genes of BrlA is another transcriptional activator encoding gene, *abaA*. When *abaA* is mutated, conidiophores resemble a mechanical calculator—an abacus—thus, the name *abaA* (Fig. 7). Genetic evidence already suggested *abaA* to be a major regulator (Clutterbuck 1969), which was proven through many molecular biological experiments with the proof of an interaction with a sequence-specific DNA-motif (5′-CATTCY-3′) (Andrianopoulos and Timberlake 1994). The protein harbors an ATTS/TEA DNA-binding motif, which is also present in a number of other transcription factors, such as the human TEF-1 or the *S. cerevisiae* Ty1 regulator TEC1 (Laloux et al. 1990; Xiao et al. 1991). Direct targets for AbaA are *brlAα*, *abaA* itself, *wetA*, and several structural genes, such as *yA*, *wA*, or *rodA*. All target genes share multiple elements with the consensus sequence 5′-CATTCY-3′ in their promoter. Overexpression of *abaA* in liquid cultures causes cessation of vegetative growth and extensive vacuolization of hyphal cells, but no conidial differentiation (Mirabito et al. 1989). Thus, earlier genes are required throughout development to ensure the correct temporal and spatial expression of later genes.

A third gene placed into the central transcriptional cascade together with *brlA* and *abaA* is *wetA*. As for *brlA* and *abaA*, mutation of the gene causes a well-defined developmental phenotype. Conidiospores are generated, as in the wild-type, but they lyze during the final stages of differentiation. This leaves a droplet in the conidiophore head and makes them look wet-white (Boylan et al. 1987). Ultrastructural analyses revealed that the cell walls of wetA mutant strains are different from wild-type and, thus, one function of WetA is the modification of the wall to gain the stability of mature conidia (Sewall et al. 1990). *wetA* Is expressed in mature conidia, in contrast to *brlA* and *abaA*, whose transcripts are only detectable in earlier developmental stages, but not in conidia (Boylan et al. 1987). WetA reinforces its own expression. Forced expression of the gene in hyphae leads to highly branched cells, again pointing to a possible effect on the remodeling of cell walls. Expression also caused the activation of several spore-specific genes as well as *wA*, whose mRNA does not occur in spores but rather in phialides. Yet, it is not clear how WetA achieves expression of development-specific genes. The WetA protein does not contain any motifs or homologues, which would allow assigning a direct DNA interaction and direct gene activation activity (Marshall and Timberlake 1991). However, homologous proteins have been found in *Penicillium chrysogenum* and, recently, in the sequencing project in *N. crassa* (Prade and Timberlake 1994 [see above]).

One interesting aspect of the central regulatory pathway is its reinforcement once development is induced. The three regulators BrlA, AbaA, and WetA, act in a positive feedback loop and thus guarantee high and quick expression required for efficient downstream gene activation (Fig. 10). There is evidence for a novel spore-specific regulator, which came from the analysis of a gene expressed in mature conidia (Stephens et al. 1999). The gene *spoC1-C1C* is a member of a cluster of 14 genes, spanning 38 kb (Gwynne et al. 1984; Timberlake and Barnard 1981). All genes are coordinately regulated, in part by a regional, position-dependent regulatory mechanism that represses expression in undifferentiated hyphae and may involve developmentally altered changes in chromatin conformation within the *spoC1* cluster (Miller et al. 1987). The biochemical function of the proteins encoded by the genes is not yet known, because deletion of the entire cluster had no discernible phenotype (Aramayo et al. 1989). Although the developmental upregulation of transcription is dependent on functional BrlA and AbaA proteins, the 5′-region of the *spoC1-C1* gene lacks response elements for either transcriptional regulator, suggesting a novel regulator downstream of the two central regulators.

Besides the transcriptional cascade of *brlA*, *abaA*, and *wetA*, several other genes have been identified that are targets of the regulators. Examples are the spore specific genes, *wA* and *yA*, which encode a polyketide synthase and a laccase, respectively, and which are both involved in synthesis of the green pigment

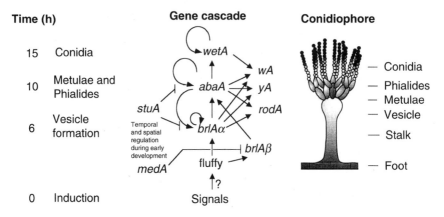

**Figure 10** Regulatory circuits of asexual reproduction. Modified from Andrianopoulos and Timberlake, 1994; Busby et al., 1996.

in the conidiospores (Aramayo and Timberlake 1990, 1993; Machida et al. 1994; Mayorga and Timberlake 1990, 1992). Other examples are two genes involved in conidiophore pigmentation, *ivoA* and *ivoB*. They encode proteins required for the synthesis of a melanin-like molecule, which colors the conidiophore brown (Birse and Clutterbuck 1990, 1991; Clutterbuck 1990). In the wild type, this color is hidden because of the highly pigmented conidia. *rodA* and *dewA*, which encode highly hydrophobic, cystein-rich secreted proteins (hydrophobins) and which are involved in spore wall formation, are other cases of regulation by the cascade (Stringer et al. 1991; Stringer and Timberlake 1994).

## 6.  The *Stunted* and *Medusa* Genes

With the transcriptional cascade of the *fluffy* genes, the *brl*, *aba*, and *wet* genes, a first concept has been established to explain the differentiation process. Some of the open questions might be explained by the genes stunted (*stuA*) and medusa (*medA*). Mutants for both genes were isolated as strains with an aberrant conidiophore morphology, but are still able to produce some viable conidiospores (Clutterbuck 1969). *Stunted* conidiophores are shorter than wild type and fail to produce metulae or phialides but, instead, generate viable conidia directly from the vesicle. Already, early expression studies showed that in *stuA* mutants, expression of several other developmental genes is altered (Zimmermann 1986). The *stuA* transcript appears already in hyphae and is significantly induced at the time where *A. nidulans* acquires developmental competence (Miller et al. 1991). During conidiation, expression is restricted to the periphery of the conidiophore vesicle, metulae and phialides (Miller et al. 1992). The *stuA* locus produces two overlapping transcripts, *stuAα* and *stuAβ*, from different promoters and is similarly complex as the *brlA* locus. *stuAα* and *stuAβ* are characterized by three common introns. Another relatively large intron of 497 bp is present in the 5′ nontranslated region of *stuAβ* (Miller et al. 1992). The complex regulation of StuA expression and its modulation of expression of *brlA* and *abaA* was nicely shown with the help of reporter constructs. StaAα and StuAβ are expressed in hyphae after acquisition of developmental competence, but *stuAα* then is further induced transcriptionally through BrlA. Both the *stuAα* and the *stuAβ* transcripts contain a long (more than 1 kb) nontranslated leader. The encoded StuA proteins are identical, but several small open reading frames in these nontranslated leaders suggest translational control—at least *stuA* translation is stimulated through a micro open reading frame located in the 5′ nontranslated leader of the *stuAα* transcript (Wu and Miller 1997). Expression of StuA is also feedback regulated (Miller 1992). The StuA proteins contain a bipartite nuclear localization signal, whose functionality was shown by fusion to GFP (Suelmann et al. 1997), and an APSES domain, which is a DNA-binding basic helix-loop-helix structure (Dutton et al. 1997). Taken together, these features, plus the developmental phenotype of *stuA* mutants, suggest StuA to be a transcriptional regulatory protein. StuA binds to MCB-like boxes in the promoters of target genes (StuA response elements) and is able to activate transcription from MCB elements in yeast. Despite the activating capacity of StuA in yeast, in *A. nidulans* it represses transcription of the *abaA* gene. However, in sexual development, StuA induces the expression of a catalase-peroxidase, showing that this transcription factor has activating and repressing properties (Scherer et al. 2002) (see Section IV.A.7). StuA response elements are found in the promoters of developmental genes, such as *brlAa* or *abaA*, but also in the promoters of genes involved in cell-cycle regulation, such as *nimE* or *nimO* (Dutton et al. 1997). In summary, StuA expression leads to a restriction of *brlA* expression to the periphery of the conidiophore vesicle, metulae, phialides, and immature conidia and *abaA* expression to metulae, phialides, and immature conidia (Aguirre 1993; Miller et al. 1992).

Conidiophores of *medusa* mutants are characterized by a delay of differentiation of phialides and conidia, which leads to branching chains of sterigmata. Frequently, secondary conidiophores are also produced. A detailed molecular analysis has still not been performed, but genetic interactions between *medA*, *brlA*, and *abaA* were studied (Busby et al. 1996). The *medA* gene is induced upon induction of development of the conidiophore and regulates the expression of *brlA* and *abaA*. MedA represses premature expression of both *brlA* transcripts during early development and down-regulates *brlAβ* during later stages. On the other hand, it is required for sufficient *abaA* expression in phialides. Interestingly, the *medA* defect can be suppressed by extra copies of *brlA*. This and other results led Busby et al. (1996) to suggest that MedA and BrlA might form a heterodimeric protein complex and, together, regulate gene activities. Both modifiers, StuA and MedA, are also required for sexual development, which is in contrast to other regulators of the central cascade (Clutterbuck 1969).

Recently, a novel regulator of the leucine zipper protein family was discovered by complementation of aconidial *dopA* mutants described earlier (Axelrod et al. 1973; Yager et al. 1982). The protein is conserved from yeast to man and was shown to be essential for viability in *S. cerevisiae*. In *A. nidulans*, *dopA* mutant strains display a somewhat pleiotropic phenotype and produce 2.5-fold less conidiophores in comparison to wild type. Furthermore, conidiophore morphology is rather rudimentary. Deletion of the gene also affects the sexual cycle (Pascon and Miller 2000) and in this point resembles the *stuA* function. The exact interaction of DopA with other regulators is not yet completely understood.

## 7. Molecular Analysis of Sexual Development

Whereas asexual development has been studied intensely over the years, the knowledge of genes and their function in the regulation of sexual development is rather limited. This might have two main reasons. The cycle requires several days to be completed in contrast to 24 hours for asexual development. This makes it more difficult to use a mutant isolation procedure, as strains have to be individually studied after mutagenesis. In contrast, for asexual mutants, hundreds of strains can be grown on one petri dish and then visually inspected for aberrant development. Nevertheless, a recent mutagenesis approach was successful, and the molecular analysis of this mutant led to the isolation of a transcriptional activator, NsdD (= never sexual development) (Chae et al. 1995). Deletion of the gene prohibited the initiation of sexual development, and overexpression induces the formation of sexual structures even in submerged cultures (Han et al. 2001).

Taking advantage of the knowledge of signaling cascades in other organisms, a homeodomain protein with similarity to Ste12 of *S. cerevisiae* was isolated from *A. nidulans*. In *S. cerevisiae*, Ste12 is involved in the mating type-controlled pheromone signaling pathway (Lengeler et al. 2000). The corresponding gene of *A. nidulans*, *steA*, is required for sexual development, although in *A. nidulans*, no such pheromone system has yet been described (Vallim et al. 2000).

The switch from vegetative growth to sexual development is regulated by environmental factors as well as through the nutritional status of the mycelium. Starvation for carbon and nitrogen favors the induction of the differentiation. Because formation of fruitbodies requires massive cell proliferation, energy for these processes has to be provided from sources other than glucose. It was shown that $\alpha$-1,3 glucan serves as the main reserve material and is accumulated during vegetative growth as a layer of the cell wall (Zonneveld 1974). After depletion of external glucose, $\alpha$-1,3 glucanase is secreted to release monosaccharides. A gene encoding an $\alpha$-1,3 glucanase was recently detected in a differential library and analyzed in detail (Wei et al. 2001). The expression of the gene was specifically observed in Hülle cells. Surprisingly, deletion of the gene did not prohibit the initiation or completion of the developmental program, suggesting that other related enzymes might substitute for the function. Indeed, several cell-wall hydrolizing enzymes are expressed before sexual development. However, as the localization studies with α-1,3 glucanase show nicely, the expression of those lytic and, thus, "dangerous" enzymes is highly spatially regulated (Fig. 11) (Prade et al. 2001). The α-1,3 glucanase gene provides a tool to study the dependence of the temporal and spatial expression on different development regulators, such as NsdD or SteA, or the isolation of novel regulators.

In another approach to isolate differentially expressed genes and use them to identify regulators, a catalase-peroxidase enzyme was characterized (Scherer et al. 2002). The enzyme copurified with a laccase, which is specifically expressed in Hülle cells (Hermann et al. 1983; Scherer and Fischer 1998). Catalase-peroxidase is not only expressed during sexual development but also upon carbon starvation. Together with three other catalases, two of which are differentially expressed during asexual development (Kawasaki and Aguirre 2001), the results demonstrate that both morphological changes and metabolic adaptations play important roles during development. A further example of how nutritional regulatory circuits are interwoven with development is the amino acid metabolism. Amino acid limitation causes a major change in amino acid metabolism, such that many biosynthetic pathways are turned on. The phenomenon is well studied in *S. cerevisiae* and named cross-pathway control. Amino acid starvation in *A. nidulans* results in similar metabolic responses and in a distinct developmental phenotype in which sexual development is initiated but not completed. The mor-

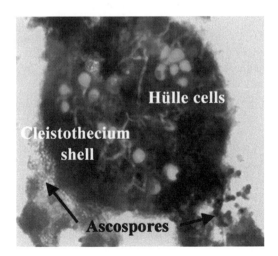

**Figure 11**  Subcellular localization of α-1,3-glucanase. The promoter of its gene was fused to sGFP and expressed in *A. nidulans* wild type. Hülle cells are brightly green fluorescent (white cells, attached to the cleisotothecium shell). From Wei et al., 2001.

phogenetic differentiation process is blocked before completion of meiosis and results in microcleistothecia filled with hyphae. It appears that the regulatory control system underlying the cross-pathway control system in *S. cerevisiae* is conserved in *A. nidulans* and is involved in the regulation of sexual development (Eckert et al. 1999; Busch et al. 2001; Hoffmann et al. 2000, 2001a,b; Strittmatter et al. 2001; Valerius et al. 2001).

Genome-wide analyses have not been performed yet with regard to the sexual cycle. However, EST sequencing identified several differentially expressed genes (Jeong et al. 2001; Lee et al. 1996; Prade et al. 2001). The information collected from the analyses of differential libraries will be implemented greatly now with the genomic DNA sequence available to the fungal community. This should allow study of the expression of hundreds of genes using cDNA arrays (Wei et al. 2001).

## B. Spore Formation in the Basidiomycetous Fungus *Coprinus cinereus*

Intracellular cytological events and signaling cascades acting in development and pathogenicity in basidiomycetes are currently best elucidated in the smut fungus *U. maydis* and the human-pathogenic yeast *C. neoformans* (for recent reviews, see Banuett 2002; Lengeler et al. 2000). Less is known in higher filamentous basidiomycetes, but sexual and asexual reproductive cycles and their environmental controls in relation are easy to follow, as they are not occuring in host organisms. The majority of edible and medicinal fungi as well as lignin-degrading and ectomycorrhizal fungi belong to the phyla of basidiomycetes (Kües and Liu 2000; Whiteford and Thurston 2000; Duplessis et al. 2001; Kothe 2001; Kües et al. 2002a). Despite their economical and ecological importance, most of the species are not (yet) accessible to classic and molecular genetic techniques, restricting such studies on fungal differentiation to a few model species, such as the homobasidiomycetes *C. cinereus* and *S. commune* (Walser et al. 2001). The developmental potential of these two species includes fruiting body development within the sexual cycle and pathways of asexual development, formation of chlamydospores in both species and, in *C. cinereus*, formation of oidia and sclerotia (Wessels 1993a; Moore 1998; Kües 2000; Kües et al. 2002a,b). *C. cinereus*, furthermore, is very special, as it is the only known organism with a naturally synchroneous meiosis. Meiosis occurs simultaneously within the $10^7$ to $10^8$ basidia present within a single fruiting body (Lu 1967), making the fungus an attractive object to study meiosis and sexual recombination (Zolan et al. 1995).

### 1. Asexual Sporulation

*Chlamydospores.*   Although mostly neglected, many higher basidiomycetes produce one or more types of mitotic spores in the homo-/monokaryotic and/or dikaryotic stage (Kendrick and Watling 1979; Kües et al. 2002b). Irregularly shaped, large chlamydospores develop in aging cultures within the sub-merged mycelium (Fig. 12). Chlamydospores in the strictest sense generate endogenously within hyphal cells after cytoplasmic condensation. These spores have a thick double-layered cell wall, one of which comes from the former hyphal cell. Others, termed blastocystes, bud off from hyphal cells that evacuate its condensed cytoplasm into the bud (Clémençon 1997; Kües et al. 2002b). In *C. cinereus*, chlamydospores occur both on the mono- and the dikaryon. However, the activated A mating-type pathway and the activated *B* mating-type pathway positively influence chlamydospore production in the dikaryon, whereas blue light has a negative effect (Kües et al. 1998a, 2002c). Chlamydospores enable the fungi to overcome adverse environmental conditions (Clémençon 1997).

*Oidia.* Oidia are small, rod-shaped, unicellular, uninuclear, haploid spores predominantly formed in the aerial mycelium (Brodie 1931, 1936; Polak et al. 1997, 2001). Oidia of *C. cinereus* easily attach to insects that distribute them onto fresh substrates (Brodie 1931). In addition of being a major distribution unit, germinating spores can fuse with hyphal cells of the same species and thus function as spermatia. Spores are able to attract hyphae of neighboring mycelia, so that these alter their growth direction (oidial homing). Fusion might also occur with hyphae of other species, with the consequence that these die because of intracellular incompatibility

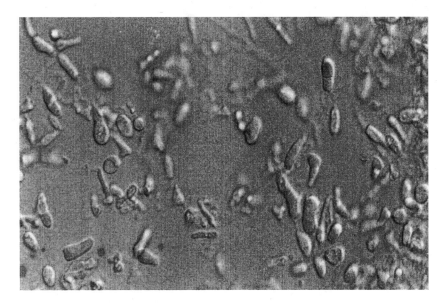

**Figure 12** Chlamydospores formed in the submerged mycelium of an aging culture of *C. cinereus* monokaryon 218.

reactions. This killing action of foreign cells by oidia gives ecological advantage for *C. cinereus* toward other fungi to develop on limited substrates (Kemp 1977).

The cytology of aerial spore formation in the mycelium of *C. cinereus* has recently been described in detail (Polak et al. 1997, 2001, Figs. 13 and 14). The process starts with the protrusion of a sporophore (the oidiophore) from a cell within an aerial hypha. There is no visible cellular differentiation to the hyphal footcell of this structure. However, a relatively defined cellular size compared to an average aerial cell suggests footcells to be physiologically predetermined (Polak et al. 2001). After a nuclear division within the footcell, one of the daughter nuclei migrates into the growing oidiophore. Subsequently, the young oidiophore separates from the footcell by septation. The oidiophore further elongates, tapering with length. When fully elongated, short hyphal segments of defined lengths (oidial hyphae) bud off one after the other from the tip of the oidiophore. Oidial hyphae each obtain one haploid nucleus from consecutive mitotic divisions within the oidiophore stemcell. Each mitosis and nuclear delivery is followed by septum formation, separating an oidial hypha from its stemcell. Another nuclear division occurs within the oidial hypha and a cross wall forms, dividing the oidial hypha in two equally sized spores of about 2 × 4–6 µm. As a final step of spore maturation, the two cells of the oidial hypha separate from each other and from

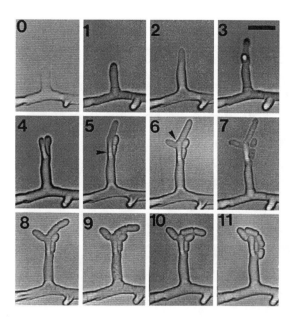

**Figure 13** Development of a simple structured oidiophore over the time on an aerial hyphae of *C. cinereus* strain AmutBmut. Numbers in the photographs correspond to hours. *Arrowheads* point to septa. From Polak et al., 1997.

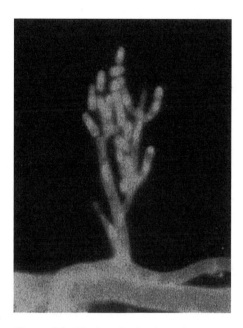

**Figure 14**  Nuclear distribution in a mature oidiophore of *C. cinereus* strain AmutBmut. Nuclei and cell walls were simultaneously stained with Hoechst 33258 and Blankophor BA 267%. Photo courtesy of E. Polak.

the oidiophore stemcell by schizolysis. Up to 200 spores collect in this way in a sticky liquid droplet secreted at the tip of the oidiophore (Polak et al. 1997).

Although the text above and Figure 14 describe the general course of oidiation, on the whole the process is not as uniform, as for example, conidiation in *Aspergillus* (IV.A.1). Between different *Coprinus* strains, within the same strain, and even on the same hypha, various subtypes of oidiophores occur next to each other. Oidiophores might be branched or not, and they might have elongated stems of one or more cells or stout stems. Oidial hyphae might directly bud off from footcells of the undifferentiated aerial hyphae and might break up into three, four, or more spores that, in some strains, vary greatly in length (Kües et al. 2002b; Polak et al. 2001). Because oidia formation involves both budding and schizolysis, the spores are classified arthroconidia (Kendrick and Watling 1979).

## 2.  Fruiting Body Development and Sexual Sporulation

*Initial Stages in Fruiting Body Development.*    Fruiting bodies are the most prominent emergent structures formed on a mycelium of a higher basidiomycete. Typically, they occur on the dikaryon under defined environmental conditions (Wessels 1993a; Moore 1998; Kües 2000; Kües and Liu 2002; Kües et al. 2002a).

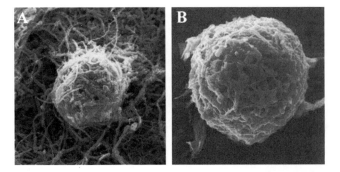

**Figure 15** Scanning electron microscopic pictures of a young (A) and a mature (B) of
*C. cinereus* strain AmutBmut. Photos courtesy of R. Boulianne.

Development starts with a dramatic change in the hyphal growth behavior. The
vegetative mycelium, consisting of a network of single hyphae of principally
unrestricted growth, starts to produce multihyphal structures—plectenchymatic
tissue that requires extensive hyphal-hyphal interactions. Usually when nutrients
are exhausted, so-called hyphal knots arise by intense ramnification at localized
areas within the vegetative mycelium. The multiple hyphal branches formed are
restricted in their tip growth. Their cells tend to inflate and finally aggregate into
complex round structures of yet undifferentiated tissue (Matthews and Niederpruem 1972, 1973; van der Valk and Marchant 1978; Liu 2001). Microscopic
studies suggest that two types of hyphal knots occur during fruiting, that is, primary and secondary hyphal knots, also referred to as primary and secondary
nodules (Clémençon 1994, 1997; Liu 2001). In *C. cinereus*, primary hyphal knot
formation occurs in the dark and is repressed by light (Kües et al. 1998a). These
primary structures are of relatively loose organization, and hyphal cells, although
already present in high density, are little aggregated (Liu 2001). When kept in
the dark, primary hyphal knots will convert into the compact multicellular resting
structures called sclerotia (Jirjis and Moore 1976; Kües et al. 1998a) (Fig. 15).
These duration bodies consist of an outer melanized rind made from small thick-
walled hyphal cells and an inner medulla formed by large irregular-shaped cells
resembling the chlamydospores (Waters et al. 1975a,b; Kües et al., 2002b; Liu
2001). However, when a light signal is given, primary hyphal knots are thought
to transform into the secondary hyphal knots (Kües 2000; Boulianne et al. 2000;
Kües et al. 2002a). Genetic evidence from mutants being blocked in both sclerotia
and fruiting body formation supports this view (Moore 1981), but cytologically,
this transition has not yet been documented for a given structure. The mature
secondary hyphal knot is a compact globular aggregate of undifferentiated hyphal
cells that are densely interwoven in a plectenchymatic tissue (Matthews and
Niederpruem 1973; Clémençon 1997; Moore 1998). As in light control of all

other developmental processes of the fungus, blue light is effective in induction of secondary hyphal knot formation. However, the light energy needed for this developmental step is comparatively low (1 s light exposure, 0.98 lx)—too much light will block secondary hyphal knot formation (Lu 1974; Kertesz-Chaloupková et al. 1998; Kües et al. 1998a). In *C. cinereus*, secondary hyphal knot formation is the first step specific to fruiting body development. The secondary hyphal knot is about 200 μm in size. Although there is no obvious cellular differentiation, a change of the wall structure has been noted at this stage of development. Chitin fibrils in cell walls of cells in the vegetative mycelium are disordered, whereas in cells of secondary hyphal knots, chitin fibrils are organized in parallel (Kamada and Tsuru 1993). At the stage of primary hyphal knot formation, expression of a galectin (Cgl2) is induced, a β-galactoside-binding lectin thought to act in hyphal aggregation. A second galectin (Cgl1) appears with formation of secondary hyphal knots right at the moment of close hyphal interactions, and the levels of both proteins rise with proceeding fruiting body development (Cooper et al. 1997; Boulianne et al. 2000; Liu 2001).

Primary and secondary hyphal knots have been studied in a number of other basidiomycetous species, including *Agaricus bisporus* and *S. commune*. The course of cytological events in hyphal knot formation appears to be like that in *C. cinereus*, although regulation by environmental signals might differ (Clémeçon 1997; Kües and Liu 2000; Kües et al. 2002a). Most of the studied species are angiocarpous or hemiangiocarpous, that is, specific fruiting body tissues, including the spore-bearing cell layers (hymenium), develop internally within the secondary hyphal knot. In contrast, *S. commune* is gymnocarpous. Its hymenium appears and develops to maturity exposed (Volz and Niederpruem 1969; Clémençon 1997). Consequently, the macroscopic steps in further development vary greatly between the species, although on the cellular levels, fewer differences might eventually be found.

*Primordia Formation and Fruiting Body Maturation.* In *C. cinereus*, tissue differentiation starts within the secondary hyphal knot (Matthews and Niederpruem 1973). Large, vacuolated cells of the outer cellular layers, the young veil cells, enclose a core of glycogen-rich, densely packed, heavily branched, short cells that gives rise to the fruiting-body stipe and cap, respectively (Fig. 16). The cytological differentiation within the aggregate becomes obvious at a size of about 0.4 to 0.7 mm. The upper third of the prosenchymal structure, the prepileus region, is the base for the upper peripheral cells (pileus trama). Underneath, a ring of cells appears in which glycogen accumulates and which mark the later position of the gills carrying the basidia. Cells in the middle part line up to form the young stipe tissue. The bottom third of the structure remains as a tightly packed prosenchym, not belonging to the stipe. At a size of 0.8 to 1.0 mm, the shape of a mushroom is clearly visible within the fruiting body primordium (for further details see Reijnders 1977, 1979; Moore 1998; Kües 2000).

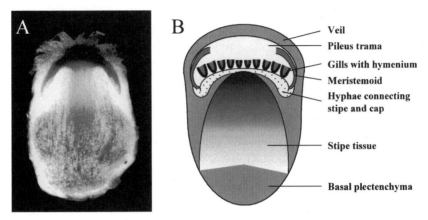

**Figure 16** Longitudinal section through a *C. cinereus* primordium before karyogamy (A) and schematic presentation of a slightly tangential section of a primordium with its main tissue organization (B). Modified after Kües (2000).

Induction of initial tissue differentiation within the secondary hyphal knot appears not to require light. However, when cultures are kept in dark after the light signal for secondary hyphal knot development was given, extensively elongated structures develop by tissue proliferation at the basal third of the secondary hyphal knot. Because they resemble etiolated potato shoots in appearance, these structures are called etiolated stipes, or darkstipes. They have only a rudimentary cap, never transform into mature fruiting bodies, but will eventually abort in favor of new secondary hyphal knot formation in case a new light signal is received (Kües 2000; Kamada 2002). To suppress etiolated stipe development and permit complete primordium and fruiting body differentiation upon secondary hyphal knot formation, additional light signals are compulsory. The first light signal induces formation of gill rudiments (Lu 1974). Programmed cell death to provide space for the developing gills has been claimed by Lu (1991) to occur in early primordial development in *C. cinereus*, but this has not been accepted by other workers in the field (Chiu and Moore 2000). However, Umar and van Griensven (1997, 1998) reported similarly the occurrence of morphogenetic cell death in fruiting body development of *A. bisporus*. Future work will have to clarify whether programmed cell death is an intrinsic factor in establishing specific tissues in the primordium.

Fruiting body development in *C. cinereus* is well adapted to the day-night light-dark rhythms. Daily light signals ensure continuation in primordial maturation and repression of etiolated stipe formation until gill tissues are fully developed with a hymenium in which the binuclear basidia are found, regularly interspersed by the larger cystesia having the task to keep opposing gills apart from each other. At this point, cellular proliferation is completed and subsequent morphological changes are based solely on extension of stipe cells and cell expansion within cap tissues (further reading in Moore 1998; Kües 2000; Liu 2001; Kües et al. 2002a). To proceed from a mature primordium to a fully developed fruiting body, karyogamy needs to be induced by light in the basidia (Lu 1974). After a period of a few hours (after ca 16 hours), meiosis follows while the primordium slowly swells and the veil ruptures. Parallel to spore production, invasion of these each by one haploid nucleus and spore maturation (Fig. 17), within another 12 hours the stipe rapidly elongates and the cap opens (Fig. 18). Because of the dark-spore color, the mature cap appears black. The mature fruiting body of *C. cinereus* is not persistent but quickly disintegrates by cap autolysis owing to the action of chitinases, proteases, and glucanases. Autolysis serves spore liberation. Spores will fall to the ground in liquid droplets that, because of the black spore color, can be used as ink, explaining the common name of the fungus: inky-cap (further reading in Moore 1998; Kües 2000; Liu 2001; Kües et al. 2002a).

Because *C. cinereus* is remarkable because it synchronizes meiosis within the basidia of a fruiting body, much attention has been given to the cytology of meiosis and meiotic recombination (Zolan et al. 1995; Kües 2000). When carefully prepared, a few gills can be removed at successive time points from a single developing mushroom without causing any apparent effects on development, and the two meiotic subdivisions can be monitored under the microscope in a regular and predictable manner (Lu 1984, 1996). Although light is not essential for mei-

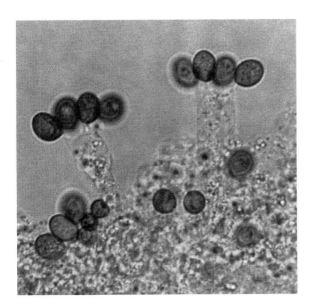

**Figure 17**   Basidia with four meiotic basidiospores from *C. cinereus* strain AmutBmut. Photo courtesy of J.D. Granado, after Kües (2000).

**Figure 18**  Mature fruiting bodies formed by the self-compatible *C. cinereus* strain AmutBmut grown on its natural substrate, horsedung. Photo courtesy of M. Aebi.

otic progression and, indeed, there are strains unaffected in meiosis by light, in others it influences the timing of meiotic events and, at high intensities, can lead to an arrest in meiosis at the stage of diplotene (Lu 2000). Most interestingly, a cell type-specific programmed cell death occurs within the basidia in case of meiotic disorders (Celerin et al. 2000; Lu et al. 2001, submitted).

## 3. Regulation of Oidia Production

Little has been unraveled on the genetic control of oidiation in *C. cinereus* (Kües 2000; Kües et al. 2002b). Oidiophore development on the monokaryon has been shown to be constitutive but renders a blue light-controlled process when the *A* mating-type pathway is active, a situation given in the dikaryon and in homokary-ons with specific mutations in the *A* mating-type locus (Polak et al. 1997; Kertesz-Chaloupková et al. 1998; Kües et al. 1998a). In the dikaryon, oidiation is only partially induced by light. The action of the *B* mating-type genes strengthens the *A*-mediated repression, in both dark and light (Kertesz-Chaloupková et al. 1998; Kües et al. 2002c), the reason why oidia formation on the dikaryon has been overlooked for decades (Brodie 1932; Kimura 1952). The oidia formed by a dikaryon are uninuclear, as are those developing on the monokaryon. However, there is usually a strong bias toward oidia production of only one nuclear type, and this bias appears to be influenced by the strength of the respective mating-type genes (Polak 1999). A similar biased production uninuclear haploid aerial spores on the dikaryon and indications for a mating-type hierarchy in this bias have been reported for other basidiomycetes (Cao et al. 2000; Kitamato et al. 2000). Thus, the presence of two genetically distinct haploid nuclei in cells of the dikaryon has important consequences for the lifestyle of the fungi: keeping two haploid nuclei together in the same cell has the advantages of genetic complementation given in the diploid situation. However, haploid nuclei can still leave a dikaryotic nuclear partnership, either to further exist independently within a monokaryotic mycelium or to enter a new nuclear association when forming a new dikaryon (Kües 2000, 2002; Kües et al. 2002b). Because maintenance of the dikaryotic situation is controlled by the mating-type genes (Sec. III.A.2), a halt in *A*-mediated repression of haploid spore formation by an environmental signal such as light is a good way to balance between the dikaryotic mycelial stage and the escape of single haploid nuclei into spores.

The gene *pccl* has been shown to bring about a light control in oidiation when mutated (Kües 2000). The gene encodes a potential HMG-box transcription factor that may act negatively in the *A* mating-type pathway or, in mutated form, bypasses the need for *A* mating-type genes in expression of dikaryon-specific mycelial characteristics (Murata et al. 1998a; Inada et al. 2001). In addition to repression of oidiation, mutations in the HMG-box of protein Pccl cause the fungus to produce *A*-regulated clamp cells and to fruit without mating (Uno and Ishikawa, 1971; Murata et al., 1998a,b; Kamada 2002). These mutations positively affect the light-controlled, mating-type regulated cAMP levels within a mycelium (Uno and Ishikawa 1982; Swamy et al. 1985a,b), which might link to the increased production of oidia in light (Kües et al. 2002b). Strikingly, increasing C (glucose) but not N (asparagine) levels negatively influence oidia production in dark and in light in the situation of an activated *A* mating-type pathway (Bottoli 2001). However, gene products acting in this nutritional control have yet to be identified. A *C. cinereus* Ras protein has been implicated in nutritional sensing (Bottoli 2001)—analysis of effects on oidiation is under way. *cac*, the gene encoding adenylate cyclase for cAMP production, is expressed in vegetative mycelium regardless of light conditions, and enzyme activity is posttranscriptionally regulated (Swamy et al. 1985b; Bottoli 2001).

Gene *clp1* coding for a novel protein of unknown function has recently been described to be essential for *A*-regulated sexual development. Mutations in the gene abolished *A*-induced clamp cell formation and, in addition, partially overcame *A*-mediated repression of oidiation. However, light still had an inducing effect on spore numbers. Forced expression of *clp1* in a homokaryon caused clamp cell formation, but effects on oidiation were not reported (Inada et al. 2001). For future definition of the molecular nature of the protein, it might be interesting to note that the Clp1 protein has low similarity to the fruiting body-specific proteins Sc7 and Scl4 from *S. commune* and to the extensin-like protein product of gene *SC25* from *Amanita muscaria* (Fig. 19; see also Sec. IV.B.5). Computer programs such as PSORT (http://psort.nibb.ac. jp/), however, predict Clp1to be a nuclear protein in contrast to Sc7, Sc 14, and SC25, which are expected to be extracellular and possibly localized within the cell walls. In accordance, Clpl appears to have a regulatory function within the A mating-type pathway acting upstream of Pcc1 (Brown and Casselton 2001; Kamada 2002).

Mutants in regulation of oidiation and in oidiophore and spore development have been generated, and the mutated genes await cloning (Polak 1999). Several of the mutant phenotypes resemble defects observed in conidiation in *A. nidulans*. For example, mutants were detected with defects in nuclear positioning as occur in *A. nidulans apsA* and *apsB* mutants (Fischer and Timberlake 1995; Polak 1999). Other mutants with stout oidiophores resemble *A. nidulans StuA* mutants that produce conidiophores with extremely shortenened stalks (Clutterbuck 1969; Polak 1999; Sec. IV.A.6). Mutants producing oidia exclusively at the aerial cells without former oidiophore formation are reminiscent of the situation where *brlA* was forcibly activated in vegetative cells (Adams et al. 1988; Polak 1999; Sec. IVA.5). Whether such phenotypic analogies between *C. cinereus* and *A. nidulans* mutants reflect genetic similarities is a question that remains to be answered in the future. No function has yet been isolated in *C. cinereus* that resembles any of those known to act in sporulation of *A. nidulans*, except of laccase genes (Bottoli et al. 1999; Yaver et al. 1999) that might not be needed for maturation of the hyaline uncolored oidia. However, in *S. commune*, an insertion of transposon Scooter interrupted a gene *thnl* for a potential signal regulator for G$\alpha$ subunits of heterotrimeric G proteins, which is 36% identical and 54% similar to the *A. nidulans flbA* gene product (Fowler and Mitton 2000). In both species, gene disruption of these functions resulted in pronounced changes in vegetative hyphal morphology (Lee and Adams 1994a; Fowler and Mitton 2000). Whether Thn 1 affects aerial sporulation as FlbA in *A. nidulans* (Lee and Adams 1994b, 1996; Sec. IV.A.4) is not possible to analyze in *S. commune*. This latter fungus lacks aerial spores but in the future, the problem might be addressed in *C cinereus*. Most interestingly, *S. commune thn1* mutants can be stimulated in production of aerial mycelium as well as in radial growth by diffusible low-molecular-weight compound(s) ($\leq$6000–8000 Dalton) provided by wild-type strains (Schuren 1999), resembling the cross-feeding effect between wild-type and *fluG* strains in *A. nidulans* (Yager et al. 1982, Sec. IV.A.4).

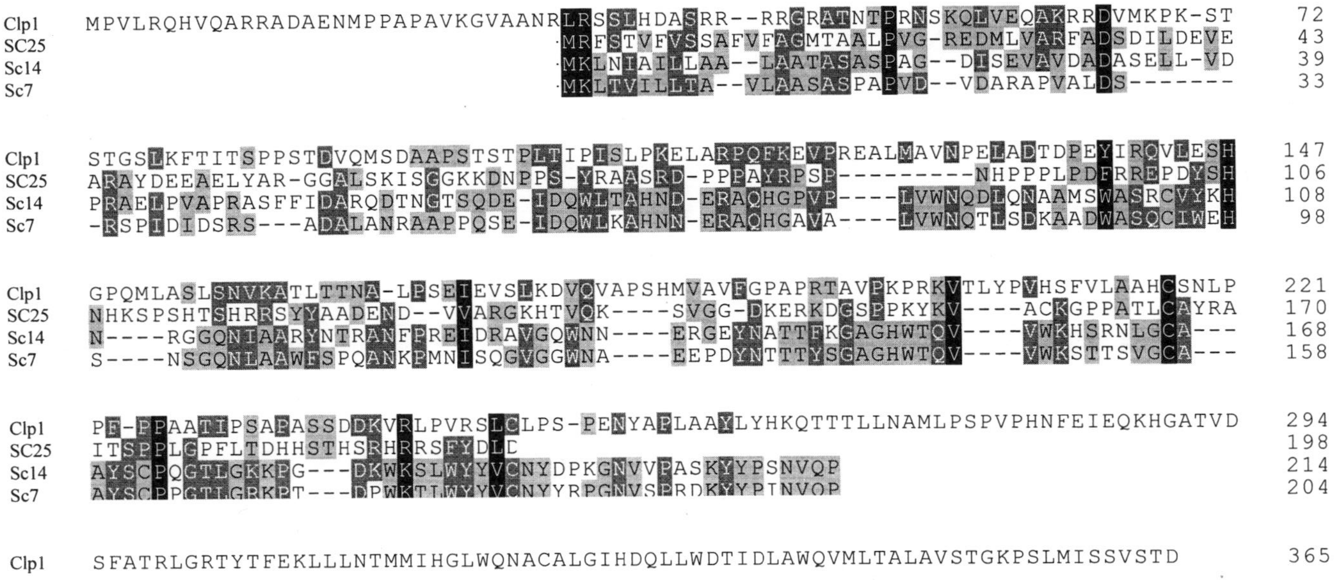

```
Clp1   MPVLRQHVQARRADAENMPPAPAVKGVAANRLRSSLHDASRR--RRGRATNTPRNSKQLVEQAKRRDVMKPK-ST   72
SC25                                 .MRFSTVFVSSAFVFAGMTAALPVG-REDMLVARFADSDILDEVE        43
Scl4                                 .MKLNIAILLAA--LAATASASPAG--DISEVAVDADASELL-VD       39
Sc7                                  .MKLTVILLTA--VLAASASPAPVD.-VDARAPVALDS------        33

Clp1   STGSLKFTITSPPSTDVQMSDAAPSTSTPLTIPISLPKELARPQFKEVPREALMAVNPELADTDPEYIRQVLESH  147
SC25   ARAYDEEAELYAR-GGALSKISGKKDNPPS-YRAASRD-PPFAYRPSP-----NHPPLPDERREPDYSH         106
Scl4   PRAELPVAPRASFFIDARQDTNGTSQDE-IDQWLTAHND-ERAQHGPVP-----LVWNQDLQNAAMSWASRCVYKH  108
Sc7    -RSPIDIDSRS---ADALANRAAPPQSE-IDQWLKAHNN-ERAQHGAVA-----LVWNQTLSDKAADWASQCIWEH   98

Clp1   GPQMLASLSNVKATLTTNA-LPSEIEVSLKDVQVAPSHMVAVEGPAPRTAVPKPRKVTLYPVHSFVLAAHCSNLP  221
SC25   NHKSPSHTSHRRSYYAADEND--VVARGKHTVQK-----SVGG-DKERKDGSPPKYKV---ACKGPPATLCAYRA   170
Scl4   N----RGGQNIAARYNTRANFPREIDRAVGQWNN----ERGEYNATTFKGAGHWTQV---VWKHSRNLGCA----   168
Sc7    S----NSGQNIAAWFSPQANKPMNISQGVGGWNA----EEPDYNTTTYSGAGHWTQV---VWKSTTSVGCA----   158

Clp1   PE-FRAATIPSAPASSDDKVRLPVRSLCLPS-PENYAPLAAYLYHKQTTLLNAMLPSPVPHNFEIEQKHGATVD   294
SC25   ITSPLGPFLTDHSTHSRHRSFYDLC                                                   198
Scl4   AYSCPQGTLGKKPG---DKWKSLWYYVCNYDPKGNVVPASKYYPSNVQP                           214
Sc7    AYSCPPGTLGRKPT---DPWKTIWYYVCNYYRPGNVSPRDKYYPTNVQP                           204

Clp1   SFATRLGRTYTFEKLLLNTMMIHGLWQNACALGIHDQLLWDTIDLAWQVMLTALAVSTGKPSLMISSVSTD     365
```

**Figure 19** Sequence alignment between Clp1 of *C. cinereus*, the extensin-like protein of SC25 of *A. muscaria*, and the PR-1 type proteins Scl4 and Sc7 from *S. commune*.

**Table 1**  Regulation of Development by the $A$ Mating-Type Pathway ($A$ on), the $B$ Mating-Type Pathway ($B$ on), Light, Temperature, and Nutrients

| Asexual development | | Sexual development | |
| --- | --- | --- | --- |
| Mycelial growth, aerial mycelium | | cAMP | |
| (Prerequisite nitrogen low) | | (Prerequisite nutrients low) | |
| $B$ on | repressed | $A$ on $+B$ on $+$ light | induced |
| | | | |
| Clamp cells | | Primary hyphal knots | |
| $A$ on | induced | (Prerequisite nutrients low, in the dark) | |
| | | $A$ on | induced |
| Clamp cell fusion | | $B$ on | induced |
| $B$ on | induced | Light | repressed |
| | | | |
| Oidia | | Secondary hyphal knots | |
| (temperature optimum 37°–42°C) | | (Prerequisite nutrients low, efficient at 25°–28°C) | |
| $A$ on | repressed | $A$ on $+$ light | induced |
| $A$ on $+$ light | induced | $A$ on $+$ $B$ on $+$ light | more efficiently induced |
| $A$ on $+$ $B$ on | more repressed | | |
| $A$ on $+$ $B$ on $+$ light | slightly induced | | |
| | | | |
| Chlamydospores and sclerotia | | Fruiting body maturation | |
| (in the dark) | | (Prerequisite nutrients low, 25°–28°C) | |
| $A$ on | induced | $B$ on $+$ light ($+$ $A$ on?) induced | |
| $B$ on | induced | | |
| Light | repressed | | |

Data were compiled from Uno and Ishikawa (1982), Swamy et al. (1985a), Kües (2000), Kües et al. (1998a,b 2002a–c), Bottoli (2001), and Muraguchi and Kamada (2002), and references therein.

## 4.  The Decision Whether to Enter the Pathway of Sexual or Asexual Sporulation

Superficially, nutritional and environmental regulation of oidia formation in the $A$-activated mycelia seems to follow that of fruiting body initiation unlike temperature control (Table 1). Light-induced oidiation occurs best at higher temperatures in contrast to fruiting that proceeds at 25° to 28°C (Lu 1974; Kües et al. 1998b; Kües 2000). Both developmental processes are favored when nutrients within the medium are low. (Bottoli 2001). Light is needed for both induction of oidiation as well as fruiting. However, prolonged light illumination and higher intensities increase asexual spore numbers but negatively influence fruiting body development (Lu 1974, 2000; Kettesz-Chaloupková et al. 1988; Kües 2000). Thus, absolute light energy is an important modifier of the intracellular signaling responses. Light is known to override $A$ mating-type-regulated developmental decisions (Kües et al. 1998), whereas $B$ mating-type genes counteract light effects in oidiation (Kertesz-Chaloupková et al. 1998; Kües et al. 2002c), but obviously not in light-induced fruiting body initiation and karyogamy (see Sec. IV.B.5). *A* and *B* mating type genes together with light regulate in the dikaryon cAMP production (Swamy et al. 1985a). Action of the *B* mating-type pathway on formation of the aerial vegetative mycelium is controlled by the available nutritions (Kües et al. 2002a,c). Mutants in gene *pcc1*, either bypassing the A mating type pathway or releasing a negative control of sexual development acting downstream of the A mating-type proteins (Murata et al. 1998a, b; Inada et al. 2001; Kamada 2002), produce increased levels of cAMP thought to be the cause of fruiting in the homokaryotic stage (Uno and Ishikawa 1982). All these observations clearly indicate that external light, temperature, and nutrient signals, in a very subtle way, act together with the products of both mating-type loci in deciding which developmental pathway will be entered. In the future, we need to define all individual components of the different signaling cascades to determine those factors in which the signaling pathways interlink. In the better understood basidiomycetous yeasts *U. maydis* and *C. neoformans*, some G proteins and various kinases acting in nutritional, temperature, and mating-type signaling pathways, as well as proteins involved in cAMP metabolism, have already been identified and more direct interactions between pathways have been deduced (Banuett 2002; Lengeler et al. 2000; Kahmann et al. 2000). Light control in signaling networks of control of development, however, is unique to *C. cinereus* and gives an extra motivation to elucidate this higher basidiomycete.

## 5. Genetic Analysis of Fruiting Body Development

Little is known about genes contributing to fruiting body initiation and formation. Est libraries and techniques tike RNA fingerprinting identified genes expressed at the onset or during fruiting in *A. bisporus* (De Groot et al. 1997; Ospina-Giraldo et al. 2000), *Lentinus edodes* (Leung et al. 2000), and *Pleurotus ostreatus* (Lee et al. 2002). Not surprisingly, several of these genes link to the metabolism of the fungi or are genes related to the cell cycle and other cellular processes not necessarily expected to be fruiting body-specific. The functions of a number of others are currently unknown, as no orthologs are present in the databases. Their specific role in the process of fruiting remains unclear up to the point when mutants have been generated from these genes. Indeed, mutants in fruiting body development can be obtained from dikaryons (Takemaru and Kamada 1969); but for the obvious reason of the two genomes presented by the two different haploid nuclei in the dikaryon, such mutants are difficult to handle.

To overcome the problem caused by the dikaryotic situation, a genetic trick can be applied. The *A* mating-type genes have been shown to induce primary and secondary hyphal knot formation in transformed heterologous *C. cinereus* monokaryons (Tymon et al. 1992; Kües et al. 1998a). Primordial development continues up the stage of karyogamy (Kües et al. 1998a). Heterologous *B* mating-type genes in transformants of monokaryons also induce primary but not secondary hyphal knot formation. However, *B* genes were found to enhance production of secondary hyphal knots in number and time when the *A* mating-type pathway was also activated. When *A*-induced primordia development is completed, heterologous *B* genes appear to induce karyogamy and, subsequently, primordia mature into fruiting bodies (Kües et al. 2002a,c). In accordance with these observations, specific mutations in the mating-type loci cause self-compatibility and result in fruiting body formation on homokaryotic mycelium (Raper et al. 1965; Swamy et al. 1984; Wessels 1993a) (Fig. 18). Similarly, mutations acting downstream in the mating-type pathways or by-passing the mating-type control can overcome self-incompatibility and the natural block in fruiting (Uno and Ishikawa 1971; Muraguchi et al. 1999). As they are homokaryotic, any of these self-compatible, fruiting-proficient mutants are ideal strains to induce and detect dominant *and* recessive mutations in fruiting body development. In *C. cinereus*, the haploid uninuclear oidia can be used for mutagenesis (Granado et al. 1997; Cummings et al. 1999; Muraguchi et al. 1999; Walser et al. 2001).

Spontaneous, UV, and REMI mutants from self-fertile homokaryons have already successfully been appointed to clone genes from *C. cinereus* acting in the process of fruiting body development (Muraguchi and Kamada, 1998, 2000; Liu 2001; Kües et al. 2002a) and meiotic spore formation (Cummings et al. 1999; Celerin et al. 2000). The mutant cloning approach detected an essential gene, *cfs1* in fruiting body initiation that encodes a product highly similar to bacterial cyclopropane fatty acid synthases (Liu 2001), a specific type of S-adenosyl-L-methionine (SAM)-dependent methyltransferases not described in eukaryotes before (Grogan and Cronan 1997). Eln2 and Ich1 are proteins that specifically act in stipe and cap tissue development, respectively (Muraguchi and Kamada 1998, 2002). Eln2 is a novel P450 cytochrome (Muraguchi and Kamada 2000) and Ich 1 has a SAM-binding domain suggesting the protein also belongs to the large family of SAM-dependent methylases (Kües 2000). From a REMI mutant with white, that is sporeless, fruiting bodies, gene *spo11* has been recovered and shown to encode a type II topoisomerase with an important role in meiosis, for example in formation of the synaptonemal complex (Cummings et al. 1999; Celerin et al. 2000; Merino et al. 2000). Many more *C. cinereus* homokaryotic mutants are now available as a reservoir for cloning in future work additional genes in the pathway of fruiting body development, meiosis, and basidiospore formation (Muraguchi et al. 1999; Kües et al., unpublished).

The usefulness of *C. cinereus* mutants for isolating genes in fruiting certainly is proven and the overall process of fruiting body development is best described for this fungus (Moore 1998; Kües 2000; Kües et al. 2002a), making it possible to relate gene functions to specific steps in fruiting (Muraguchi and Kamada 1998, 2000; Liu 2001; Kamada 2002). Moreover, the number of active *Coprinus* labs currently overweights those of *S. commune*, promising that scientific progress will proceed faster for the first organism. However, there is currently one advantage to *S. commune* compared to *C. cinereus*: knocking out genes becomes routine in *S. commune*. Gene replacement in this fungus occurred in the majority of cases in around 2% of all transformants (Robertson et al. 1995; van Wetter et al. 2000a; Lengeler and Kothe 1999a), but in some cases it has been much higher (8%: Horton et al. 1999; 33%: Marion et al. 1995; 50%: Lengeler and Kothe 1999b). In *C. cinereus*, so far there are only a few reports on gene replacements: replacement of *A* mating-type DNA was observed in less than 1% of transformants (Pardo 1995) and gene *trp1* had been replaced in 5% to 8% of transformants (Binninger et al. 1991). Gene *cfs1* appears to integrate in 0.2% of all cases at its natural locus (Liu 2001) and a knock-out has been reported for gene *hmg1* for a HMG-box transcription factor but without data of knock-out frequencies (Aime and Casselton 2002). Thus, establishing an efficient and reliable gene-knockout protocol for *C. cinereus* is the major problem to be solved in the future. In *C. neoformans*, this has recently been achieved by changing from classic protoplast transformation to a ballistic transformation system (Davidson et al. 2000). Possibly, the type (mitotic or meiotic spore, or hyphal cell) and age of cells and the cell cycle in general next to positional effects in the chromosomal DNA will play a considerable role in succeeding in obtaining knock-outs. Successful application of RNA interference in silencing gene functions in the basidiomycetous yeast *C. neoformans* (Liu et al. 2002), raises hope for establishing alternative methods.

One of the genes that has been knocked out in *S. commune* is *frt1* (Horton et al. 1999). This gene was fortuitously isolated when screening for *B* mating genes by its ability to induce fruiting in heterologous homokaryotic hosts not carrying the cloned *frt1* allele (Horton and Raper 1991). So far, the mechanism of this fruiting-inductive reaction is not clear. A prominent motif in Frt1 is a P-loop important in many nucleotide-binding proteins (Horton and Raper 1995). Deletion of the gene causes expression of dikaryon-specific genes in the homokaryon. For fruiting in the dikaryon, the gene is dispensable, although it has a positive effect also on the dikaryon (Horton and Raper 1995; Horton et al. 1999).

Isolation of highly expressed cDNAs from fruiting bodies of *S. commune* (Mulder and Wessels 1986) started research on the continually increasing family of fungal hydrophobins (Wessels 1997). Within the inner tissues of the fruiting body, the dikaryon-specific hydrophobin Sc4 is highly expressed where it lines gas channels and prevents these from filling with water under wet conditions (Lugones et al. 1999; van Wetter et al. 2000a). Such coating of fruiting body tissues apparently contributes to maintaining the size and shape of the mushroom (Hershko and Nussinovitch 1998). Sc3 is another protein of the family that covers aerial hyphae of both mono- and dikaryons in the form of an amphipathic rodlet layer, gives the vegetative aerial mycelium its hydrophilic character, and enables it to attach to hydrophobic surfaces (Wösten et al. 1993, 1994a,b). Within the fruiting body, Sc3 is expressed by hyphae covering the outer surface of the structure (Ásgeirsdottir et al. 1995). The hydrophilic side of assembled Sc3 is more hydrophilic than that of Sc4, and this appears to be the reason that Sc4 can only partially substitute Sc3 in function (van Wetter et al. 2000a). The role of two other dikaryon-specific hydrophobins, Sc1 and Sc6, in fruiting body development still has to be elucidated (Wessels et al. 1995). Multiple hydrophobins have also been shown to contribute to the shape of fruiting bodies of *A. bisporus* (Lugones et al. 1999; De Groot et al. 1999). Not surprisingly, the interesting features of these proteins and the importance in construction of fruiting bodies prompted the isolation of respective fruiting body-specific genes from a number of other basidiomycetes, including the edible fungi *P. ostreatus* and *L. edodes* (Ásgeirsdottir et al. 1997; Santos and Labarère 1999; Ng et al. 2000; Tagu et al. 2001). In *C. cinereus*, such proteins are, however, currently only known from the vegetative mycelium of the monokaryon (Ásgeirsdottir et al. 1995). Because the mature *C. cinereus* fruiting body quickly disintegrates by autolysis, there is a special stimulus to study hydrophobins in the fruiting body of this species.

Another class of highly expressed fruiting body specific proteins is presented through Sc7 and Sc14 of *S. commune* (Mulder and Wessels 1986). Antibodies showed the proteins to be secreted in accordance with the presence of an N-terminal secretion signal and the proteins appear to loosely interact with the fungal cell walls (Schuren et al. 1993). There is no defined function yet to Sc7 and Sc14, but they relate to the PR-1 type proteins, a group of specific pathogen-induced plant proteins whose biological role is also still unknown (van Loon and van Strien 1999). As mentioned before (Sec. IV.B.3), Sc7 and Sc14 have weak sequence homology to Clp1 of *C. cinereus* and SC25 of *Amanita* (Fig. 19). Expression of the latter has been shown to be induced in fruiting bodies and in ectomycorrhiza (Nehls et al. 1999).

## 6.  Proteins in Hyphal Aggregation

Hyphal aggregation is an essential part in the process of fruiting body development (Boulianne et al. 2000), as well as in lichen formation (Oberwinkler 2001), and in the course of plant root colonization in basidiomycetous ectomycorrhizal interactions—for example, in formation of the pseudoparenchymatous hyphal mantle that covers the root. Moreover, in ectomycorrhizza, fungal cells and cells of the plant symbiotic partner need to recognize and interact with each other (F. Martin et al. 1999; Bonfante 2001; Duplessis et al. 2002). Thus, extracellular proteins are expected to play an important role in the differentiation of fruiting bodies, lichen formation, and ectomycorrhizza development. For example, lectins have been implicated in hyphal aggregation in fruiting body formation of *C. cinereus* and other basidiomycetes (Wang et al. 1998; Boulianne et al. 2000) and in photobiont interaction in lichens (Elifio et al. 2000). Lectins from *Lactarius deterrimus* fruiting bodies are also expressed during ectomycorrhizza formation and might have an additional role in the fungal-plant contact (Giollant et al. 1993). Furthermore, hydrophobins are found within fruiting bodies (Sec. IV.B.5) and in the fungal tissues of basidiomycetous (as well as ascomycetous) lichens (Scherrer et al. 2000, 2002; Trembley et al. 2002a,b) and ectomycorrhizal interactions (Tagu et al., 1998, 2001; Duplessis et al. 2001, Mankel et al. 2002). It will be interesting to establish whether, within a given species, the same hydrophobins will be present in fruiting bodies and, respectively, in the lichen thallus or in the ectomycorrhizal fungal mass. Hydrophobins in the vegetative mycelium and in fruiting bodies tend to be distinct from each other (Mulder and Wessels 1986; Ásgeirsdottir et al. 1998; Ng et al. 2000). Like lectins, hydrophobins have been implicated in hyphal-hyphal interactions (Wessels 1993b). The hyphal cell wall structure drastically changes in fruiting body formation (Kamada and Tsuru 1993; Kamada 1994), as well as with ectomycorrhiza formation (F. Martin et al. 1999; Duplessis et al. 2002). A recent study showed hydrophobins to influence cell wall composition, namely the level of chitin-linked alkali-resistant glucans and the amount of soluble mucilage consisting of 1,3-β-glucans (van Wetter et al. 2000b). Another study suggests an interaction of hydrophobins with the soluble 1,3-β-glucan Schizophyllan of *S. commune* (G.G. Martin et al. 1999, 2000). Lectins from fruiting bodies of different basidiomycetes vary in their preference for binding sugar residues (Cooper et al. 1997; Wang et al. 1998), but respective ligands present within fungal cell walls still have to be identified.

More parallels to the two processes of fruiting body development and ectomycorrhizza formation might exist. In a search for extracellular cell wall proteins, a small alanine-rich protein has been isolated from *Pisolithus tinctorius*, whose expression is several-fold up-regulated during mycorrhizzal development. Although secreted, the protein is lacking a typical N-terminal secretion signal (F. Martin et al. 1999; Laurent et al. 1999). Likewise, the galectins found in cell walls and the extracellular matrix of fruiting bodies of *C. cinereus* have no typical secretion signals (Cooper et al. 1997; Boulianne et al. 2000). In a yeast *sec18* mutant, being defective in secretion by the endoplasmatic reticulum, the *Coprinus* galectins have been shown to be secreted by a novel, yet unidentified pathway (Boulianne et al. 2000). Classic secretion in filamentous fungi goes through the hyphal tip (Sec. II). This might not be the case for nonclassic secretion. In *C. cinereus*, galectins are found expressed in the outer stipe and cap tissues of the fruiting body (Boulianne et al. 2000). These tissues need to resist very strong tension forces during stipe elongation and cap expansion. Stipe elongation and cap expansion is by extension growth of existing cells and occurs over the whole cellular surface (Gooday 1975; Kamada 1994; Moore 1998). Most efficient would be to secrete the components required for cell wall extension over the whole cellular surface. In *C. cinereus*, galectins are found intracellularly in bodies that appear to interact with the whole outer cell membrane, suggesting a possible role for these bodies in secretion, possibly using the whole cellular surface (Boulianne et al. 2000). Further studies on galectin secretion in *C. cinereus* promise to bring

exciting findings on a new secretory pathway that also might be active in ectomycorrhizal species in mycorrhiza formation.


## V.  FUTURE PERSPECTIVES

The compilation of genetic studies on developmental processes in filamentous
fungi shows that of all, conidiation in *A. nidulans* is best understood. There are
two main rationales for studying conidiation in *A. nidulans* in such detail. One
is to understand the process of conidium formation in this mold and, thus, broaden
our knowledge of fungal biology. A second is the possible discovery or better
understanding of general regulatory principles and their contribution to morphological changes. The analysis of conidiation in *A. nidulans* has revealed a basic
knowledge of the regulatory circuits underlying the developmental process. However, many interesting questions need to be solved, such as which other components are required and how gene activity is modulated and fine tuned during
differentiation to achieve the structural and physiological changes in the conidiophore. Several regulatory principles well established in other systems have not
yet been studied during development. One example is the modulation of gene
activity through chromatin changes, as they were observed in the nitrogen metabolism regulatory system of *A. nidulans* (Muro-Pator et al. 1999) or in the regulation of pathogenicity of *U. maydis* (Quadbeck-Seeger et al. 2000) and appear to
play a role in the transition from the syncytical to true cellular stages during
sexual reproduction in *P. anserina* (Berteaux-Lecellier et al. 1998). First indications for such a regulatory mechanism during conidiation of *A. nidulans* came
from the analysis of the *spoC1* cluster, which is positionally regulated (Miller et
al. 1987). Another example is the modulation of the activity of regulatory proteins
through their subcellular localization. A shuttle of a transcription factor between
the nucleus and the cytoplasm has been established for the regulation of acetate
utilization in *A. nidulans* and for several other systems in a variety of eukaryotes
(Komeili and O'Shea 2000). Most likely, this regulatory principle also plays a
role during conidiation of *A. nidulans*, and it will be the challenge of future
research to identify those regulatory circuits. There is also evidence that "two-
component" systems, which are well-known signal transduction pathways in prokaryotes, also operate in *A. nidulans* in conidiation. A detailed analysis of the
link to the known regulatory pathways and the exact role of this two-component
system have not yet been established (Appleyard et al. 2000).

How widely distributed is the developmental regulation revealed in *A. nidulans* among other fungi? Studies are rather limited and the few published examples do not answer this question satisfactorily. The *brlA* gene, which does not
have a homologue in *S. cerevisiae*, was identified in *Aspergillus oryzae*, and
functional analysis revealed a conserved function (Yamada et al. 1999). The same

holds true for the *wetA* gene in *P. chrysogenum* (Prade and Timberlake 1994). The AbaA homologue in *S. cerevisiae*, TEC1, regulates pseudohyphal development (Gavrias et al. 1996) and the one in *Candida albicans* temperature-controlled hyphal growth and virulence (Schweizer et al. 2000). The pathogenic and dimorphic growing *Penicillium marneffei* grows filamentous and reproduces with conidiophore-borne asexual spores at 25° C, and is fission-yeast-like at 37° C. The conidiation phenotype of a *P. marneffei abaA* mutant strain at 25° C is similar to the phenotype described in *A. nidulans*, and *abaA* genes are interchangeable between the two species (Borneman et al. 2000). These results indicate that the central regulatory pathway of *A. nidulans* seems to be conserved among filamentous ascomycetes, and some regulators are also involved in developmental processes in yeasts. However, the cases of *P. marneffei* and *C. albicans* clearly emphasize that different fungi need to be studied, because in these pathogenic fungi, sensing of the temperature or other host-specific environmental conditions must trigger the developmental cascade defined by the regulators.

Another example for a species-specific modulation of the regulatory cascade is *fibD*. This gene is well characterized in *A. nidulans* as a typical *fluffy* gene (see Sec. IV.A.4), but deletion of the gene in *N. crassa* revealed no evidence for an important developmental role, although the gene could complement the defect in *A. nidulans* (Shen et al. 1998). Because the *fluffy* genes are required early during development, perhaps for the integration of environmental signals, this situation might indicate the differences in the processing of these initial signals. Seeing such differences within the ascomycetes, how will it be in the basidiomycetes? The example of the *thn* gene in *S. commune* (Fowler and Mitton 2000) promises that more elements of the regulatory cascade of *A. nidulans* conidiation will be discovered in the future. Understanding the genetic basis of asexual spore development in basidiomycetes is only starting. Currently, only the negative link to the *A* mating-type genes has firmly been established, and, in turn, a control by blue light (Kües et al. 1998a). The blue-light signaling pathway in *C. cinereus*—will it differ completely or only in the receptor molecule from the red-light driven pathway in *A. nidulans*? How similar is the blue-light signaling pathway in *N. crassa*?

The transcriptional regulation of asexual development in *A. nidulans* nicely demonstrates the potential of transcriptional profiling to improve the knowledge of differentiation. However, recent studies indicate that besides the induction of gene activities during the course of development, those required during vegetative growth are modulated during conidiation to meet specific requirements in certain developmental stages. One example is the regulation of the cell cycle and the regulation of nuclear migration during conidiogenesis (Fischer and Timberlake 1995; Schier et al. 2000; Ye et al. 1999). Likewise, modulation of such generally required gene activities will be part of sexual development (Graia et al. 2000), an assumption also supported, for example, by results from Est libraries in the ascomycete *N. crassa* (Nelson et al. 1997; Braun et al. 2000), as well as from basidiomycetes (Leung et al. 2000; Ospina-Giraldo et al. 2000). The mating-type genes, standing at the beginning of sexual development, are now relatively well understood within several species (Sec. III.A), but beyond that, what we know is very limited. However, study of sexual development by molecular means in both asco- and basidiomycetes has lately gotten more attention, following complementation of interesting mutants with specific defects in sexual development. The functions unravelled so far are solitary cases but with time, a picture will emerge about their specific position in the broader context of other genes. Interestingly, several of the currently identified genes encode proteins linked to lipid metabolism. There is the gene *cfs1* for a cyclopropane fatty acid synthase in *C. cinereus* (Liu 2001). Failure in formation of cytosolic acetyl-CoA (mainly used for lipid and sterol biosynthesis) by genes *acl1* and *acl2* for two subunits of ATP citrate lyase blocks fruiting body maturation in *S. macrospora* (Nowrousian et al. 1999, 2000). Gene *car1* in *P. anserina* encodes a peroxisomal protein found to be essential for karyogamy required for sexual sporulation (Berteaux-Lecellier et al. 1995). Lipid metabolism in developmental processes is rather novel, be it for providing the high energy amounts needed or for new signaling pathways acting on membranes (Huber et al. 2002; Ruprich-Robert et al. 2002; Liu 2001). Nevertheless a need for lipid functions is not unique to sexual development. For example, mutations *eel* (chain-elongation) and *chol-1* (choline-requirer) causing defects in lipid synthesis, affect the period and temperature compensation of the circadian rhythm in *N. crassa* (Lakin-Thomas and Brody 2000; Ramsdale and Lakin-Thomas 2000).

After the increase in studies, for example on cell cycle, budding, and pseudohyphal growth in *S. cerevisiae* after the publication of the complete yeast genome, we are far away from knowing what makes a fungus filamentous. Estimates in *N. crassa* expect 1.5 to 2.2 more genes to be present in this filamentous ascomycete than in the yeast (Kupfer et al. 1997; Nelson et al. 1997) and about 66% of all genes (so-called orphan genes) cannot be assigned to functions already known from other organisms; for comparison, more than 57% of predicted genes from *S. cerevisiae* have clear homologs in the databases (Braun et al. 2000). Is a bulk of these responsible for the hyphal growth form? Also, for sexual and asexual development, which in filamentous fungi both are more complex than in *S. cerevisiae*? When comparing genomes of different filamentous ascomycetes, will these be very different or will there be only small differences that account for species-specific characteristics? How similar are asco- and basidiomycetes and, finally, how about zygomycetes and chytridiomycetes? When comparing to other eukaryotes, will we unravel why several of the fungi have heterokaryotic growth phases, and why their nuclei do not fuse directly upon cellular fusion as it is regular in higher eukaryotes?

Many other fascinating questions have been addressed in this chapter, for example, the various systems of vegetative incompatibilities and the repeated discovery of programmed cell death in hyphal interactions and developmental processes. Studying these and other cellular events in filamentous fungi will help us understand the general concepts in living cells and will bring into focus the individual facets.

## ACKNOWLEDGMENTS

The authors are very grateful to the members of their groups for their published and unpublished work and would like to thank C. d'Enfert, L. Yager, and K.S. Chae for communicating unpublished results, M. Aebi, R. Boulianne, J. Granado, E. Polak, and R. Prade for supplying photographs. The work of RF was supported by the DFG, the Philipps-University of Marburg, and the Max-Planck-Institute of Marburg. UK acknowledges support by the Swiss National Science Foundation (grants 31-46'940.96, 31-46'940.96/2, and 31-59157.99), by the ETH Zürich, and the DBU (Deutsche Bundesstiftung Umwelt).

## REFERENCES

TH Adams, MT Boylan, WE Timberlake. *brlA* is necessary and sufficient to direct conidiophore development in *Aspergillus nidulans*. Cell 54:353–362, 1988.

TH Adams, H Deising, WE Timberlake. *brlA* requires both zinc fingers to induce development. Mol Cell Biol 10:1815–1817, 1990.

TH Adams, WA Hide, LN Yager, BN Lee. Isolation of a gene required for programmed initiation of development by *Aspergillus nidulans*. Mol Cell Biol 12: 3827–3833, 1992.

TH Adams, JK Wieser, J-H Yu. Asexual sporulation in *Aspergillus nidulans*. Microbiol Mol Biol Rev 62:35–54, 1998.

J Aguirre. Spatial and temporal controls of the *Aspergillus brlA* developmental regulatory gene. Mol Microbiol 8:211–218, 1993.

MC Aime, LA Casselton. *hmgl* encodes an HMG domain protein implicated in nuclear migration in *Coprinus cinereus*. Abstract lo-6. ECFG6, Pisa 6-9 April 2002, p. 24.

T Aimi, Y Yotsutani, T Moringa. Cytological analysis of anastomoses and vegetative incompatibility reactions in *Heliobasidium monpa*. Curr Microbiol 44:148–152, 2002.

AM Ainsworth, JR Breeching, SJ Broxholme, BA Hunt, ADM Rayner, PT Scard. Complex outcome of reciprocal exchange of nuclear DNA between two members of the basidiomycete genus *Stereum*. J Gen Microbiol 138:1147–1157, 1992.

C Alberti-Segui, F Dietrich, R Altmann-Johl, D Hoepfner, P Philippsen. Cytoplasmic dynein is required to oppose the force that moves nuclei towards the hyphal tip in the filamentous ascomycete *Ashbya gossypii*. J Cell Sci 114:975–986, 2001.

JA Alspaugh, LM Cavallo, JR Perfect, J Heitmann. RAS1 regulates filamentation, mating and growth at high temperature of *Cryptococcus neoformans*. Mot Microbiol 36: 352–365, 2000.

B Andrews, V Measday. The cyclin family of budding yeast: Abundant use of a good idea. Trends Genet 14:66–72, 1998.

A Andrianopoulos, WE Timberlake. The *Aspergillus nidulans abaA* gene encodes a transcriptional activator that acts as a genetic switch to control development. Mol Cell Biol 14:2503–2515, 1994.

MVCL Appleyard, WL McPheat, MJR Stark. A novel "two-component" protein containing histidine kinase and response regulator domains required for sporulation in *Aspergillus nidulans*. Curr Genet 37:364–372, 2000.

R Aramayo, TH Adams, WE Timberlake. A large cluster of highly expressed genes is dispensable for growth and development in *Aspergillus nidulans*. Genetics 122: 65–71,1989.

R Aramayo, WE Timberlake. Sequence and molecular structure of the *Aspergillus nidulans yA* (lactase I) gene. Nucl Acids Res 18:3415, 1990.

R Aramayo, WE Timberlake. The *Aspergillus nidulans yA* gene is regulated by abaA. EMBO J 12:2039–2048, 1993.

S Arnaise, R Debuchy, M Picard. What is a bona fide mating-type gene? Internuclear complementation of *mat* mutants in *Podospora anserina*. Mol Gen Genet 256:169–178, 1997.

S Arnaise, D Zickler, S Le Bilcot, C Poisier, R Debuchy. Mutations in mating-type genes of the heterothallic fungus *Podospora anserina* lead to self-fertility. Genetics 159: 545–556, 2001.

G Arpaia, F Cerri, S Baima, G Macino. Involvement of protein kinase C in the response of *Neurospora crassa* to blue light. Mol Gen Genet 262:314–322,1999.

Y Asada, C Yue, J Wu, G-P Shen, CP Novotny, RC Ullrich. *Schizophyllum commune* Aα mating-type proteins, Y and Z, form complexes in all combinations in vitro. Genetics 147:117–123, 1997.

RN Asante-Owusu, AH Banham, HU Böhnert, EJC Mellor, LA Casselton. Heterodimerization between two classes of homeodomain proteins in the mushroom *Coprinus cinereus* brings together potential DNA-binding and activation domains. Gene 172: 25–31, 1996.

SA Ásgeirsdottir, MA van Wetter, JGH Wessels. Differential expression of genes under control of the mating-type genes in the secondary mycelium of *Schizophyllum commune*. Microbiology 141:1281–1288, 1995.

SA Ásgeirsdottir, OMH de Vries, JGH Wessels. Identification of three differentially expressed hydrophobins in *Pleurotus ostreatus* (oyster mushroom). Microbiology 144:2961–2969, 1998.

SA Ásgeirsdottir, JR Halsall, LA Casselton. Expression of two closely linked hydrophobin genes of *Coprinus cinereus* is monokaryon-specific and down-regulated by the *oid-1* mutation. Fungal Genet Biol 22:54–63, 1997.

DE Axelrod, M Grealt, M Pastushok Gene control of developmental competence in *Aspergillus nidulans*. Dev Biol 34:9–15, 1973.

Y Ayad-Durieux, P Knechtle, S Goff, F Dietrich. P Philippsen. A PAK-like protein kinase is required for maturation of young hyphae and septation in the filamentous ascomycete *Ashbya gossypii*. J Cell Sci 113:4563–4575, 2000.

G Bakkeren, JW Kronstad. Conservation of the *b* mating-type gene complex among bipolar and tetrapolar smut fungi. Plant Cell 5, 123–136, 1993.

G Bakkeren, JW Kronstad. Linkage of mating-type loci distinguishes bipolar from tetrapolar mating in basidiomycetous smut fungi. Proc Natl Acad Sci USA 91: 7085–7089,1994.

P Ballario, G Macino. White collar proteins: PASsing the light signal in *Neurospora crassa*. Trends Microbiol 5:458–462, 1997.

P Ballario, C Talora, D Galli, H Linden, G Macino. Roles in dimerization and blue light photoresponse of the PAS and LOV domains of *Neurospora crassa* white collar proteins. Mol Microbiol 29:719–729, 1998.

AH Banham, RN Asante-Owusu, B Göttgens, SAJ Thompson, CS Kingsnorth, EJC Mellor, LA Casselton. An N-terminal dimerization domain permits homeodomain proteins to choose compatible partners and initiate sexual development in the mushroom *Coprinus cinereus*. Plant Cell 7:773–783, 1995.

F Banuett. Genetics of *Ustilago maydis*, a fungal pathogen that induces tumors in maize. Ann Rev Genet 29:179–208, 1995.

F Banuett. Pathogenic development of *Ustilago maydis*: A progression of morphological transitions that results in tumor formation and teliospore formation. In: HD Osiewacz, ed. Molecular Biology of Fungal Development. New York: Marcel Dekker, 2002, pp 349–399.

S Bartnicki-Garcia. Hyphal tip growth: Outstanding questions. In: HD Osiewacz, ed. Molecular Biology of Fungal Development. New York: Marcel Dekker, 2002, pp 29–58.

CR Benjamin. Ascocarps of *Aspergillus* and *Penicillium*. Mycologia 47:669–687, 1955.

V Berteaux-Lecellier, M Picard, C Thompson-Coffe, D Zickler, A Panvier-Adoutte, JM Simonet. A nonmammalian homolog of the *paf1* gene (Zellweger syndrome) discovered as a gene involved in caryogamy in the fungus *Podospora anserina*. Cell 81:1043–1051, 1995.

V Berteaux-Lecellier, D Zickler, R Debuchy, A Panvier-Adoutte, C Thompson-Coffe, M Picard. A homologue of the yeast *SHE4* gene is essential for the transition between the syncytical and cellular stages during sexual reproduction of the fungus *Podospora anserina*. EMBO J 17:1248–1258, 1998.

DM Binninger, L Le Chevanton, C Skrzynia, CD Shubkin, PJ Pukkila. Targeted transformation in *Coprinus cinereus*. Mol Gen Genet 227:245–251, 1991.

B Birren, G Fink, E Lander. Fungal genome initiative. White paper developed by the fungal research community. February 8, 2002. http://www-genome.wi.mit.edu/seq/fgi/FGI_whitepaper_Feb8.pdf.

CE Birse, AJ Clutterbuck. N-Acetyl-6-hydroxytryptophan oxidase, a developmentally controlled phenol oxidase from *Aspergillus nidulans*. J Gen Microbiol 136:1725–1730, 1990.

CE Birse, AJ Clutterbuck. Isolation and developmentally regulated expression of an *Aspergillus nidulans* phenol oxidase-encoding gene, *ivoB*. Gene 98:69–76,1991.

M Bölker. Sex and crime: Heterotrimeric G-proteins in fungal mating and pathogenesis. Fungal Genet Biol 25:143–156, 1998.

M Bölker, M Urban, R Kahmann. The *a* mating type locus of *U. maydis* specifies cell signaling components. Cell 68:441–450, 1992.

P Bonfante. At the interface between mycorrhizal fungi and plants: The structural organization of cell wall, plasma membrane and cytoskeleton. In: B Hock, ed. The Mycota. Vol. IX. Fungal Associations. Berlin, Germany: Springer, 2001, pp 45–62.

AR Borneman, MJ Hynes, A Andrianopoulos. The *abaA* homologue of *Penicillium marneffei* participates in two developmental programmes: Conidiation and dimorphic growth. Mol Microbiol 38:1034–1047, 2000.

APF Bottoli. Metabolic and environmental control of development in *Coprinus cinereus*. Ph.D. thesis, ETH Zurich, Zurich, Switzerland, 2001.

APF Bottoli, K Kertesz-Chaloupková, RP Boulianne, JD Granado, M Aebi, U Kües. Rapid isolation of genes from an indexed genomic library of *C. cinereus* in a novel *pab1*$^+$ cosmid. J Microbiol Meth 35:129–141, 1999.

RP Boulianne, Y Liu, BC Lu, M Aebi, U Kües. Fruiting body development in *Coprinus cinereus*: Regulated expression of two galectins secreted by a non-classical pathway. Microbiology 146:1841–1853, 2000.

MT Boylan, PM Mirabito, CE Willett, CR Zimmerman, WE Timberlake. Isolation and physical characterization of three essential conidiation genes from *Aspergillus nidulans*. Mol Cell Biol 7:3113–3118, 1987.

A Brachmann, G Weinzierl, J Kämper, R Kahmann. Identification of genes in the bW/bE regulatory cascade in *Ustilago maydis*. Mol Microbiol 42:1047–1063, 2001.

EL Braun, AL Halpern, MA Nelson, Natvig DO. Large-scale comparison of fungal sequence information: Mechanisms of innovation in *Neurospora crassa* and gene loss in *Saccharomyces cerevisiae*. Genome Res 10:416–430, 2000.

HJ Brodie. The oidia of *Coprinus lagopus* and their relation with insects. Ann Bot 45: 315–344, 1931.

HJ Brodie. Oidial mycelia and the diploidization process in *Coprinus lagopus*. Ann Bot 46:727–732, 1932.

HJ Brodie. The occurrence and function of oidia in the hymenomycetes. Am J Bot 23: 309–327, 1936.

AJ Brown, LA Casselton. Mating in mushrooms: Increasing the chances but prolonging the affair. Trends Genet 17:393–400, 2001.

AHR Buller. The biological significance of conjugate nuclei in *Coprinus lagopus* and other basidiomycetes. Nature 126:686–689, 1930.

TM Busby, KY Miller, BL Miller. Suppression and enhancement of the *Aspergillus nidulans medusa* mutation by altered dosage of the *bristle* and *stunted* genes. Genetics 143:155–163, 1996.

S Busch, G Hoffmann, O Valerius, K Starke, K Düvel, GH Braus. Regulation of the *Aspergillus nidulans hisB* gene by histidine starvation. Curr Genet 38:314–322, 2001.

HJ Bussink, SA Osmani. A cyclin-dependent kinase family member (PHOA) is required to link developmental fate to environmental conditions in *Aspergillus nidulans*. EMBO J 17:3990–4003, 1998.

AM Calvo, RA Wilson, JW Bok, NP Keller. Relationship between secondary metabolism and fungal development. Microbiol Mol Biol Rev 66:447–459, 2002.

MA Calvo, M Agut. Observation of Woronin bodies in *Arthrinium aureum* by scanning electron microscopy. Mycopathologia 153:137–139, 2002.

H Cao, H Yamamoto, Y Kitamoto. Genetic study on monokaryotic oidium formation from dikaryotic mycelia in a basidiomycete, *Pholiota nameko*. Mushroom Sci 15:259–265, 2000.

LA Casselton, U Kües. Mating-type genes in homobasidiomycetes. In: JGH Wessels, F Meinhardt, eds. The Mycota. Vol. I. Growth, Differentiation and Sexuality. Berlin, Germany: Springer, 1994, pp 307–322.

LA Casselton, NS Olesnicky. Molecular genetics of mating recognition in basidiomycete fungi. Microbiol Mol Biol Rev 62:55–70, 1998.

LA Casselton, M Zolan. The art and design of genetic screens: Filamentous fungi. Nat Rev Genet 3:683–697, 2002.

M Celerin, ST Merino, JE Stone, AM Menzie, ME Zolan. Multiple roles of Spo11 in meiotic.chromosome behavior. EMBO J 19:2739–2750, 2000.

K-S Chae, JH Kim, Y Choi, DM Han, K-Y Jahng. Isolation and characterization of a genomic DNA fragment complementing an *nsdD* mutation of *Aspergillus nidulans*. Mol Cells 5:146–150, 1995.

SP Champe, AE El-Zayat. Isolation of a sexual sporulation hormone from *Aspergillus nidulans*. J Bacteriol 171:3982–3988, 1989.

SP Champe, P Rao, A Chang. An endogenous inducer of sexual development in *Aspergillus nidulans*. J Gen Microbiol 133:1383–1387, 1987.

YC Chang, WE Timberlake. Identification of *Aspergillus brlA* response elements (BREs) by genetic selection in yeast. Genetics 133:29–38, 1992.

SW Chiu, D Moore. Programmed cell death is not involved in initiation of the gill cavity of *Coprinus cinereus*: A study using morphological mutants. Mushroom Sci 15:115–120, 2000.

CR Cisar, DO TeBeest. Mating system of the filamentous ascomycete, *Glomerella cingulata*. Curr Genet 35:127–133, 1999.

H Clémençon. Der Nodulus and die Ontogenese während der frühen Fruchtkörperentwicklung von *Psilocybe cyanescens*. Z Mykol 60:49–68, 1994.

H Clémençon. Anatomie der Hymenomyceten. Eine Einführung in die Cytologie und Plectologie der Krustenpilze, Porlinge, Keulenpilze, Leistlinge, Blätterpilze und Röhrlinge. (Anatomy of the Hymenomycetes. An Introduction to the Cytology and Plectology of Crust Fungi, Bracket Fungi, Club Fungi, Chantarelles, Agarics and Boletes.) Teufen, Switzerland: F Flück-Wirth, 1997.

AJ Clutterbuck. A mutational analysis of conidial development in *Aspergillus nidulans*. Genetics 63:317–327, 1969.

AJ Clutterbuck. The genetics of conidiophore pigmentation in *Aspergillus nidulans*. J Gen Microbiol 136:1731–1738, 1990.

DNW Cooper, RP Boulianne, S Charlton. E Farrell, A Sucher, BC Lu. Fungal galectins: Sequence and specificity of two isolectins from *Coprinus cinereus*. J Biol Chem 272:1514–1521, 1997.

E Coppin, R Debuchy. Co-expression of the mating-type genes involved in internuclear recognition is lethal in *Podospora anserina*. Genetics 155:657–669, 2000.

E Coppin, R Debuchy, S Arnaise, M Picard. Mating-types and sexual development in filamentous ascomycetes. Microbiol Mol Biol Rev 61:411–428, 1997.

V Coustou-Linares, ML Maddelein, J Begueret, SJ Saupe. In vivo aggregation of the HET-s prion protein of the fungus *Podospora anserina*. Mol Microbiol 42:1325–1335, 2001.

SK Crosthwaite, JC Dunlap, JJ Loros. *Neurospora wc-1* and *wc-2*: Transcription, photoresponses, and the origins of circadian rhythmicity. Science 276:763–769, 1997.

WJ Cummings, M Celerin, J Crodian, LK Brumick, ME Zolan. Insertional mutagenesis in *Coprinus cinereus*: Use of a dominant selectable marker to generate tagged, sporulation-defective mutants. Curr Genet 36:371–382, 1999.

RC Davidson, M Cristina Cruz, RAL Sia, B Allen, JA Alspaugh, J Heitman. Gene disruption by biolistic transformation in serotype D strains of *Cryptococcus neoformans*. Fungal Genet Biol 29:38–48, 2000.

R Debuchy. Internuclear recognition: A possible connection between euascomycetes and homobasidiomycetes. Fungal Genet Biol 27:218–223, 1999.

PWJ De Groot, PJ Schaap, LJLD Van Griensven, J Visser. Isolation of developmentally regulated genes from the edible mushroom *Agaricus bisporus*. Microbiology 143: 1993–2001, 1997.

PWJ De Groot, RTP Roeven, LJLD Van Griensven, J Visser, PJ Schaap. Different temporal and spatial expression of two hydrophobin-encoding genes of the edible mushroom *Agaricus bisporus*. Microbiology 145:1105–1113, 1999.

FS Dietrich, S Voegeli, T Gaffney, C Mohr, C Rebischung, R Wing, S Goff, P Philippsen. Gene map of chromosome I of Ashbya gossypii. Fungal Genetics Newsletter 46 (Supplement):127, 1999.

C Dietzel, J Kurjan. Pheromonal regulation and sequence of the *Saccharomyces cerevisiae* SST2 gene: A model for desensitization to pheromone. Mol Cell Biol 7:4169–4177, 1987.

KM Druey, KY Blumer, VH Kang, JH Kehrl. Inhibition of G-protein mediated MAP kinase activation by a new mammalian gene family. Nature 379:742–746, 1996.

S Duplessis, C Sorin, C Voiblet, B Palin, F Martin, D Tagu. Cloning and expression analysis of a new hydrophobin cDNA from the ectomycorrhizal basidiomycete *Pisolithus tinctorius*. Curr Genet 39:335–339, 2001.

S Duplessis, D Tagu, F Martin. Living together underground: A molecular glimpse of ectomycorrhizal symbiosis. In: HD Osiewacz, ed. Molecular Biology of Fungal Development. New York: Marcel Dekker, 2002, pp 297–324.

JR Dutton, S Johns, BL Miller. StuAp is a sequence-specific transcription factor that regulates developmental complexity in *Aspergillus nidulans*. EMBO J 16:5710–5721, 1997.

SE Eckert, B Hoffmann, C Wanke, GH Braus. Sexual development of *Aspergillus nidulans* in tryptophan auxotrophic strains. Arch Microbiol 172:157–166, 1999.

E Eidam. Zur Kenntniss der Entwicklung bei den Ascomyceten. III *Sterigmatocystis nidulans*. Beitr Biol Pfl 3:392–411, 1883.

SL Elifio, MDCC Da Solva, M Iacomini, PAJ Gorin. A lectin from the lichenized basidiomycete *Dictyonema glabratum*. New Phytol 148:327–334, 2000.

E Espagne, P Balhadere, ML Penin, C Barreau, B Turcq. HET-E and HET-D belong to a new subfamily of WD40 proteins involved in vegetative incompatibility specificity in the fungus *Podospora anserina*. Genetics 161:71–81, 2002.

K Esser. Kryptogamen. Cyanobakterien, Algen, Pilze, Flechten: Praktikum und Lehrbuch. 3rd ed. Berlin, Germany: Springer; 2000.

K Esser, R Blaich. Heterogenic incompatibility in fungi. In: JGH Wessels, F Meinhardt, eds. The Mycota. Vol. I. Growth, differentiation and sexuality. Berlin, Germany: Springer, 1994, pp 211–232.

M Evangelista, K Blundell, MS Longtine, CJ Chow, N Adames, JR Pringle, M Peter, C Boone. Bnilp, a yeast formin linking cdc42p and the actin cytoskeleton during polarized morphogenesis. Science 276:118–122, 1997.

AVB Ferreira, Z An, RL Metzenberg, NL Glass. Characterization of *mat A-2, mat A-3* and Δ*matA* mating-type mutants of *Neurospora crassa*. Genetics 148:1069–1079, 1998.

S Fillinger, MK Chaveroche, K Shimizu, N Keller, C d'Enfert. cAMP and ras signalling independently control spore germination in the filamentous fungus *Aspergillus nidulans*. Mol Microbiol 44, 1001–1016, 2002.

R Fischer, WE Timberlake. *Aspergillus nidulans apsA* (anucleate primary sterigmata) encodes a coiled-coil protein necessary for nuclear positioning and completion of asexual development. J Cell Biol 128:485–498, 1995.

TJ Fowler, MF Mitton. Scooter, a new active transposon in *Schizophyllum commune*, has interrupted two genes regulating signal transduction. Genetics 156:1585–1594, 2000.

TJ Fowler, SM DeSimone, MF Mitton, J Kurjan, CA Raper. Multiple sex pheromones and receptors of a mushroom-producing fungus elicit mating in yeast. Mol Biol Cell 10:2559–2572, 1999.

TJ Fowler, MF Mitton, LA Vaillancourt, CA Raper. Changes in mate recognition through alterations of pheromones and receptors in the multisexual mushroom fungus *Schizophyllum commune*. Genetics 158:1491–1503, 2001.

P Franken, N Requena. Molecular approaches to arbuscular mycorrhiza functioning. In: B Hock, ed. The Mycota. Vol. IX. Fungal associations. Berlin: Springer, 2001, pp 19–28.

AC Froehlich, Y Liu, JJ Loros, JC Dunlap. White-collar-1, a circadian blue light photoreceptor, binding to the *frequency* promoter. Science 297:815–819, 2002.

V Gavrias, A Andrianopoulos, CJ Gimeno, WE Timberlake. *Sacchromyces cerevisiae TEC1* is required for pseudohyphal growth. Mol Microbiol 19:1255–1263, 1996.

B Gillissen, J Bergemann, C Sandmann, B Schroer, M Bölker, R Kahmann. A two-component regulatory system for self/non-self recognition in *Ustilago maydis*. Cell 68:647–657, 1992.

M Giollant, J Guillot, M Damez, M Dusser, P Didier, E Didier. Characterization of a lectin from *Lactarius deterrimus*—Research on the possible involvement of the fungal lectin in recognition between mushroom and spruce during early stages of mycorrhizae formation. Plant Physiol 101:513–522, 1993.

NL Glass, DJ Jacobson, PKT Shiu. The genetics of hyphal fusion and vegetative incompatibility in filamentous ascomycete fungi. Ann Rev Genet 34:165–186, 2000.

A Goffeau, BG Barrell, H Bussey, RW Davis, B Dujon. H Feldmann, F Galibert, JD Hoheisel, C Jacq, M Johnston, EJ Louis, HW Mewes, Y Murakami, P Philippsen, H Tettelin, SG Oliver. Life with 6000 genes. Science 274:563–567, 1996.

S Gola, J Hegner, E Kothe. Chimeric pheromone receptors in the basidiomycete. *Schizophyllum commune*. Fungal Genet Biol 30:191–196, 2000.

GW Gooday. The control of differentiation in fruit bodies of *Coprinus cinereus* Rep Tottori Mycol Inst 12:151–160, 1975.

F Graia, V Berteaux-Lecellier, D Zickler, M Picard. *ami1*, an orthologue of the *Aspergillus nidulans apsA* gene, is involved in nuclear migration events throughout the life cycle of *Podospora anserina*. Genetics 155:633–646, 2000.

JD Granado, K Kertesz-Chaloupková, M Aebi, U Kües. Restriction enzyme-mediated DNA integration in *Coprinus cinereus*. Mol Gen Genet 256:28–36, 1997.

GW Griffith, MS Stark, AJ Clutterbuck. Wild-type and mutant alleles of the *Aspergillus nidulans* developmental regulator gene *brlA*: Correlation of variant sites with protein function. Mol Gen Genet 262:892–897, 1999.

DW Grogan, JE Cronan Jr. Cyclopropane ring formation in membrane lipids of bacteria. Microbiol Mol Biol Rev 61:429–441, 1997.

DI Gwynne, BL Miller, KY Miller, WE Timberlake. Structure and regulated expression of the *SpoC1* gene cluster from *Aspergillus nidulans*. J Mol Biol 180:91–109, 1984.

JR Halsall, MI Milner, LA Casselton. Three subfamilies of pheromone and receptor genes generate multiple B mating type specificities in the mushroom *Coprinus cinereus*. Genetics 154:1115–1123, 2000.

K-H Han, K-Y Han, J-H Chu, K-S Chae, K-Y Jahng, D-M Han. The *nsdD* gene encodes a putative GATA-type transcription of *Aspergillus nidulans*. Mol Microbiol 141: 299–309, 2001.

S Han, J Navarro, RA Greve, TH Adams. Translational repression of *brlA* expression prevents premature development in *Aspergillus*. EMBO J 12:2449–2457, 1993.

FM Harold. In pursuit of the whole hypha. Fungal Genet Biol 27:128–133, 1999.

DL Hawksworth, PM Kirk, BC Sutton, DN Pegler. Dictionary of the Fungi. 8th ed. Cambridge, UK: University Press, 1995.

Q He, P Cheng, Y Yang, L Wang, KH Gardner, Y Liu. White collar-1, a DNA binding transscription factor and a light sensor. Science 297:840–843, 2002.

J Hegner, C Siebert-Bartholmei, E Kothe. Ligand recognition in multiallelic pheromone receptors from the basidiomycete *Schizophyllum commune* studied in yeast. Fungal Genet Biol 26:190–197, 1999.

G Hennebert, LS Pascal, M Cosyns. Interactions d'incompatibilité entre homocaryons bifactoriel *Lenzites betulinus*. Une phéromone de rèpulsion sexuelle? Cryptogam Mycol 15:83–116, 1994.

TE Hermann, MB Kurtz, SP Champe. Laccase localized in Hulle cells and cleistothecial primordia of *Aspergillus nidulans*. J Bacteriol 154:955–964, 1983.

V Hershko, A Nussinovitch. Relationships between hydrocolloid coating and mushroom structure. J Agricul Food Chem 46:2988–2997, 1998.

JK Hicks, J-H Yu, NP Keller, TH Adams. *Aspergillus* sporulation and mycotoxin production both require inactivation of the FadA Gα protein-dependent signaling pathway. EMBO J 16:4916–4923, 1997.

SJ Hiscock, U Kües. Cellular and molecular mechanisms of sexual incompatibility in plants and fungi. Int Rev Cytol 193:165–295, 1999.

B Hoffmann, C Wanke, KS LaPaglia, GH Braus. c-Jun and RACK1 homologues regulate a control point for sexual development in *Aspergillus nidulans*. Mol Microbiol 37: 28–41, 2000.

B Hoffmann, SE Eckert, S Krappmann, GH Braus. Sexual diploids of *Aspergillus nidulans* do not form by random fusion of nuclei in the heterokaryon. Genetics 157:141–147, 2001a.

B Hoffmann, O Valerius, M Andermann, GH Braus. Transcriptional autoregulation and inhibition of mRNA translation of amino acid regulator gene *cpcA* of filamentous fungus *Aspergillus nidulans*. Mol Biol Cell 12:2846–2857, 2001b.

JS Horton, CA Raper. A mushroom-inducing DNA-sequence isolated from the basidiomycete, *Schizophyllum commune*. Genetics 129:707–716, 1991.

JS Horton, CA Raper. The mushroom inducing gene *frtl* of *Schizophyllum commune* encodes a putative nucleotide-binding protein. Mol Gen Genet 247:358–366, 1995.

JS Horton, GE Palmer, Smith WJ. Regulation of dikaryon-expressed genes by FRT1 in the basidiomycete *Schizophyllum commune*. Fungal Genet Biol 26:33–47, 1999.

SMFE Huber, F Lottspeich, J Kämper. A gene that encodes a product with similarity to dioxygenases is highly expressed in teliospores of *Ustilago maydis*. Mol Genet Genom 267:757–771, 2002.

K Inada, Y Morimoto, T Arima, Y Murata, T Kamada. The *clp1* gene of the mushroom *Coprinus cinereus* is essential for A-regulated sexual development. Genetics 157: 133–140, 2001.

M Iwasa, S Tanabe, T Kamada. The two nuclei in the dikaryon of the homobasidiomycete *Coprinus Cinereus* change position after each conjugate division. Fungal Genet Biol 23:110–116, 1998.

S Jacobsen, M Wittig, S Pöggeler. Interaction between mating-type proteins from the homothallic fungus *Sordaria macrospora*. Curr Genet 41:150–158, 2002.

G Jedd, N-H Chua. A new self-assembled peroxisomal vesicle required for efficient resealing of the plasma membrane. NCB 2:226–231, 2000.

H-Y Jeong, B-B Cho, K-Y Han, J Kim, DM Han, K-Y Jahng, K-S Chae. Differential expression of house-keeping genes of *Aspergillus nidulans* during sexual development. Gene 262:215–219, 2001.

RI Jirjis, D Moore. Involvement of glycogen in morphogenesis of *Coprinus cinereus*. J Gen Microbiol 95:348–352, 1976.

AD Johnson. Molecular mechanisms of cell-type determination in budding yeast. Curr Opin Genet Dev 5:552–558, 1995.

IL Johnstone, SG Hughes, AJ Clutterbuck. Cloning an *Aspergillus nidulans* developmental gene by transformation. EMBO J 4:1307–1311, 1985.

E Käfer. Origins of translocations in *Aspergillus nidulans*. Genetics 52:217–232, 1965.

E Käfer. Meiotic and mitotic recombination in *Aspergillus* and its chromosomal aberrations. Adv Genet 19:33–131, 1977.

R Kahmann, C Basse, M Feldbrügge. Fungal-plant signalling in the *Ustilago maydis* maize pathosystem. Curr Opin Microbiol 2:647–650, 1999.

R Kahmann, G Steinberg, C Basse, M Feldbrügge, J Kämper. *Ustilago maydis*, the causative agent of corn and disease. In: JW Kronstad, ed. Fungal Pathology. Dordrecht: Kluwer Academic Publishers, 2000, pp 347–372.

T Kamada. Stipe elongation in fruit bodies. In: JGH Wessels, F Meinhard, eds. The Mycota. Vol I. Growth, Differentiation and Sexuality. Berlin, Germany: Springer, 1994, pp 367–380.

T Kamada. Molecular genetics of sexual development in the mushroom *Coprinus cinereus*. BioEssays 24:449–459, 2002.

T Kamada, M Tsuru. The onset of the helical arrangement of chitin microfibrils in fruit-body development of *Coprinus cinereus*. Mycol Res 97:884–888, 1993.

SGW Kaminskyj. Septum position is marked at the tip of *Aspergillus nidulans* hyphae. Fungal Genet Biol 31:105–114, 2000.

SGW Kaminskyj, JE Hamer. hyp loci control cell pattern formation in the vegetative mycelium of *Aspergillus nidulans*. Genetics 148:669–680, 1998.

J Kämper, M Reichmann, T Romeis, M Bölker, R Kahmann. Multiallelic recognition, nonself-dependent dimerization of the bE and bW homeodomain proteins in *Ustilago maydis*. Cell 81:73–83, 1995.

J Kämper, G Weinzierl, A Brachmann, M Feldbrügge, C Basse, G Steinberg, R Kahmann, G Friedrich, V Vollenbroich, E Koopmann, I Häuser-Hahn, D Nennstiel, K Sievert, R Suelmann, M Vaupel, C Aichinger, R Ebbert, B Leuthner, B Jaitner, V Li, P Schreier, T Schlüter, D Schütte, H Kranz, J Heinrich, G Kurapkat, M Arenz, H Voss. The *Ustilago maydis* sequencing project. Fungal Genet Newsl 48 (Suppl): 169, 2001.

JC Kapteyn, E van den Ende, FM Klis. The contribution of cell wall proteins to the organization of the yeast cell wall. Biochim Biophys Acta 1426:373–383, 1999.

M Karos, R Fischer. *hymA* (Hypha-like metulae), a new developmental mutant of *Aspergillus nidulans*. Microbiology 142:3211–3218, 1996.

L Kawasaki, J Aguirre. Multiple catalase genes are differentially regulated in *Aspergillus nidulans*. J Bacteriol 183:1434–1440, 2001.

AM Kays, PS Rowlry, RA Baasiri, KA Borkovich. Regulation of conidiation and adenylyl cyclase levels by the Gα protein GNA-3 in *Neurospora crassa*. Mol Cell Biol 20: 7693–7705, 2000.

M Kelly, J Burke, M Smith, A Klar, D Beach. Four mating-type genes control sexual differentiation in the fission yeast. EMBO J 7:1537–1547, 1988.

ROF Kemp. Oidial homing and the taxonomy and speciation of basidiomycetes with special reference to the genus *Coprinus*. In: H Clémençon, ed. The Species Concept in Hymenomycetes. Vaduz, Switzerland: Cramer, 1977, pp 259–276.

B Kendrick, R Watling. Mitospores in basidiomycetes. In: B Kendrick, ed. The Whole Fungus. Ottawa, Canada: National Museums of Canada, 1979, pp 473–546.

K Kertesz-Chaloupková, PJ Walser, M Aebi, U Kües. Blue light overrides repression of asexual sporulation by mating type genes in the basidiomycete *Coprinus cinereus*. Fungal Genet Biol 23:18–33, 1998.

S Kim, K-Y Han, K-J Kim, D-M Han, K-Y Jahng, K-S Chae. The *veA* gene activates sexual development in *Aspergillus nidulans*. Fungal Genet Biol 37: 72–80, 2002.

K Kimura. Studies on the sex of *Coprinus macrorhizus* Rea f. *microsporus* Hongo. I. Introductory experiments. Biol J Okayama Univ 1:72–79, 1952.

Y Kitamoto, M Shishida, H Yamamoto, K Tatkeo, P Masuda. Nuclear selection in oidium formation from dikaryotic mycelia of *Flammulina velutipes*. Mycoscience 41:417–424, 2000.

A Komeili, EK O'Shea. Nuclear transport and transcription. Curr Opin Cell Biol 12: 355–360, 2000.

E Kothe. Mating types and pheromone recognition in the homobasidiomycete *Schizophyllum commune*. Fungal Genet Biol 27:146–152, 1999.

E Kothe. Mating type genes for basidiomycete strain improvement in mushroom farming. Appl Microbiol Biotechnol 56:602–612, 2001.

GO Kothe, SJ Free. The isolation and characterization of *nrc-1* and *nrc-2*, two genes encoding protein kinases that control growth and development in *Neurospora crassa*. Genetics 149:117–130, 1998.

JW Kronstad, C Staben. Mating type in filamentous fungi. Ann Rev Genet 31:245–276, 1997.

M Krüger, N Sievers, R Fischer. Molekularbiologie der Sporenträgerentwicklung des Schimmelpilzes *Aspergillus nidulans*. Biol in unserer Zeit 6:375–382, 1997.

U Kües. Life history and developmental processes in the basidiomycete *Coprinus cinereus*. Microbiol Mol Biol Rev 64:316–353, 2000.

U Kües. Sexuelle and asexuelle Fortpflanzung bei einem Pilz mit 12000 Geschlechtern. [Sexual and asexual reproduction of a fungus with 12000 sexes.] Vierteljahrsschrift der Naturforschenden Gesellschaft in Zürich 147:23–34, 2002.

U Kües, LA Casselton. Homeodomains and regulation of sexual development in basidiomycetes. Trends Genet 8:154–155, 1992.

U Kües, LA Casselton. The origin of multiple mating types in mushrooms. J Cell Sci 104: 227–230, 1993.

U Kües, Y Liu. Fruiting body production in basidiomycetes. Appl Microbiol Biotechnol 54:141–152, 2000.

U Kües, WVJ Richardson, AM Tymon, ES Mutasa, B Göttgens, S Gaubatz, A Gregoriades, LA Casselton. The combination of dissimilar alleles of the $A\alpha$ and $A\beta$ gene complexes, whose proteins contain homeodomain motifs, determine sexual development in the mushroom Coprinus cinereus. Genes Dev 6:568–577, 1992.

U Kües, RN Asante-Owusu, ES Mutasa, AM Tymon, EH Pardo, SF O'Shea, LA Casselton. Two classes of homeodomain proteins specifiy the multiple A mating types of the mushroom *Coprinus cinereus*. Plant Cell 6:1467–1475, 1994a.

U Kües, B Göttgens, R Stratmann, WVJ Richardson, SF O'Shea, LA Casselton. A chimeric homeodomain protein causes self-compatibility and constitutive sexual development in the mushroom *Coprinus cinereus*. EMBO J 13: 4054–4059, 1994b.

U Kües, JD Granado, R Hermann, RP Boulianne, K Kertesz-Chaloupková, M Aebi. The A mating type and blue light regulate all known differentiation processes in the basidiomycete *Coprinus cinereus*. Mot Gen Genet 260:81–91, 1998a.

U Kües, JD Granado, K Kertesz-Chaloupková, PJ Walser, M Hollenstein, E Polak, Y Liu, RP Boulianne, APF Bottoli, M Aebi. Mating types and light are major regulators in *Coprinus cinereus*. In: LJLD Van Griensven, J Visser, eds. Proceedings of the Fourth Meeting on the Genetics and Cellular Biology of Basidiomycetes. Horst, Netherlands: Mushroom Experimental Station, 1998b, pp 113–118.

U Kües, TY James, R Vilgalys, MP Challen. The chromosomal region containing *pab-1*, *mip*, and the *A* mating type locus of the secondarily homothallic homobasidiomycete *Coprinus bilanatus*. Curr Genet 294:16–24, 2001.

U Kües, M Kilnzler, APF Bottoti, PJ Walser, JD Granado, Y Liu, RC Bertossa, D Ciardo, P-H Clergeot, S Loos, G Ruprich-Robert, M Aebi. Mushroom development in higher basidiomycetes: Implications for human and animal health. In: RKS Kushwaha, ed. Fungi in Human and Animal Health. Jodhpur, India: Scientific Publishers (India), 2002a, in press.

U Kües, E Polak, M Hollenstein, APF Bottoli, PJ Walser, RP Boulianne, M Aebi. Vegetative development in *Coprinus cinereus*. In: HD Osiewacz, ed. Molecular Biology of Fungal Development. New York: Marcel Dekker, 2002b, pp 133–164.

U Kües, PJ Walser, MJ Klaus, M Aebi. Influence of activated *A* and *B* mating type pathways on developmental processes in the basidiomycete *Coprinus cinereus*. Mol Genet Genom 268:262–271, 2002.

A Kumata, T Takehara, S Aono. Changes of culture properties and colony types in the process of consecutive subculturing of *Pholiota nameko* stock on potato-dextrose-agar plate media. J Japan Wood Res Soc 42:101–104, 1996.

H Kunitomo, T Higuchi, Y Iino, M Yamamoto. A zinc-finger protein, rst2p, regulates transcription of the fission yeast *ste11(+)* gene, which encodes a pivotal transcription factor for sexual development. Mol Biol Chem 11:3205–3217, 2000.

DM Kupfer, CA Reese, SW Clifton, BA Roe, RA Prade. Multicellular ascomycetous fungal genomes contain more than 8000 genes. Fungal Genet Biol 21:364–372, 1997.

PL Lakin-Thomas, S Brody. Circadian rhythms in *Neurospora crassa*: Lipid deficiencies restore robust rhythmicity to null frequency and white-collar mutants. Proc Natl Acad Sci USA 97:256–261, 2000.

I Laloux, E Dubois, M Dewerchin, E Jacobs. TEC1, a gene involved in the activation of Ty1 and Ty1-mediated gene expression in *Saccharomyces cerevisiae*: Cloning and molecular analysis. Mol Cell Biol 10:3541–3550, 1990.

P Laurent, C Voiblet, D Tagu, D de Carvalho, U Nehls, R De Bellis, R Balestrini, G Bauw, P Bonfante, F Martin. A novel class of ectomycorrhiza-regulated cell wall polypeptides in *Pisolithus tinctorius*. Mol Plant Microbe Interact 12:862–871, 1999.

BN Lee, TH Adams. The *Aspergillus nidulans fluG* gene is required for production of an extracellular developmental signal and is related to prokaryotic glutamine synthetase I. Genes Dev 8:641–651, 1994a.

BN Lee, TH Adams. Overexpression of *flbA*, an early regulator of *Aspergillus* asexual sporulation, leads to activation of *brlA* and premature initiation of development. Mol Microbiol 14:323–334, 1994b.

BN Lee, TH Adams. *fluG* and *flbA* function independently to initiate conidiophore development in *Aspergillus nidulans* through *brlA* β activation. EMBO J 15:299–309, 1996.

DW Lee, SH Lee, H-A Hwang, JH Kim, K-S Chae. Quantitative analysis of gene expression in sexual structures of *Aspergillus nidulans* by sequencing of 3′-directed cDNA clones. FEMS Microbiol Letters 138:71–76, 1996.

N Lee, G Bakkeren, K Wong, JE Sherwood, JW Kronstad. The mating-type and pathogenicity locus of the fungus *Ustilago hordei* spans a 500-kb region. Proc Natl Acad Sci USA 96:15026–15031, 1999.

SH Lee, BG Kim, KJ Kim, JS Lee, DW Yun, JH Hahn, GH Kim, KH Lee, DS Suh, ST Kwon, CS Lee, YB Yoo. Comparative analysis of sequences expressed during the liquid-cultured mycelia and fruit body stages of *Pleurotus ostreatus*. Fungal Genet Biol 35:115–134, 2002.

K Lengeler, E Kothe. Identification and characterization of *brt1*, a gene down-regulated during *B*-regulated development in *Schizophyllum commune*. Curr Gen 35:551–556, 1999a.

K Lengeler, E Kothe. Mated: A putative peptide transporter of *Schizophyllium commune* expressed in dikaryons. Curr Genet 36:159–164, 1999b.

KB Lengeler, RC Davidson, C D'Souza, T Harashima, W-C Shen, P Wang, X Pan, M Waugh, J Heitman. Signal transduction cascades regulating fungal development and virulence. Microbiol Mol Biol Rev 64:746–785, 2000.

JF Leslie. Fungal vegetative compatibility. Ann Rev Phytopathol 31:127–150, 1993.

GSW Leung, M Zhang, WJ Xie, HS Kwan. Identification by RNA fingerprinting of genes differentially expressed during the development of the basidiomycete *Lentinula edodes*. Mol Gen Genet 262:977–990, 2000.

H Linden. A white collar protein senses blue light. Science 297:777–778, 2002.

H Linden, P Ballario, G Macino. Blue light regulation in *Neurospora crassa*. Fungal Genet Biol 22:141–150, 1997.

H Liu, TR Cottrel, LM Pierini, WE Goldman, TL Doering. RNA interference in the pathogenic fungus *Cryptococcus neoformans*. Genetics 160:463–470, 2002.

Y Liu. Fruiting body initiation in the basidiomycete *Coprinus cinereus*. PhD thesis, ETH Zurich, Zurich, Switzerland, 2001.

G Loubradou, A Brachmann, M Feldbrügge, R Kahmann. A homologue of the transcriptional repressor Ssn6p antagonizes cAMP signalling in *Ustilago maydis*. Mol Microbiol 40:719–730, 2001.

BC Lu. Meiosis in *Coprinus lagopus*: A comparative study with light and electron microscopy. J Cell Sci 2:529–536, 1967.

BC Lu. Meiosis in *Coprinus*. V. The role of light on basidiocarp initiation, meiosis, and hymenium differentiation in *Coprinus lagopus*. Can J Bot 52:299–305, 1974.

BC Lu. The cellular program for the formation and dissolution of the synaptonemal complex in *Coprinus*. J Cell Sci 67:25–43, 1984.

BC Lu. Cell degeneration and gill remodeling during basidiocarp development in the fungus *Coprinus cinereus*. Can J Bot 69:1161–1169, 1991.

BCK Lu. Chromosomes, mitosis, and meiosis. In: CJ Bos, ed. Fungal Genetics. Principles and Practice. New York Marcel Dekker, Inc., 1996, pp 119–176.

BC Lu. The control of meiosis progression in the fungus *Coprinus cinereus* by light/dark cycles. Fungal Genet Biol 31:33–41, 2000.

BC Lu, N Gallo, U Kües. Apoptosis occurs in the basidia of mutants of *Coprinus cinereus*. Fungal Genet Newsl 48 (Suppl):49, 2001.

LG Lugones, HAB Wösten, KU Birkenkamp, KA Sjollema, J Zagers, JGH Wessels. Hydrophobins line air channels in fruiting bodies of *Schizophyllum commune* and *Agaricus bisporus*. Mycol Res 103:635–640, 1999.

YH Luo, RC Ullrich, CP Novotny. Only one of the paired Schizophyllium commune Aα mating-type putative homeobox genes encodes a homeodomain essential for Aα regulated development. Mot Gen Genet 244:318–324, 1995.

S Machida, Y Itho, H Kishida, T Higasa, M Saito. Localization of chitin synthase in *Absidia glauca* studied by immunoelectron microscopy: Application of cryo-ultra-microtomy. Biosci Biotech Biochem 58:1983–1989, 1994.

ML Maddelein, S Dos Reis, S Duvezin-Caubet, B Coulary-Salin, SJ Saupe. Amyloid aggregates of the HET-s prion protein are infectious. Proc Natl Acad Sci USA 99: 7402–7407, 2002.

Y Magae, C Novotny, R Ullrich. Interaction of the Aα Y mating-type and Z mating-type homeodomain proteins of *Schizophyllum commune* detected by the two-hybrid system. Biochem Biophys Res Commun 211:1071–1076, 1995.

A Mankel, K Krause, E Kothe. Identification of a hydrophobin gene that is developmentally regulated in the ectomycorrhizal fungus *Tricholoma terreum*. App Env Microbiol 68:1408–1413, 2002.

JF Marhoul, TH Adams. Identification of developmental regulatory genes in *Aspergillus nidulans* by overexpression. Genetics 139:537–547, 1995.

JF Marhoul, TH Adams. *Aspergillus fabM* encodes an essential product that is related to poly(A)-binding proteins and activates development when overexpressed. Genetics 144:1463–1470, 1996.

AL Marion, KA Bartholomew KA, J Wu, HL Yang, CP Novotny, RC Ullrich. The Aα mating-type locus of *Schizophyllum*; Structure and function of gene X. Curr Genet 29:143–149, 1995.

MA Marshall, WE Timberlake. *Aspergillus nidulans wetA* activates spore-specific gene expression. Mol Cell Biol 11:55–62, 1991.

F Martin, P Lauren, D de Carvalho, C Voiblet, R Balestrini, P Bonfante, D Tagu. Cell wall proteins of the ectomycorrhizal basidiomycete *Pisolithus tinctorius*: Identification, function, and expression in symbiosis. Fungal Genet Biol 27:161–174, 1999.

GG Martin, GC Cannon, CL McCormick. Adsorption of a fungal hydrophobin onto surfaces as mediated by the associated polysaccharide schizophyllan. Biopolymers 49:621–633, 1999.

GG Martin, GC Cannon, CL McCormick. Sc3p hydrophobin organization in aqueous media and assembly onto surfaces as mediated by the associated polysaccharide schizophyllan. Biomacromolecules 1:49–60, 2000.

SD Martinelli, AJ Clutterbuck. A quantitative survey of conidiation mutants in *Aspergillus nidulans*. J Gen Microbiol 69:261–268, 1971.

TR Matthews, DJ Niederpruem. Differentiation in *Coprinus lagopus*. I. Control of fruiting and cytology of initial events. Arch Mikrobiol 87:257–268,1972.

TR Matthews, DJ Niederpruem. Differentiation in *Coprinus lagopus*. II. Histology and ultrastructural aspects of developing primordia. Arch Mikrobiol 88:169–180, 1973.

G May. Somatic incompatibility and individualism in the coprophilus basidiomycete *Coprinus cinereus*. Trans Br Mycol Soc 91:443–451, 1988.

ME Mayorga, WE Timberlake. Isolation and molecular characterization of the *Aspergillus nidulans wA* gene. Genetics 126:73–79, 1990.

ME Mayorga, WE Timberlake. The developmentally regulated *Aspergillus nidulans wA* gene encodes a polypeptide homologous to polyketide and fatty acid synthases. Mol Gen Genet 235:205–212, 1992.

P Mazur, K Nakanishi. Enantioselective synthesis of PsiAβ, a sporogenic metabolite of *Aspergillus nidulans*. J Org Chem 57:1047–1051, 1992.

P Mazur, HV Meyers, K Nakanishi. Structural elucidation of sporogenic fatty acid metabolites from *Aspergillus nidulans*. Tetrahedron Lett 31:3837–3840, 1990.

CA McGoldrick, C Gruver, GS May. *myoA* of *Aspergillus nidulans* encodes an essential myosin I required for secretion and polarized growth. J Cell Biol 128:577–587, 1995.

V Measday, L Moore, R Retnakaraan, J Lee, M Donoviel, AM Neiman, B Andrews. A family of cyclin-like proteins that interact with the Pho85 cyclin-dependent kinase. Mot Cell Biol 17:1212–1223, 1997.

V Measday, H McBride, J Moffat, D Stillman, B Andrews. Interactions between Pho85 cyclin-dependent kinase complexes and the Swi5 transcription factor in budding yeast Mol microbiol 35:825–834, 2000.

ST Merino, WJ Cummings, SN Acharya, ME Zolan. Replication-independent early meiotic requirement for Spoll and Rad50. Proc Nati Acad Sci USA 97:10477–10482, 2000.

RL Metzenberg, NL Glass. Mating and mating strategies in *Neurospora*. BioEssays 12: 53–59, 1990.

BL Miller, KY Miller, KA Roberti, WE Timberlake. Position-dependent and -independent mechanisms regulate cell-specific expression of the *SpoC1* gene cluster of *A. nidulans*. Mol Cell Biol 7:427–434, 1987.

KY Miller, TM Toennis, TH Adams, BL Miller. Isolation and transcriptional characterization of a morphological modifier: The *Aspergillus nidulans* stunted (stuA) gene. Mol Gen Genet 227:285–292, 1991.

KY Miller, J Wu, BL Miller. StuA is required for cell pattern formation in *Aspergillus*. Genes Dev 6:1770–1782, 1992

CW Mims, EA Richardson, WE Timberlake. Ultrastructural analysis of conidiophore development in the fungus *Aspergillus nidulans* using freeze-substitution. Protoplasma 44:132–141, 1988.

PM Mirabito, TH Adams, WE Timberlake. Interactions of three sequentially expressed genes control temporal and spatial specificity in *Aspergillus* development. Cell 57: 859–868, 1989.

N Mochizuki, M Yamamoto. Reduction in the intracellular cAMP level triggers initiation of sexual development in fission yeast. Mol Gen Genet 233:17–24, 1992.

M Momany, EA Richardson, C Van Sickle, G Jedd. Mapping Woronin body position in *Aspergillus nidulans*. Mycologia 94:260–266, 2002.

JL Mooney, LN Yager. Light is required for conidiation in *Aspergillus nidulans*. Genes Dev 4:1473–1482, 1990.

D Moore. Developmental genetics of *Coprinus cinereus*: genetic evidence that carpophores and sclerotia share a common pathway of initiation. Curr Genet 3:145–150, 1981.

D Moore. Fungal Morphogenesis. Cambridge, UK: Cambridge University Press, 1998.

RT Moore. The challenge of the dolipore/parenthosome septum. In: D Moore, LA Casselton, DA Wood, JC Frankland, eds. Developmental Biology of Higher Fungi. Cambridge, UK: Cambridge University Press, 1985, pp 175–212.

HU Mösch, E Kubler, S Krappmann, GR Fink, GH Braus. Crosstalk between the Ras2p-controlled mitogen-activated protein kinase and CAMP pathways during invasive growth of *Saccharomyces cerevisiae*. Mol Cell Biol 10:1325–1335, 1999.

CA Muirhead, NL Glass, M Slatkin. Multilocus self-recognition systems in fungi as a cause of trans-species polymorphism. Genetics 161:633–641, 2002.

GH Mulder, JGH Wessels. Molecular cloning of RNA species differentially expressed in monokaryons and dikaryons of *Schizophyllum commune* in relation to fruiting. Exp Mycol 10:214–227, 1986.

H Muraguchi, T Kamada. The *ich1* gene of the mushroom *Coprinus cinereus* is essential for pileus formation in fruiting. Development 125:3133–3141, 1998.

H Muraguchi, T Kamada. A mutation in the *eln2* gene encoding a cytochrome P450 of *Coprinus cinereus* affects mushroom morphogenesis. Fungal Genet Biol 29:49–59, 2000.

H Muraguchi, T Kamada. A distinctive phenotype in the common-*A* heterokaryon of *Coprinus cinereus*. Mycoscience 43:77–79, 2002.

H Muraguchi, T Takemaru, T Kamada. Isolation and characterization of developmental variants in fruiting using a homokaryotic fruiting strain of *Coprinus cinereus*. Mycoscience 40:227–233, 1999.

Y Murata, M Fujii, ME Zolan, T Kamada. Molecular analysis of *pcc1*, a gene that leads to A-regulated sexual morphogenesis in *Coprinus cinereus*. Genetics 149:1753–1761, 1998a.

Y Murata, M Fujii, ME Zolan, T Kamada. Mutations in clamp-cell formation, the A-regulated sexual morphogenesis, in homobasidiomycetes. In: LJLD Van Griensven, J Visser, eds. Proceedings of the Fourth Meeting on the Genetics and Cellular Biology of Basidiomycetes. Horst, The Netherlands: Mushroom Experimental Station, 1998b, pp 143–146.

MI Muro-Pator, R Gonzalez, J Strauss, F Narendja, C Scazzocchio. The GATA factor AreA is essential for chromatin remodelling in a eukaryotic bidirectional promoter, EMBO J 15:1584–1597, 1999.

U Nehls, S Mikolajewski, M Ecke, R Hampp. Identification and expression analysis of two fungal cDNAs regulated by ectomycorrbiza and fruit body formation. New Phytol 144:195–202, 1999.

MA Nelson, S Kang, El Braun, ME Crawford, PL Dolan, J Leonard, J Mitchell, AM Armijo, L Bean, E Blueyes, T Cushing, A Errett, M Fleharty, M Gorman, K Judson, R Miller, J Ortega, I Pavlova, J Perea, S Todisco, R Trujillo, J Valentine, A Wells, M Werner-Washburne, S Yazzie, DO Natvig. Expressed sequences from conidial, mycelial and sexual stages of *Neurospora crassa*. Fungal Genet Biol 21: 348–363,1997.

WL Ng, TP Ng, HS Kwam. Cloning and characterization of two hydrophobin genes differentially expressed during fruit body development in *Lentinula edodes*. FEMS Microbiol Lett 185:139–145, 2000.

M Nowrousian, S Masloff S Pöggeler, U Kück Cell differentiation during sexual development of the fungus *Sordaria macrospora* requires ATP citrate lyase activity. Mol Cell Biol 19:450–460, 1999.

M Nowrousian, U Kück, K Loser, KM Weltring. The fungal *acl1* and *acl2* genes encode two polypeptides with homology to the N- and C-terminal parts of the animal ATP citrate lyase polypeptide. Cuff Genet 37:189–193, 2000.

F Oberwinkler. Basidiolichens. In: The Mycota. Vol IX. Fungal Associations. B Hock, ed. Berlin, Germany: Springer, 2001, pp 211–228.

NS Olesnicky, AJ Brown, SJ Dowell, LA Casselton. A constitutively active G protein coupled receptor causes mating self-compatibility. EMBO J 18:2756–2763, 1999.

NS Olesnicky, AJ Brown, Y Honda, SL Dyos, SJ Dowell, LA Casselton. Self-compatible B mutants in *Coprinus* with altered pheromone-receptor specificities. Genetics 156: 1025–1033, 2000.

PTP Oliver. Conidiophore and spore development in *Aspergillus nidulans*. J Gen Microbiol 73:45–54, 1972.

SF O'Shea, PT Chaure, JR Halsall, NS Olesnicky, A Leibbrandt, IF Connerton, LA Casselton. A large pheromone and receptor gene complex determines multiple B mating type specificities in *Coprinus cinereus*. Genetics 148:1081–1090, 1998.

N Osherov, G May. Conidial germination in *Aspergillus nidulans* requires RAS signaling and protein synthesis. Genetics 155:647–656, 2000.

MD Ospina-Giraldo, PD Collopy, CP Romaine, DJ Royse. Classification of sequences expressed during the primordial and basidiome stages of the cultivated mushroom *Agaricus bisporus*. Fungal Genet Biol 29:81–94, 2000.

EH Pardo. Organisation of the A mating type locus of *Coprinus cinereus*. PhD thesis, University of Oxford, Oxford, UK, 1995.

EH Pardo, SF O'Shea, LA Casselton. Multiple versions of the A mating type locus of *Coprinus cinereus* are generated by three paralogous pairs of multiallelic homeobox genes. Genetics 144:87–94, 1995.

M Pardo, M Ward, S Bains, M Molina, W Blackstock, C Gil, C Nombela. A proteomic approach for the study of *Saccharomyces cerevisiae* cell wall biogenesis. Electrophoresis 21:3396–3410, 2000

RC Pascon, BL Miller. Morphogenesis in *Aspergillus nidulans* requires Dopey (DopA), a member of a novel family of leucine zipper-like proteins conserved from yeast to humans. Mol Microbiol 36:1250–1264, 2000.

Philley ML, C Staben. Functional analyses of the *Neurospora crassa* MT a-1 mating type polypeptide. Genetics 137:722–751, 1994.

S Pöggeler. Phylogenetic relationships between mating-type sequences from homothallic and heterothallic ascomycetes. Curr Genet 36:222–231, 1999.

S Pöggeler. Two pheromone precursor genes are transcriptionally expressed in the homothallic ascomycete *Sordaria macrospora*. Curr Genet 37:403–411, 2000.

S Pöggeler. Mating-type genes for classical strain improvements of ascomycetes. Appl Microbiol Biotechnol 56:589–601, 2001.

S Pöggeler, U Kück. Comparative analysis of mating-type loci from *Neurospora crassa* and *Sordaria macrospora*: Identification of novel transcribed ORFs. Mol Gen Genet 263:292–301, 2000.

S Pöggeler, U Kück. Identification of transcriptionally expressed pheromone receptor genes in filamentous ascomycetes. Gene 280:9–17, 2001.

S Pöggeler, S Risch, U Kück, HD Osiewacz. Mating-type genes from the homothallic fungus *Sordaria macrospora* are functionally expressed in a heterothallic ascomycete. Genetics 147:567–580,1997.

E Polak. Asexual Sporulation in the Basidiomycete *Coprinus cinereus*. PhD thesis, ETH Zurich, Zurich, Switzerland, 1999.

E Polak, R Hermann, U Kües, M Aebi. Asexual sporulation in *Coprinus cinereus*: Structure and development of oidiophores and oidia in an *Amut Bmut* homokaryon. Fungal Genet Biol 22:112–126, 1997.

E Polak, M Aebi, U Kües. Morphological variations in oidium formation in the basidiomycete *Coprinus cinereus*. Mycol Res 105:603–610, 2001.

G Pontecorvo, JA Roper, LM Hemmons, KD MacDonald, AWJ Bufton. The genetics of *Aspergillus nidulans*. Adv Genet 5:141–238, 1953.

RA Prade. The reliability of the *Aspergillus nidulans* physical map. Fungal Genet Biol 29:175–185, 2000.

R Prade, WE Timberlake. The *Aspergillus nidulans brlA* regulatory locus consists of two overlapping transcription units that are individually required for conidiophore development. EMBO J 12:2439–2447, 1993.

RA Prade, WE Timberlake. The *Penicillium chrysogenum* and *Aspergillus nidulans wetA* developmental regulatory genes are functionally equivalent. Mol Gen Genet 244: 539–547, 1994.

RA Prade, P Ayoubi, S Krishnan, S Macwana, H Russell. Accumulation of stress and inducer-dependent plant-cell-wall-degrading enzymes during asexual development in *Aspergillus nidulans*. Genetics 157:957–967, 2001.

C Quadbeck-Seeger, G Wanner, S Huber, R Kahmann, J Kämper. A protein with similarity to the human retinoblastoma binding protein 2 acts specifically as a repressor for genes regulated by the b mating type locus in *Ustilago maydis*. Mol Microbiol 38: 154–166, 2000.

M Ramsdale, PL Lakin-Thomas. sn-1,2-Diacylglycerol levels in the fungus *Neurospora crassa* display circadian rhythmicity. J Biol Chem 275:27541–27550, 2000.

JR Raper. Genetics of Sexuality in Higher Fungi. New York: Ronald Press, 1966.

JR Raper, DH Boyd, CA Raper. Primary and secondary mutations at the incompatibility loci in *Schizophyllum*. Proc Natl Acad Sci USA 53:1324–1332, 1965.

M Raudaskoski. The relationship between *B*-mating-type genes and nuclear migration in *Schizophyllum commune*. Fungal Genet Biol 24:207–227, 1998.

ADM Rayner. The challenge of the individualistic mycelium. Mycologia 83:48–71, 1991.

AFM Reinjnders. The histogenesis of bulb and trama tissue of the higher basidiomycetes and its phylogenetic implications. Perssonia 9:329–362, 1977.

AFM Reinjnders. On the origin of specialized trama types in the Agaricales. Mycol Res 97:257–268, 1979.

CG Reynaga-Peñ, G Gierz, S Bartnicki-Garcia. Analysis of the role of the Spitzenkörper in fungal morphogenesis by computer simulation of apical branching in *Aspergillus niger*. Proc Natl Acad Sci USA 94:9096–9101, 1997.

M Riquelme, CG Reqnaga-Pena, G Gierz, S Bartnicki-Garcia. What determines growth direction in fungal hyphae? Fungal Genet Biol 24:101–109, 1998.

M Riquelme, G Gierz, S Bartnicki-Garcia. Dynein and dynactin deficiencies affect the formation and function of the Spitzenkörper and distort hyphal morphogenesis of *Neurospora crassa*. Microbiology 146:1743–1752, 2000.

CI Robertson, KA Bartholomew, CP Novotny, RC Ullrich. Deletion of the *Schizophyllum commune A*α locus: The roles of Aα Y and Z mating-type genes. Genetics 144: 1437–1444, 1996.

CI Robertson, AM Kende, K Toenjes, CP Novotny, RC Ullrich. Evidence for interaction of *Schizophyllum commune* Y mating-type proteins in vivo. Genetics 160:1461–1467, 2002.

H Röhr, U Kües, U Stahl. Organelle DNA of plants and fungi: Inheritance and recombination. Prog Bot 58:307–351, 1998.

T Romeis, A Brachmann, R Kahmann, J Kämper. Identification of a target gene for the bE-bW homeodomain protein complex in *Ustilago maydis*. Mol Microbiol 37:54–66, 2000.

S Rosén, J-H Yu, TH Adams. The *Aspergillus nidulans sfaD* gene encodes a G protein β subunit that is required for normal growth and repression of sporulation. EMBO J 18:5592–5600, 1999.

G Ruprich-Robert, V Berteaux-Lecellier, D Zickler, A Panvier-Adoutte, M Picard. Identification of six loci in which mutations partially restore peroxisome biogenesis and/or alleviate the metabolic defect of pex2 mutants in *Podospora*. Genetics 161:1089–1099, 2002.

C Santos, J Labarére. Aa-Pri2, a single copy gene from *Agrocybe aegerita*, specifically expressed during friuting initiation, encodes a hydrophobin with a leucine-zipper domain. Curr Genet 35:564–570, 1999.

SJ Saupe. Molecular genetics of heterokaryon incompatibility in filamentous ascomycetes. Microbiol Mol Biol Rev 64:489–502, 2000.

SJ Saupe, C Clavé, J Begueret. Vegetative incompatibility in filamentous fungi: *Podospora* and *Neurospora* provide some clues. Curr Opin Microbiol 3:608–612, 2000.

M Scherer, R Fischer. Purification and characterization of lactase II of *Aspergillus nidulans*. Arch Microbiol 170:78–84, 1998.

M Scherer, H. Wei, R Liese, R Fischer. *Aspergillus nidulans* catalase-peroxidase (*cpeA*) is transcriptionally induced during sexual development through the APSES-transcription factor StuA Euk Cell 1:725–735, 2002.

S Scherrer, OHM de Vries, R Dudler, JGH Wessels, R Honegger. Interfacial self-assembly of fungal hydrophobins of the lichen-forming ascomycetes *Xanthoria parietina* and *X. ectaneoides*. Fungal Genet Biol 30, 81–93, 2000.

S Scherrer, A Haisch, R Honegger. Characterization and expression of XPH1, the hydrophobin gene of the lichen-forming ascomycete *Xanthoria parietina*. New Phytol 154:175–184, 2002.

N Schier, R Liese, R Fischer. A pcl-like cyclin of *Aspergillus nidulans* is transcriptionally activated by developmental regulators and is involved in sporulation. Mol Cell Biol 21:4075–4088, 2001.

R Schlesinger, R Kahmann, J Kämper. The homeodomain of the heterodimeric bE and bW proteins of *Ustilago maydis* are both critical for function. Mol Gen Genet 254:514–519, 1997.

FHJ Schuren. Atypical interactions between *thn* and wild type mycelia of *Schizophyllum commune*. Mycol Res 103:1540–1544, 1999.

FHJ Schuren, SA Ásgeirsdottir, EM Kothe, JM Scheer, JGH Wessels. The *Sc7/Sc14* gene family of *Schizophyllum commune* codes for extracellular proteins specifically expressed during fruit-body formation. J Gen Microbiol 139:2083–2090, 1993.

A Schweizer, S Rupp, BN Taylor, M Rollinghoff, K Schroppel. The TEA/ATTS transcription factor CaTec1p regulates hyphal development and virulence in Candida albicans. Mol Microbiol 38:435–445, 2000.

S Seiler, FE Nargang, G Steinberg, M Schliwa. Kinesin is essential for cell morphogenesis and polarized secretion in *Neurospora crassa*. EMBO J 16:3025–3034, 1997.

TC Sewall, CW Mims, WE Timberlake. Conidium differentiation in *Aspergillus nidulans* wild-type and wet-white (*wet*) mutant strains. Dev Biol 138:499–508, 1990.

G-P Shen, D-C Park, RC Ullrich, CP Novotny. Cloning and characterization of a *Schizophyllum* gene with *ABG* mating-type activity. Curr Genet 29:136–142, 1996.

WC Shen, J Wieser, TH Adams, DJ Ebbole. The *Neurospora rca-1* gene complements an *Aspergillus flbD* sporulation mutant but has no identifiable role in *Neurospora* sporulation. Genetics 148:1031–1041, 1998.

WC Shen, P Bobrowicz, DJ Ebbole. Isolation of pheromone precursor genes of *Magnaporthe* grisea. Fungal Genet Biol 27:253–263, 1999.

VA Shepherd, DA Orlovich, AE Ashford. Cell-to-cell transport via motile tubules in growing hyphae of a fungus. J Cell Sci 105:1173–1178, 1993.

PKT Shiu, NL Glass. Cell and nuclear recognition mechanisms mediated by type in filamentous ascomycetes. Curr Opin Microbiol 3:183–188, 2000.

JH Sietsma, HAB Wösten, JGH Wessels. Cell wall growth and protein secretion in fungi. Can J Bot 73 (Suppl, 1):S388–S395, 1995.

N Sievers, M Krüger, R Fischer. Kreuzung von *Aspergillus nidulans*. Biol in unserer Zeit 6:383–388, 1997.

I Skromne, O Sánchez, J Aguirre. Starvation stress modulates the expression of the *Aspergillus nidulans brlA* regulatory gene. Microbiology 141:21–28, 1995.

T Som, VSR Kolaparthi. Developmental decisions in *Aspergillus nidulans* are modulated by ras activity. Mol Cell Biol 14:5333–5348, 1994.

CA Specht, MM Stankis, L Giasson, CP Novotny, RC Ullrich. Functional analysis of the homeodomain related proteins of the Aα locus of *Schizophyllum commune*. Proc Natl Acad Sci USA 89:7174–7178, 1992.

CA Specht, MM Stankis, CP Novotny, Ullrich RC. Mapping the heterogeneous DNA region that determines the nine Aα mating-type specificities of *Schizophyllum commune*. Genetics 137:709–714, 1994.

T Spellig, M Bölker, F Lottspeich, RW Frank, R Kahmann. Pheromones trigger filamentous growth in *Ustilago maydis*. EMBO J 13:1620–1627, 1994.

A Spit, RH Hyland, ECJ Mellor, LA Casselton. A role for heterodimerization in nuclear localisation of a homeodomain protein. Proc Natl Acad Sci USA 95: 6228–6233, 1998.

MM Stankis, CA Specht, H Yang, L Giasson, RC Ullrich, CP Novotny. The Aα mating type locus of *Schizophyllum commune* encodes two dissimilar multiallelic homeodomain proteins. Proc Natl Acad Sci USA 89:7169–7173, 1992.

KE Stephens, KY Miller, BL Miller. Functional analysis of DNA sequences required for conidium-specific expression of the SpoC1-C1C gene of *Aspergillus nidulans*. Fungal Genet Biol 23:231–242, 1999.

MA Stringer, RA Dean, TC Sewall, WE Timberlake. *Rodletless*, a new *Aspergillus* developmental mutant induced by directed gene inactivation. Genes Dev 5:1161–1171, 1991.

MA Stringer, WE Timberlake. dewA encodes a fungal hydrophobin component of the *Aspergillus* spore wall. Mol Microbiol 16:33–44, 1994.

AW Strittmatter, S Irninger, GH Braus. Induction of *jlbA* mRNA synthesis for a putative bZIP protein of *Aspergillus nidulans* by amino acid starvation. Curr Genet 39:327–334, 2001.

R Suelmann, N Sievers, R Fischer. Nuclear traffic in fungal hyphae: In vivo study of nuclear migration and positioning in *Aspergillus nidulans*. Mol Microbiol 25: 757–769. http://www.blacksci.co.uk/products/elecpro/anims/mole254.gif, 1997.

S Swamy, I Uno, T Ishikawa. Morphogenetic effects of mutations at the A and B incompatibility factors in *Coprinus cinereus*. J Gen Microbiol 130:3219–3224, 1984.

S Swamy, I Uno, T Ishikawa. Regulation of cyclic AMP metabolism by the incompatibility factors in *Coprinus cinereus*. J Gen Microbiol 131:3211–3217, 1985a

S Swamy, I Uno, T Ishikawa. Regulation of cyclic AMP-dependent phosphorylation of cellular proteins by the incompatibility factors in *Coprinus cinereus*. J Gen Microbiol 31:339–346, 1985b.

A Tag, J Hicks, G Garifullina, C Ake, TD Phillips, M Beremand, N Keller. G-protein signalling mediates differential production of toxic secondary metabolites. Mol Microbiol 38:658–665, 2000.

D Tagu, I Kottke, F Martin. Hydrophobins in ectomycorrhizal symbiosis: Hypothesis. Symbiosis 25:5–18, 1998.

D Tagu, R De Bellis, R Balestrini, OMH De Vries, G Piccoli, V Stocchi, P Bonfante, F Martin. Immunolocalization of hydrophobin HYDPt-1 from the ectomycorrhizal basidiomycete *Pisolithus tinctorius* during colonization of *Eucalyptus globulus* roots. New Phytol. 149:127–135, 2001.

T Takemaru, T Kamada. The induction of morphogenetic variations in *Coprinus* basidiocarps by UV irradiation. Rep Tottori Mycol Inst 7:71–77, 1969.

C Talora, L Franchi, H Linden, P Ballario, G Macino. Role of a white collar-1-white collar-2 complex in blue-light signal transduction. EMBO J 18:4961–4968, 1999.

K Tenney, I Hunt, J Sweigard, JI Pounder, C McClain, EM Bowman, BJ Bowman. *Hex-I*, a gene unique to filamentous fungi, encodes the major protein of the Woronin body and functions as a plug for septal pores. Fungal Genet Biol 31:205–217, 2000.

WE Timberlake, EC Barnard. Organization of a gene cluster expressed specifically in the asexual spores of *A. nidulans*. Cell 26:29–37, 1981.

WE Timberlake, MA Marshall. Genetic engineering of filamentous fungi. Science 244: 1313–1317, 1989.

ML Trembley, C Ringli, R Honegger. Hydrophobins DGH1, DGH2, and DGH3 in the lichen-forming basidiomycete *Dictyonema glabratum*. Fungal Genet Biol 35:247–259, 2002a.

ML Trembley, C Ringli, R Honegger. Differential expression of hydrophobin DGH1, DGH2 and DGH3 and immunolocalization of DGH1 in strata of the lichenized basidiocarp of *Dictyonema glabratum*. New Phytol 154:185–195, 2002b.

BG Turgeon. Application of mating-type gene technology to problems in fungal biology. Ann Rev Phytopathol 36:115–137, 1998.

BG Turgeon. OC Yoder. Proposed nomenclature for mating type genes of filamentous ascomycetes. Fungal Genet Biol 31:1–5, 2000.

BG Turgeon, H Bohlmann, LM Ciufetti, SK Christiansen, G Yang, W Schäfer, OC Yoder. Cloning and analysis of the mating-type genes from *Cochliobolus heterostrophus*. Mol Gen Genet 238:270–284, 1993.

AM Tymon, U Kües, WVJ Richardson, LA Casselton. A fungal mating type protein that regulates sexual and asexual development contains a POU-related domain. EMBO J 11:1805–1813, 1992.

MH Umar, LJLD van Griensven. Morphogenetic cell death in developing primordia of *Agaricus bisporus*. Mycologia 89:274–277, 1997.

MH Umar, LJLD van Griensven. The role of morphogenetic cell death in the histogenesis of the mycelial cord of *Agaricus bisporus* and in the development of macrofungi. Mycol Res 102:719–735, 1998.

I Uno, T Ishikawa. Chemical and genetical control of induction of monokaryotic fruitine bodies in *Coprinus macrorhizus*. Mol Gen Genet 113:228–239,1971.

I Uno, T Ishikawa. Biochemical and genetic studies on the initial events of fruitbody formation. In: K Wells, EK Wells (eds). Basidium and Basidiocarp. Evolution, Cytology, Function and Development. New York: Springer. 1982, pp 113–123.

M Urban, R Kahmann, M Bölker. The biallelic *a* mating type locus of *Ustilago maydis*: remnants of an additional pheromone gene indicate evolution from a multiallelic ancestor. Mol Gen Genet 250:414–420, 1996.

LJ Vaillancourt, M Raudaskoski, CA Specht, CA Raper. Multiple genes encoding phero-mones and a pheromone receptor define the Bβ1 mating-type specificity in *Schizophyllum commune*. Genetics 146:541–551, 1997.

O Valerius, O Draht, E Kübler K Adler, B Hoffmann, GH Braus. Regulation of hisHF transcription of *Aspergillus nidulans* by adenine and amino acid limitation. Fungal Genet Biol 32:21–31, 2001.

MA Vallim, KY Miller, BL Miller. *Aspergillus* SteA (Sterile 12-like) is a homeodomain-$C_2/H_2$-$Zn^{+2}$ finger transcription factor required for sexual reproduction. Mol Micro-biol 36:290–301, 2000.

P van der Valk, R Marchant. Ultrastructure in fruit-body primordial of the basidiomycetes *Schizophyllum commune* and *Coprinus cinereus*. Protoplasma 95:57–72, 1978.

WJ van Heeckeren, DR Dorris, K Struhl. The mating-type proteins of fission yeast induce meiosis by directly activating mei3 transcription. Mol Cell Biol 18:7317–7326, 1998.

LC van Loon, EA van Strien. The families of pathogenesis-related proteins, their activities, and comparative analysis of PR-1 type proteins. Physiol Mol Plant Pathol 55:85–97, 1999.

M-A van Wetter, HAB Wösten, JGH Wessels. SC3 and SC4 hydrophobins have distinct roles in formation of aerial structures in dikaryons of *Schizophyllum commune*. Mol Microbiol 36:201–210, 2000a.

M-A van Wetter, HAB Wösten, JH Sietsma, JGH Wessels. Hydrophobin gene expression affects hyphal wall composition in *Schizophyllum commune*. Fungal Genet Biol 31:99–104, 2000b.

PA Volz, DJ Niederpruem. Dikaryotic fruiting in *Schizophyllum commune* Fr.: Morphol-ogy of the developing basidiocarp. Arch Mikrobiol 68:246–258, 1969.

PJ Walser, M Hollenstein, M Klaus, U Kües. Genetic analysis of basidiomycete fungi. In: NJ Talbot, ed. Molecular and Cell Biology of Filamentous Fungi: A Practical Approach. Oxford, UK: Oxford University Press, 2001, pp 59–90.

HX Wang, TB Ng, VEC Ooi. Lectins from mushrooms. Mycol Res 102:897–906; 1998.

H Waters, RD Butler, D Moore. Structure of aerial and submerged sclerotia of *Co-prinus lagopus*. New Phytol 74:199–205, 1975a.

H Waters, D Moore, RD Butler, Morphogenesis of aerial sclerotia of *Coprinus lagopus*. New Phytol 74:207–213, 1975b.

J Webster. Introduction to Fungi. Cambridge, UK: Cambridge University Press, 1980.

H Wei, M Scherer, A Singh, R Liese, R Fischer. *Aspergillus nidulans* alpha-1,3 glucanase (mutanase), mutA, is expressed during sexual development and mobilises mutan. Fung Genet Biol 34:217–227, 2001.

H Wei, N Requena, R. Fischer. The MAPKK-kinase SteC is essential for hyphal fusion and sexual development in the homothallic fungus *Aspergillus nidulans*. Mol Microbiol, submitted, 2003.

J Wendland, P Philippsen. Determination of cell polarity in germinated spores and hyphal tips of the filamentous ascomycete *Ashbya gossypii* requires a rhoGAP homolog. J Cell Sci 113:1611–1621, 2000.

J Wendland, P Philippsen. Cell polarity and hyphal morphogenesis are controlled by multiple Rho-Protein modules in the filamentous fungus *Ashbya gossypii*. Genetics 157: 601–610, 2001.

J Wendland, LJ Vaillancourt, J Hegner, KB Lengeler, KJ Laddison, CA Specht, CA Raper, E Kothe. The mating-type locus B$\alpha$1 of *Schizophyllum commune* contains a pheromone receptor gene and putative pheromone genes. EMBO J 14:5271–5278,1995.

JGH Wessels. Fruiting in the higher fungi. Adv Microb Physiol 34:147–202, 1993a.

JGH Wessels. Wall growth, protein excretion and morphogenesis in fungi. New Phytol 123:397–413, 1993b.

JGH Wessels. Hydrophobins: Proteins that change the nature of the fungal surface. Adv Microb Physiol 38:1–45, 1997.

JGH Wessels. Fungi in their own rights. Fungal Genet Biol 27:134–145, 1999.

JGH Wessels, SA Ásgeirsdottir, KU Birkenkamp, OMH de Vries, LG Lugones, JMJ Scheer, FHJ Schuren, TA Schuurs, MA van Wetter, HAB Wösten. Genetic regulation of emergent growth in *Schizophyllum commune*. Can J Bot 73 (Suppl1): S273– S281, 1995.

JR Whiteford, Thurston CF The molecular genetics of cultivated mushrooms. Adv Microb Physiol 42:1–23, 2000.

HLK Whitehouse. Multiple allelomorph heterothallism in the fungi. New Phytol 48:212– 244, 1949.

J Wieser, TH Adams. *flbD* encodes a myb-like DNA-binding protein that coordinates initiation of *Aspergillus nidulans* conidiophore development. Genes Dev 9:491– 502, 1995.

J Wieser, BN Lee, JW Fondon III, TH Adams. Genetic requirements for initiating asexual development in *Aspergillus nidulans*. Curr Genet 27:62–69, 1994.

M Willer, L Hoffman, U Styrarsdottir, R Egel, J Davey, O Nielsen. Two-step activation of meiosis by the *mat1* locus in *Schizosaccharomyces pombe*. Mol Cell Biol 15: 4964–4970, 1995.

J Wöstemeyer, A Wöstemeyer, A Burmester, K Czempinski. Relationship between sexual processes and parasitic interactions in the host-pathogen system *Absidia glauca-Parasitella parasitica*. Can J Bot 73 (Suppl. 1):S243–S250, 1995.

HAB Wösten, OMH de Vries, JGH Wessels. Interfacial self-assembly of a fungal hydrophobin into a rodlet layer. Plant Cell 5:1567–1574, 1993.

HAS Wösten, SA Ásgeirsdottir, JH Krook, JHH Drenth, JGH Wessels. The fungal hydrophobin Sc3p self-assembles at the surface of aerial hyphae as a protein membrane constituting the hydrophobic rodlet layer. Eur J Cell Biol 63:122–129, 1994a.

HAB Wösten, FHJ Schuren, JGH Wessels. Interfacial self-assembly of a hydrophobin into an amphipathic membrane mediates fungal attachment to hydrophobic surfaces. EMBO J 13:5848–5854, 1994b.

J Wu, BL Miller. *Aspergillus* asexual reproduction and sexual reproduction are differentially affected by transcriptional and translational mechanisms regulating *stunted* gene expression. Mol Cell Biol 17:6191–6201, 1997.

JH Xiao, I Davidson, H Matthes, J-M Garnier, P Chambon. Cloning, expression, and transcriptional properties of the human enhancer factor TEF-1. Cell 65:551–568, 1991.

LN Yager, MB Kurtz, S Champe. Temperature-shift analysis of conidial development in *Aspergillus nidulans*. Dev Biol 93:92–103, 1982.

LN Yager, HO Lee, DL Nagle, JE Zimmermann. Analysis of *fluG* mutations that affect light-dependent conidiation in *Aspergillus nidulans*. Genetics 149:1777–1786, 1998.

O Yamada, BR Lee, K Bomi, Y Iimura. Cloning and functional analysis of the *Aspergillus oryzae* conidiation regulator gene *brlA* by its disruption and mis-scheduled expression. J Biosci Bioeng 87:424–429, 1999.

DS Yaver, MD Overjero, F Xu, BA Nelson, KM Brown, T Halkier, S Bernauer, SH Brown, S Kauppinen. Molecular characterization of laccase genes from the basidiomycete *Coprinus cinereus* and heterologous expression of the laccase lcc1. Appl Environ Microbiol 65:4943–4948, 1999.

XS Ye, S-L Lee, TD Wolkow, S-L McGuire, JE Hamer, GC Wood, SA Osmani. Interaction between developmental and cell cycle regulators is required for morphogenesis in *Aspergillus nidulans*. EMBO J 18:6994–7001, 1999.

MM Yelton, JE Hamer, WE Timberlake. Transformation of *Aspergillus nidulans* by using a *trpC* plasmid. Proc Natl Acad Sci USA 81:1470–1474, 1984.

J-H Yu, J Wieser, TH Adams. The *Aspergillus* F1bA RGS domain protein antagonizes G-protein signaling to block proliferation and allow development. EMBO J 15:5184–5190, 1996.

SH Yun, ML Berbee, OC Yoder, BG Turgeon. Evolution of the fungal self-fertile reproductive life style from self-sterile ancestors. Proc Natl Acad Sci USA 96:5592–5597, 1999.

L Zhang, RA Baasiri, KK van Alfen. Viral expression of the fungal pheromone-precursor gene expression. Mol Cell Biol 18:953–959, 1998.

D Zickler, S Arnaise, E Coppin, R Debuchy, M Picard. Altered mating-type identity in the fungus *Podospora anserina* leads to selfish nuclei, uniparental progeny, and haploid meiosis. Genetics 140:493–503, 1995.

CR Zimmermann. A molecular genetic analysis of developmental gene regulation in *Aspergillus nidulans*. PhD thesis, University of California, Davis, 1986.

ME Zolan, NY Stassen, MA Ramesh, BC Lu, G Valentine. Meiotic mutants and DNA repair genes of *Coprinus cinereus*. Can J Bot 73 (Suppl 1):S226–S223, 1995.

BJM Zonneveld. The significance of α-1,3-glucan of the cell wall and α-1,3-glucanase for cleistothecium development. Biochim Biophys Acta 273:174–187, 1972.

BJM Zonneveld. α-1,3 Glucan synthesis is correlated with α-1,3 glucanase synthesis, conidiation and fructification in morphogenetic mutants of Aspergillus nidulans. J Gen Microbiol 81:445–451, 1974.

BJM Zonneveld. The effect of glucose and manganese on adenosine-3′,5′-monophosphate levels during growth and differentiation of *Aspergillus nidulans*. Arch Microbiol 108:41–44, 1976.

BJM Zonneveld. Biochemistry and ultrastructure of sexual development in *Aspergillus*. In: JE Smith and JA Pateman (eds.). Genetics and Physiology of *Aspergillus*. London: Academic Press, 1977, pp. 58–80.

# 3

# Multiple GATA Transcription Factors Control Distinct Regulatory Circuits and Cellular Activities in *Neurospora*

**George A. Marzluf**
*The Ohio State University, Columbus, Ohio, U.S.A.*

## I. INTRODUCTION

GATA factors are sequence-specific DNA-binding proteins that possess one or two Cys2/Cys2-type zinc fingers and are widely distributed in eukaryotic organisms from yeasts to higher plants and mammals. The regulatory functions performed by GATA factors are diverse, ranging from controlling transcription of sets of downstream genes involved in metabolic pathways, cellular activities, to governing cell differentiation and development (1–3). In the case of mammals, mice and humans possess at least six different GATA factors, each of which displays a tissue-specific pattern of expression and which controls different aspects of cellular differentiation and development (3, 4). GATA-1, the best studied members of this family, is responsible for development of the erythropoietic cell lineage (5), whereas GATA-2 and GATA-3 control early developmental steps of the central nervous system (6). GATA-4 and GATA-6 are involved in the gastrulation step of early embryonic development (6) and of cardiac-specific gene expression (7). The mammalian GATA factors all contain two tandem Cys2/Cys2-type zinc fingers. The C-terminal finger is responsible for specific DNA binding, whereas the N-terminal finger assists but is not required for DNA binding, but is required for protein-protein interactions with important coactivators such as recognition by GATA-1 of FOG-1 (8) and FOG-2 (9) or of GATA-2 with PML (10). GATA factors have also been implicated as agents for chromatin remodeling, a critical step in gene activation (11, 12). Most GATA factors act in a positive fashion to turn on the expression of downstream genes; however, certain ones, such as the well-studied yeast DAL80 protein, act negatively to repress gene expression (13, 14).

In yeasts and filamentous fungi, most, but not all, GATA factors have only a single Cys2/Cys2-type-4 zinc finger and are involved in regulation of the expression of genes encoding catabolic enzymes (1). The three well-studied model fungi, *Aspergillus nidulans, Saccharomyces cerevisiae*, and *Neurospora crassa* all contain multiple GATA proteins. Five GATA factors are found in the yeast, *S. cerevisiae*. These include Gln3p (15) and Nil1/Gat1p (16), which are positive regulators of nitrogen metabolism, and Dal80p (14) and Nil2p (17), which also regulate nitrogen catabolic genes, but in a negative fashion. Ash1p is a daughter cell-specific GATA factor that acts to inhibit a mating type switch in daughter cells in yeast (18).

The filamentous fungus, *N. crassa*, contains at least 5 GATA factors, NIT2, WC1, WC2, SRE, and ASD4 (14,19–22). These GATA factors display similar DNA-binding activity, each containing one or two Cys2/Cys2 zinc finger motifs, each with a loop of either 17 or 18 amino acid residues (Fig. 1). Moreover, because the *Neurospora* genomic sequence is now available, no additional proteins within this family remain to be discovered. NIT2, the first GATA factor to be identified in *Neurospora*, is a global regulator of nitrogen metabolism and serves to activate a large number of genes that encode enzymes of multiple pathways for nitrogen source utilization (14). Numerous other fungi, including yeast,

```
GATA1    CTNCQTTTTTLWRRNASGD-PVCNACGL-YYKLHQVNRPLTMRKDGIQ
GLN3     CFNCKTFKTPLWRRSPEGN-TLCNACGL-FQKLHGTMRPLSLKSDVIK
AREA     CTNCFTQTTPLWRRNPEGQ-PLCNACGL-FLKLHGVVRPLSLKTDVIK
NIT2     CTNCFTQTTPLWRRNPDGQ-PLCNACGL-FLKLHGVVRPLSLKTDVIK
ASD4     CQNCATSTTPLWRRDEMGQ-VLCNACGL-FLKLHGRPRPISLKTDVIK
SRE-C    CQNCGTTITPLWRRDEAGH-TICNACGL-YYKLHGVHRPVTMKKAIIK
SRE-N    CSNCGTTHTPLWRRSPQGA-IICNACGL-YLKARNAARPANIRRPPSV
WC1      CANCHTRNTPEWRRGPSGNRDLCNSCGLRWAKQTGRVSPRTSSRGGNG
WC2      CTDCGTLDSPEWRKGPSGPKTLCNACGLRWAKKEKKKNANNNNNGGGI
```

**Figure 1** Amino acid sequence of the zinc finger region of various GATA factors. GATA1, human (carboxy-terminal finger), GLN3, yeast nitrogen regulatory protein, AREA, nitrogen regulatory protein of *Aspergillus nidulans*. NIT2, ASD4, SRE, WC1, and WC2 are all GATA factors of *Neurospora crassa*. The carboxy terminal (C) and amino terminal (N) fingers of SRE are both shown. These GATA factors each have a central loop of 17 residues between their zinc fingers, except that WC1 and WC2 contain 18 amino acids in this region. Conserved residues at the Cys2/Cys2 fingers and an internal PLWRR sequence are in bold.

*A. nidulans, Penicillium chrysogenum, Magnaporte grisea*, and *Gibberella fuji-kuroi* possess a homologous GATA factor that functions in nitrogen control (23–26). In fact, most fungal species probably have a similar GATA-type nitrogen regulatory protein. Neurospora preferentially uses ammonia, glutamine, and glutamate, and precludes the use of alternative nitrogen sources, such as nitrate or purines, when sufficient levels of primary nitrogen sources are available, a feature commonly known as nitrogen catabolite repression. Upon nitrogen derepression, when the favored nitrogen sources are absent or severely limited, NIT2 activates the expression of genes encoding enzymes for use of secondary nitrogen sources. The NIT2 protein binds specifically to GATA elements in the promoter of nitrogen catabolic genes (27–29), but also participates in protein-protein interactions that are critical for proper regulatory responses. Establishment of nitrogen repression depends on the direct binding of the negative-acting nitrogen repressor protein, NMR, to two regions of NIT2, a motif within the zinc finger domain and a sequence of approximately 12 amino acids at its carboxy terminus (30). On the other hand, upon derepression and nitrate induction, NIT2 interacts with a pathway-specific factor, NIT4, to activate expression of nitrate assimilatory genes, *nit-3* and *nit-6* (31). The nit-2 mRNA and NIT2 protein are both relatively stable and not subject to rapid turnover during nitrogen repression or derepression (32). Thus, because the NIT2 protein is present regardless of the nitrogen status of the cells, its regulatory function must somehow cycle between active and inactive states, for example, by post-translational modification, although no information of potential controls is yet available. As described below, the *Neurospora* WC1 GATA factor is subject to light-induced phosphorylation (33).

The brilliant pioneering work by Macino and his colleagues (19, 20, 33) has demonstrated that light induction of gene expression in *Neurospora* centers on GATA proteins. Two GATA factors, white collar-1 (WC1) and white collar-2 (WC2), are both required in the regulatory response to blue light and are responsible for turning on the expression of light-regulated genes such as the albino genes, *al-1, al-2,* and *al-3*, conidiation-specific genes *con-9* and *con-10*, and the clock-controlled genes *ccg-1*, and *ccg-2/eas* (19, 20). White collar-1 and WC2 also are essential for the circadian rhythm system of *Neurospora* (34). In addition to the zinc finger domain, WC1 and WC2 both possess PAS domains, motifs that are well known to promote protein-protein interactions. White collar-1 and WC2 form homodimers and heterodimers in vitro (35). A WC1-WC2 heterodimer, designated as a white collar complex (WCC), is found in vivo (33). Expression of the *wc-1+* gene is induced by light and requires functional WC1 and WC2 products, that is, it is subject to induction and to positive autogenous regulation (19). Wild-type *Neurospora* cells in the dark possess the white collar complex, that is, WC1/WC2 heterodimers. Upon illumination with blue light, expression of the various light-regulated genes, including *wc-1+*, is turned on. Moreover, when illuminated, the WC1 protein of the complex becomes hyperphosphorylated and then is degraded. In contrast, WC2 is stable and not subject to modification. The cells maintain a WCC by association of the WC2 subunit with newly synthesized WC1 (33). One particularly intriguing finding is that expression of the various light-inducible genes in vivo appears to require activation by the WCl-WC2 heterodimer (the WCC), even though WC1 and WC2 alone will bind to the *al-3* promoter in vitro (19, 20, 33), demonstrating the importance of protein-protein interactions in gene regulation. Identity of the *Neurospora* blue-light photoreceptor remains elusive, although Macino and colleagues (33) have postulated that an additional motif, a LOV domain, that is present in WC1 might represent a flavin-binding site, which would imply that the WCC transcription complex perceives and is directly controlled by light. Finally, WC1 and WC2 both play fundamental roles in maintenance of circadian rhythm in *Neurospora*. White collar-1 is essential for the light-induced resetting of the circadian cycle, by induction of frq transcription, a major component of the clock (36). The frequency protein, FRQ, in turn positively activates WC1 synthesis (36). Interestingly, although *wc-1+* mRNA levels remain constant, the WC1 protein shows a dramatic rhythmic variation during the circadian cycle, suggesting it is subject to translational control or regulated turnover (36). The presence of two feedback loops that involve FRQ and WC1 are visualized to promote stability in the *Neurospora* circadian system (34, 36).

Iron is an essential element and an important component of many enzymes in all living organisms. Microorganisms, including the fungi, have developed strategies to acquire iron, while at the same time protecting themselves from the potential toxic affects that would occur if iron were in excess. The primary strategy used by *Neurospora* and other fungi to acquire iron involves the production of small molecular weight compounds known as siderophores. Siderophores, cyclic hexapeptides, are synthesized during conditions of iron limitation and secreted into the environment where they can sequester ferric iron and allow its uptake (21, 37). Pathogenic microorganisms appear to use siderophores to acquire iron from their unwilling hosts. *Neurospora* elaborates four different siderophores, coprogen, coprogen-B, ferrichrome, and ferricrocin, and their synthesis is tightly regulated, depending on the level of iron available to the cells. A recently discovered GATA factor, SRE (for siderophore regulation), controls siderophore formation and iron acquisition in *Neurospora*. SRE is unusual in that it contains two zinc finger motifs and because it acts in a negative fashion to prevent expression of the siderophore biosynthetic pathway during conditions of iron sufficiency. Most fungal GATA factors have only a single zinc finger motif and act in a positive fashion. SRE is a homolog of URBS1 of *Ustilago maydis* and of SREA of *A. nidulans* (37, 38). URBS1 was the first iron regulatory protein of this class to be described in the excellent work of Leong and her coworkers (37).

The *Neurospora* SRE GATA factor is composed of 587 amino acids and shows sequence-specific binding to DNA fragments containing GATA elements

(21, 39). The cellular content of *sre+* mRNA occurs at the same level whether the cells are growing in conditions of excess iron or extremely limited iron (21). It remains unknown whether the SRE protein is expressed constitutively or that its synthesis may be subject to translational control. An interesting sequence of amino acids between the two zinc finger motifs is conserved in URBS1, SRE, and SREA. This region, which contains four cysteine residues, might constitute a binding site for an iron-sulfur cluster, which suggests the possibility that their function is regulated by sensing iron similar to that found with the iron binding protein of mammalian cells (40). Indeed the spectrum of SRE is indicative of the presence of an iron-sulfur cluster (K. Harrison, personal communication). A mutant strain in which the sre gene has been "knocked out" by the Rip procedure shows partial constitutive expression of siderophores and of ornithine-N5-oxygenase, the first enzyme of the siderophore synthetic pathway (21, 39). However, the finding that the loss of control is only partial implies that additional positive or negative factors beyond SRE regulate siderophore synthesis. Additional research is needed to identify the entire set of regulatory proteins that interact to control siderophore biosynthesis and iron homeostasis. Other unknown proteins may undertake protein-protein interactions with SRE to fully repress the pathway, and in the absence of SRE, they may be able to partially prevent siderophore synthesis.

The most recently discovered GATA factor of *Neurospora*, ASD4, contains a single type 4 finger motif and displays sequence-specific binding to DNA fragments containing GATA core elements (22). The amino acid sequence of ASD4 showed similarities to that of the yeast Dal8Op and the *Penicillium* NREB, both of which appear to act solely in nitrogen regulation (13, 41). However, an *Asd-4* loss-of-function mutant obtained by the Rip procedure (42) displayed a normal expression level and regulation of *nit-3*, the highly controlled nitrate reductase structural gene, and showed wild-type growth on various secondary nitrogen sources. Furthermore, Northern blot analysis showed that various nitrogen sources did not affect *Asd-4+* mRNA levels, nor was its expression altered in an *nit-2* mutant, implying that *Asd-4+* itself, unlike *DAL80* and *nreb*, is not subject to nitrogen control. It may be significant that the ASD4 protein possesses an acidic carboxyl-terminal region that is not present in NREB nor Dal80p. ASD4 does not function in nitrogen control but instead appears to play a critical regulatory role in development of asci and ascospores, structures that result from a sexual cross (22).

*Neurospora crassa* has both a vegetative and sexual cycle. Complex morphological changes take place during a sexual cross that result in formation of specialized maternal structures, the perithecia, within which asci develop, each containing eight haploid ascospores. It was surprising to find that when crossed with wild-type, the *Asd-4* mutant acts as a dominant allele and causes severe defects in ascus and ascospore development (22). Crosses of wild type with the

Asd-4 mutant resulted in fruiting bodies that appeared macroscopically to be normal, but that were completely devoid of asci and ascospores. Moreover, this failure of culmination of the normal sexual cycle occurred whether the *Asd-4* mutant or the wild type served as the maternal parent. When the wild-type *Asd-4+* gene was transformed into the mutant, normal ascospore development was regained in crosses. The fact that the ASD4 protein shows specific DNA binding and is a member of the well-studied GATA family of transcription factors suggests that it plays a central regulatory role in development of asci and ascospores. GATA factors are known to control the sexual cycle or reproductive stages of several higher organisms, for example, in development of the egg yolk protein in *Drosophila* (43) and a maternal factor in Xenopus (44). Several mutants of *Neurospora* that cause a failure of ascus or ascospore development have been described, but their molecular basis is largely unknown. The *asd-1* mutant is recessive and in a homozygous cross, *asd-1* × *asd-1*, large numbers of asci form, but they are completely devoid of ascospores (45,46). The *asd-1+* gene has been isolated and encodes a protein that appears to have rhamnogalacturonase enzyme activity (45). Downstream genes that may be controlled by the ASD4 GATA factor have not yet been identified; however, because *asd-1+* is clearly involved in ascospore development, it might be subject to Asd-4+ regulation.

As described above, *Neurospora* possesses at least five distinct GATA factors, each appearing to regulate a unique set of downstream genes, whose promoters contain one or more GATA core elements. However, these various GATA factors show overlapping DNA binding in vitro, such that it is not possible to show selective affinity for promoter fragments for their set of regulated genes. Thus, NIT2 binds to fragments of the nit-3 promoter which it controls but also binds very well to *al-3* promoter DNA fragments, which it does not regulate (Fig. 2). Similarly, WC1 binds to genes it controls (*al-3*) but also to others (*nit-3*), which it does not regulate. ASD4 binds to al-3 and nit-3 promoters, neither of which it controls. Moreover, the promoters of some genes, for example, *cys-14+*, contain GATA elements, even though they do not appear to respond to any GATA factor. Thus, DNA-binding specificity alone is not sufficient to explain the precise regulatory responses elicited by the various GATA factors.

A major factor in determining the regulatory specificity of GATA factors now appears to be critical protein-protein interactions with other regulatory proteins in the context of a target promoter (8, 9, 31). Expression of *nit-3*, which encodes nitrate reductase, requires both the global-acting NIT2 GATA factor and the pathway-specific NIT4 protein, as well as conditions of nitrogen source derepression and induction via inorganic nitrate (47). Both NIT2 and NIT4 are DNA-binding proteins and each recognizes specific elements in the nit-3 promoter (28,29). In addition, a well-defined protein-protein interaction occurs between NIT4 and NIT2 as recognized by affinity chromatography (31). Significantly, NIT4 interacts with NIT2 but not with any of the other *Neurospora* GATA

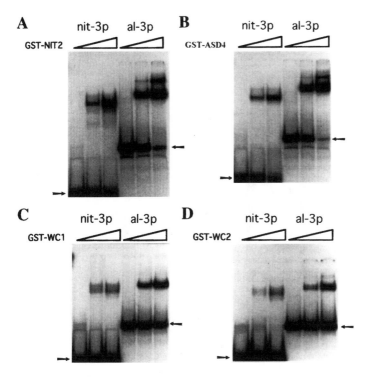

**Figure 2** Mobility shifts to examine DNA binding by *Neurospora* GATA factors to fragments containing the *nit-3* promoter (nit-3p) or the albino-3 promoter (al-3p). DNA binding by (A) GST-NIT2 fusion protein, (B) GST ASD4, (C) GST WC1, (D) GST WC2. *Arrows* indicate the position of each free DNA fragment. These GATA factors each bind both DNA fragments whether or not they regulate the respective genes in vivo.

factors, such as WC1, WC2, or ASD4 (31). It is instructive to inquire whether the NIT2-NIT4 interaction is functional, that is, is it required for the controlled high-level expression of *nit-3*? To address this question, site-directed mutants of NIT2 were obtained that retain wild-type DNA binding but are partially deficient in the protein-protein interaction with NIT4. These NIT2 mutant proteins are substantially less able to turn on nitrate reductase gene expression, displaying only approximately 15% of that obtained with the wild-type NIT2 protein (Fig. 3). A NIT2 mutant that completely lacked DNA-binding activity but interacted fully with NIT4 showed a very weak expression (4% of the wild-type level) of *nit-3*. Thus, these results imply that NIT2 and NIT4 must both function in sequence-specific DNA binding and must participate in a defined protein-protein interaction with each other to fully activate the nitrate reductase structural gene.

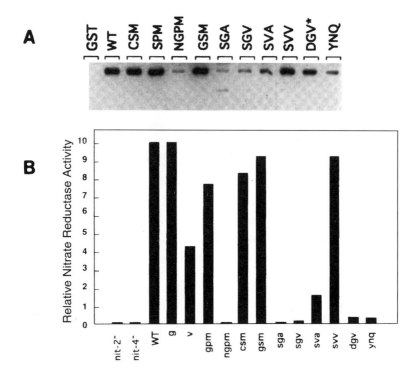

**Figure 3** (A), Affinity chromatography "pull down" assays to determine binding of wild type and various mutant NIT2 proteins to a GST-NIT4 fusion protein, with GST alone serving as a negative control. Several of the NIT2 mutant proteins, CSM, SPM, GSM, SVV, and DGV bind to NIT4 with the wild-type affinity, whereas others, such as SGA and SVA, show significantly reduced NIT4 binding. (B), In vivo activation of nitrate reductase gene expression by each of the NIT2 proteins. All NIT2 proteins that show a deficient binding to NIT4 only poorly activate the *nit-3* gene. NIT2 mutant protein SVA shows completely normal DNA binding, reduced interaction with NIT4, and only approximately 15% of the normal level of nitrate reductase expression. Mutant DGV shows wild-type interaction with NIT4 but completely lacks any DNA-binding activity and still is able to weakly activate *nit-3* (4% of wild-type level), presumably solely because of its protein-protein interaction with NIT4 at the promoter.

The NIT2 GATA factor is essential for the expression of genes that encode nitrogen catabolic enzymes of multiple other nitrogen catabolic pathways, such as in purine utilization or amino acid metabolism (1). One is tempted to predict that NIT2 also interacts with other pathway-specific regulatory proteins that are essential for the utilization of other secondary nitrogen sources. NIT2 also is involved in two protein-protein interactions with the negative-acting protein,

NMR, both of which are required to establish nitrogen catabolite repression (30). The AREA protein of *Aspergillus* also may functionally interact with multiple positive and negative control proteins, including NMRA and pathway-specific factors such as UAY and PRNA, which control, respectively, purine and proline catabolic genes (48–51). NIT2 and AREA are remarkable regulatory proteins in view of the multiple catabolic pathways they control and the apparent large number of significant protein interactions in which each of them participates. Other GATA factors such as WC1 may also represent extremely versatile proteins. Much additional work is needed to explore the exciting possibility that the regulatory specificity and multiple responses obtained with each GATA factor of *Neurospora* and of other fungal species is dependent on DNA-binding activity, potential posttranslational modifications, and specific protein-protein interactions with a unique spectrum of additional control proteins.

## ACKNOWLEDGMENTS

Research in the authors laboratory is supported by National Institutes of Health grant GM-23367. I thank Ying-Hui Fu, Bo Feng, Hubert Haas, Fei Gao, Xiaodong Xiao, Hongoo Pan, Ying Tao, Kelly Harrison, Liwei Zhou, Peter Rubinelli, Tawei Liu, and Xiaokui Mo for their valuable contributions to our research program.

## REFERENCES

1.  GA Marzluf. Genetic regulation of nitrogen metabolism in the fungi. Microbiol Mol Biol Rev 61:17–32, 1997.
2.  SH Orkin. Hematopoiesis: How does it happen? Curr Opin Genet Dev 7:870–877, 1997.
3.  T Mohun, D Sparrow. GATA-binding transcription factors in hematopoietic cells. Curr Opin Genet Dev 7:628–633, 1997.
4.  J Nardelli, D Thiesson, Y Fujiwara, FY Tsai, SH Orkin. Expression and genetic interaction of transcription factors GATA-2 and GATA-3 during development of the mouse central nervous system. Dev Biol 210:305–321, 1999.
5.  SH Orkin GATA-binding transcription factors in hematopoietic cells. Blood 80: 575–581, 1992.
6.  K Zaret. Developmental competence of the gut endoderm: Genetic potentiation by GATA and HNBF3/fork bead proteins. Dev Biol 209:1–10, 1999.
7.  F Charron, P Paradis, O Bronchain, G Nemer, M Nemer. Cooperative interaction between GATA-4 and GATA-6 regulates myocaridal gene expression. Mol Cell Biol 19:4355–4365, 1999.

8.  AP Tsang, JE Visvader, CA Turner, Y Fujiwara, C Yu, MY Weiss, M Crossley, SH Orkin. FOG, a multitype zinc finger protein, acts as a cofactor for transcription factor GATA-1 in erythroid and megakaryocytic differentiation. Cell 90:109–119, 1997.

9.  SG Tevosian, AE Deconninck, M Tanaka, M Schinke, SH Litovsky, S Izumo, Y Fujiwara, SH Orkin. FOG-2, a cofactor for GATA transcription factors, is essential for heat morphogenesis and development of coronary vessels from epicardium. Cell 101:729–739, 2000.

10. S Tsuzuki, M Towatari, H Saito, T Enver. Potentiation of GATA-2 activity through interactions with the promyelocytic leukemia protein (PML) and the t(15:17)-generated PML-retinoic acid receptor protein. Mol Cell Biol 20:6276–6286,2000.

11. J Boyes, J Omichinski, D Clark, M Pikaart, G Felsenfeld. Perturbation of nucleosome structure by erythroid transcription factor GATA-1. J Mol Biol 279:529–544, 1998.

12. GA Blobel, T Nakajima, R Eckner, M Montminy, SH Orkin. CREB-binding protein cooperates with transcription factor GATA-1 and is required for erythroid differentiation. Proc Natl Acad Sci USA 95:2061–2066, 1998.

13. TS Cunningham, TC Cooper. Expression of the DAL80 gene, whose product is homologous to the GATA factors and is a negative regulator of multiple nitrogen catabolic genes in *Saccharomyces cerevisiae*, is sensitive to nitrogen catabolite repression. Mol Cell Biol 11:6205–6215, 1991.

14. TS Cunningham, TG Cooper. The *Saccharomyces cerevisiae* DAL80 repressor protein binds to multiple copies of GATAA-containing sequences (URSgata). J Bacteriol 175:5851–5861, 1993.

15. PL Minehart, B Magasanik. Sequence and expression of GLN3, a positive nitrogen regulatory gene of *Saccharomyces cerevisiae* encoding a protein with a putative zinc finger DNA-binding domain. Mol Cell Biol. 12:6216–6226, 1991.

16. M Stanbrough, DW Rowen, B Magasanik. Role of the GATA factors Gln3p and Nil1p of *Saccharomyces cerevisiae* in the expression of nitrogen-regulated genes. Proc Natl Acad Sci USA 92:9450–9454, 1995.

17. DW Rowen, N Esiobu, B Magasanik. Role of GATA factor Nil2p in nitrogen regulation of gene expression in *Saccharomyces cerevisiae*. J Bacteriol 179:3761–3766,1997.

18. N Bobola, RP Jansen, TH Shin, K Nasmyth. Asymmetric accumulation of Ash1p in postanaphase nuclei depends on a myosin and restricts yeast mating-type switching to mother cells. Cell 84:699–709, 1996.

19. P Ballario, P Vittorioso, A Magrelli, C Talora, A Cabibbo, G Macino. White collar-1, a central regulatory of blue light responses in *Neurospora*, is a zinc finger protein. EMBO J 15:1650–1657, 1996.

20. H Linden, G Macino. White collar 2, a partner in blue light signal transduction, controlling expression of light-regulated genes in *Neurospora crassa*. EMBO J 16: 98–107, 1997.

21. LW Zhou, H Haas, GA Marzluf. Isolation and characterization of a new gene, sre, which encodes a GATA-type regulatory protein that controls iron transport in *Neurospora crassa*. Mol Gen Genet 259:532–540, 1998.

22. B Feng, H Haas, GA Marzluf. ASD4, a new GATA factor of *Neurospora crassa*,

displays sequence-specific DNA binding and functions in ascus and ascospore development. Biochemistry 39:11065–11073, 2000.

23. MX Caddick, HN Arst, LH Taylor, RI Johnson, AG Brownlee. Cloning of the regulatory gene areA mediating nitrogen metabolite repression in *Aspergillus nidulans*. EMBO J 5:1087–1090, 1986.

24. H Haas, B Bauer, B Redl, G Stoffler, GA Marzluf. Molecular cloning and analysis of nre, the major nitrogen regulatory gene of *Penicillium chrysogenum* Curr Genet 27:150–158, 1995.

25. E Froeliger, B Carpenter. NUT-1, a major nitrogen regulator in *Magnaporthe grisea* is dispensable for pathogenicity. Mol Gen Genet 251:647–656, 1996.

26. B Tudzynski, V Homann, B Feng, GA Marzluf. Isolation, characterization and disruption of the areA nitrogen regulatory gene of *Gibberella fujikuroi*. Mol Gen Genet 261:106–114, 1999.

27. TY Chiang, GA Marzluf. DNA recognition by the NIT2 nitrogen regulatory protein: Importance of the number, spacing, and orientation of GATA core elements and their flanking sequences upon NIT2 binding. Biochemistry 33: 576–582, 1994.

28. TY Chiang, GA Marzluf. Binding affinity and functional significance of NIT2 and NIT4 binding sites in the promoter of the highly regulated *nit-3* gene, which encodes nitrate reductase in *Neurospora crassa*. J Bacteriol 177:6093–6099, 1995.

29. Y Tao, GA Marzluf. Analysis of a distal cluster of binding elements and other unusual features of the promoter of the highly regulated *nit-3* gene of *Neurospora crassa*. Biochemistry 37:11136–11142, 1998.

30. HG Pan, B Feng, GA Marzluf. Two distinct protein-protein interactions between the NIT2 and NMR regulatory proteins are required to establish nitrogen metabolite repression in *Neurospora crassa*. Mol Microbiol 26:721–729, 1997.

31. B Feng, GA Marzluf. Interaction between major nitrogen regulatory protein NIT2 and pathway-specific regulatory factor NIT4 is required for their synergistic activation of gene expression in *Neurospora crassa*. Mol Cell Biol 18:3983–3990,1998.

32. Y Tao, GA Marzluf. The NIT2 nitrogen regulatory protein of Neurospora: Expression and stability of nit-2 mRNA and protein. Curr Genet 36:153–158, 1999.

33. C Talora, L Franchi, H Linden, P Ballario, G Macino. Role of a white collar-1-white collar-2 complex in blue-light signal transduction. EMBO J 18:4961–4968, 1999.

34. JC Dunlap. Molecular bases for circadian clocks. Cell 96:271–290, 1999.

35. P Ballario, C Talora, C Galli, H Linden, G Macino. Roles in dimerization and blue light photoresponse of the PAS and LOV domains of *Neurospora crassa* white collar proteins. Mol Microbiol 29:719–729, 1998.

36. K Lee, JJ Loros, JC Dunlap. Interconnected feedback loops in the *Neurospora* circadian system. Science 289:107–110, 2000.

37. B Mei, AD Budde, SA Leong. sid1, A gene initiating siderophore biosynthesis in *Ustilago maydis*: Molecular characterization, regulation by iron, and role in phytopathogenicity. Proc Natl Acad Sci USA 90:903–907, 1993.

38. H Haas, I Zadra, G Stoffler, K Angermayr. The *Aspergillus nidulans* GATA factor SREA is involved in regulation of siderophore biosynthesis and control of iron uptake. J Biol Chem 274:4613–4619, 1999.

39. LW Zhou, GA Marzluf. Functional analysis of the two zinc fingers of SRE, a

GATA-type factor that negatively regulates siderophore synthesis in *Neurospora crassa*. Biochemistry 38:4335–4341, 1999.

40.  JP Basilion, TA Rouault, CM Massinople, RD Klausner. The iron-responsive element-binding protein: Localization of the RNA-binding site to the aconitase active-site cleft. Proc Natl Acad Sci USA 91:574-578, 1994.

41.  H Haas, K Angermayr, G Stoffler. Overexpression of nreB, a new GATA factor-encoding gene of *Penicillium chrysogenum*, leads to repression of the nitrate assimilatory cluster. J Biol Chem 272:22576–22582, 1997.

42.  EU Selker, PW Garrett. DNA sequence duplications trigger gene inactivation in *Neurospora crassa*. Proc Natl Acad Sci USA 85:6870–6874, 1988.

43.  M Lossky, OC Wensink. Regulation of *Drosophila* yolk protein genes by an ovary-specific GATA factor. Mol Cell Biol 15:6943–6952, 1995.

44.  GA Partington, D Bertwistle, RH Nicolas, WJ Kee, LA Pizzey, RK Patient. GATA-2 is a maternal transcription factor present in *Xenopus oocystes* as a nuclear complex which is maintained throughout early development. Dev Biol 181:144–155, 1997.

45.  MA Nelson, ST Merino, RL Metzenberg. A putative rhamnogalacturonase required for sexual development of *Neurospora crassa*. Genetics 146:531–540, 1997.

46.  MA Nelson, RL Metzenberg. Sexual development genes of *Neurospora crassa*. Genetics 132:149–162, 1992.

47.  YH Fu, GA Marzluf. Molecular cloning and analysis of the regulation of nit-3, the structural gene for nitrate reductase in *Neurospora crassa*. Proc Natl Acad Sci USA 84:8243–8247, 1987.

48.  T Suárez N Oestreicher, MA Peñalva, C Scazzocchio. Molecular cloning of the uaY regulatory gene of *Aspergillus nidulans* reveals a favoured region for DNA insertions. Mol Gen Genet 230:369–375, 1991.

49.  T Suárez, MV de Queiroz, N Oestreicher, C Scazzocchio. The sequence and binding specificity of UaY, the specific regulator of the purine utilization pathway in *Aspergillus nidulans*, suggest an evolutionary relationship with the PPR1 protein of *Saccharomyces cerevisiae*. EMBO J 14:1453–1467, 1995.

50.  EP Hull, PM Green, HN Arst, C Scazzocchio. Cloning and characterization of the L-proline catabolism gene cluster of *Aspergillus nidulans*. Mol Microbiol 3:553–560, 1989.

51.  V Sophianopoulou, T Suárez, G Diallinas, C Scazzocchio. Operator derepressed mutations in the proline utilisation gene cluster of *Aspergillus nidulans*. Mol Gen Genet 236:209–213, 1993.

# 4

# Molecular Genetics of Metabolite Production by Industrial Filamentous Fungi

**Christian P. Kubicek**
*Institute for Chemical Engineering, Vienna University of Technology, Vienna, Austria*

## I. INTRODUCTION

Fungi have been at the forefront of developments in microbiology since the early years of this century because of their vital role in agriculture and as producers of medically and nutritionally valuable compounds. Today, filamentous fungi are classic participants in many fermentation processes, and the production of extracellular enzymes, secondary metabolites, and organic chemicals are biotechnological fields in which fungi compete successfully with bacteria or cell cultures. The latter two areas, summarized here as "metabolite production," and in some cases initiated at least 50 years ago, have given rise to some of the most efficient fermentation processes (citric acid, penicillin), and have significantly contributed to the development of bioengineering principles for cultivating fungi in fermenters. In addition, some of the fungi involved in industrial metabolite production (*Aspergillus niger, Penicillium chrysogenum*) were among the first for which molecular genetic methods were developed. Nevertheless, genomic approaches have not yet been applied for any of these fungi (although such an initiative on *A. niger* has recently been started in Europe). Legislative regulations, particularly in the food and feed area, are certainly a major reason for this fact.

This chapter will present an overview of the molecular genetic information that has become available during the past years on the production of several, but not all of the metabolites produced by filamentous fungi on a commercial scale.

Because of the long industrial tradition of many of these products, they are produced by strains that had been screened for and breeded by classic empirical methods, and consequently most of the molecular genetic information available is only from strains used by academic laboratories. Furthermore, production of some industrially produced metabolites will not be dealt with (i.e., itaconic acid, malic acid, some minor secondary metabolites) because of the total lack of molecular genetic information.

## II.  ORGANIC ACIDS

### A.  Citric Acid

Citric acid (2-hydroxy-propane-1,2,3-tricarboxylic acid) was first produced from lemons by an Italian cartell, but the discovery of its accumulation by *A. niger* in the early 1920s led to the rapid development of a fermentation process that 15 years later accounted for more then 95% of the world's production of citric acid. With its 400,000 tons per year worldwide, it is one of the major fungal fermentation products.

Industrial citric acid-producing strains of this fungus are among the most secretely kept organisms in biotechnology, including knowledge of the strategy used for their isolation during strain selection and improvement. Also, information on genomic approaches has been kept secret. Although there has recently been a European initiative toward sequencing the genome of *A. niger*, public awareness and concerns over recombinant products in the food industry until today would strongly argue against the use of recombinant mutants of *A. niger* for citric acid fermentation. Consequently, most molecular genetic studies in this area have been merely academic, and their major breakthroughs will be illustrated below.

The biochemical pathway of citric acid formation involves glycolytic catabolism of glucose to 2 mol of pyruvate, and their subsequent conversion to the precursors of citrate, oxaloacetate and acetyl-CoA (Fig. 1). A key in this process is the use of 1 mol of pyruvate as an acceptor for the carbon dioxide (released during the formation of acetyl-CoA) to form oxaloacetate. If citrate would be formed only from reactions within the tricarboxylic acid cycle, 2 mol of $CO_2$ would be lost, and consequently only two-thirds of the carbon of glucose would accumulate as citric acid, that is, 0.70 kg/kg sugar. Practical yields, however, are much higher and only possible by the synthesis of oxaloacetate by carbon dioxide fixation by pyruvate carboxylase. Of additional importance is the fact that pyruvate carboxylase of *A. niger* is localized in the cytosol, and the oxaloacetate formed is further converted to malate by the cytosolic malate dehydrogenase, thereby also regenerating 50% of the glycolytically produced NADH. This provision of cytosolic malate as "endproduct" of glycolysis is of utmost importance

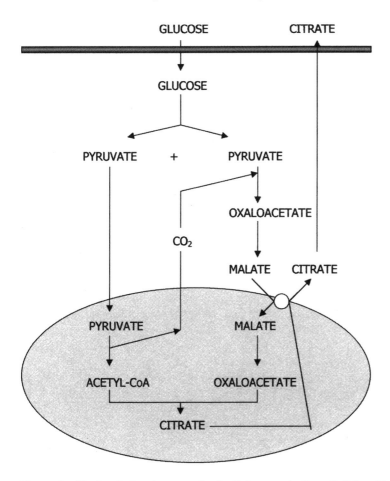

**Figure 1** Biochemical pathways and subcellular organization of citric acid accumulation by *Aspergillus niger*. The *shaded area* indicates the mitochondrium, and the *open circle* suggests the tricarboxylate/dicarboxylate transporter. The *shaded bar* represents the cytoplasma membrane.

to citric acid overflow, because it is the cosubstrate of the mitochondrial tricarboxylic acid carrier in eukaryotes. Consequently, glycolysis and its regulation are of major importance for citric acid accumulation. The current knowledge has been reviewed recently (1–3).

Based on the importance of glycolysis for citric acid accumulation, some attempts have been made toward an improvement of the rate of citric acid biosynthesis by glycolytic gene amplification. Ruijter et al. (4) have, selectively and in combination, amplified the genes encoding phosphofructokinase 1 (*pfkA*) and pyruvate kinase (*pkiA*), but the rates of citrate accumulation by the moderately citric acid-producing strain used (N400) were not increased. Similar findings were also obtained in strains bearing multiple copies of the *citA* (citrate synthase) gene (5). Torres et al. (6), using the biochemical system theory and a constrained linear optimization method, calculated that the activities ($V_{max}$) of at least seven glycolytic enzymes must be simultaneously increased to obtain an increased citric acid accumulation rate, which explains why these strategies failed. Clearly, such an increase can only be achieved by appropriate manipulation of the transcription factors regulating the genes encoding the enzymes for citric acid biosynthesis.

Unfortunately, transcriptional regulation of glycolytic genes has not yet been studied in sufficient detail in *A. niger* nor in any related fungus. In *Saccharomyces cerevisiae*, GCR1 and GCR2 cooperate in a transcriptional activation complex (7, 8), and further factors are also involved (RAP1, REB1, ABF1) (9–12). It is intriguing that the above-named binding sites have so far not been detected in the 5′-noncoding sequences of the few glycolytic genes studied in *A. niger* or the close relative *A. nidulans* (2). Transcriptional regulation of glyceraldehyde-3-phosphate dehydrogenase (13, 14) and of 3-phosphoglycerate kinase (15, 16) have been studied in some detail in the closely related fungus *A. nidulans*. A sequence, located between $-161$ and $-120$, in the *pgkA* promoter was shown to be essential for expression of the respective gene. It consists of two nonoverlapping octamer sequences that match 7 out of 8 nucleotides to the higher eukaryotic consensus ATGCAAAT (17). Punt et al. (13, 14) identified a "glycolytic box" as responsible for transcription. A 24-bp region, which shares 60% similarity with the glycolytic box, is also present at $-638$ and $-488$ of the *pgkA* promoter. However, its deletion from the *pgkA* promoter produced no effect on *pgkA* transcription.

The fact that amplification of glycolytic genes did not lead to an increase in the rate of citric acid accumulation by *A. niger* indicates the operation of very tight fine control on the enzymes involved. Fine control of citric acid accumulation has been a major target of investigation since the early 1980s, and has been reviewed recently (1–3). Torres (18, 19) used these data to conclude from theoretical calculations that a major part of the actual control of citric acid production must occur at hexose uptake, phosphorylation step, or both. As trehalose-6-phosphate is a strong inhibitor of *A. niger* hexokinase (20), citrate accumulation was investigated in mutants in which the constitutively expressed trehalose-6-phosphate synthase gene *tpsA* was knocked out in mutants bearing multiple copies of *tpsA* (21, 22). In fact, the knock-out mutants produced citric acid at increased rates, whereas the rate was decreased in the multicopy mutants (21). These data indicate that the cellular level of trehalose-6-phosphate is involved in the regulation of the flux from glucose to citric acid.

Other means for improvement of citric acid production would be the elimination of formation of production of unwanted byproducts. Oxalic acid is such a byproduct, which is accumulated under certain conditions, such as high pH and low sugar concentration. Biochemical studies have shown that the biosynthesis of oxalate occurs by cytosolic oxaloacetate hydrolysis (23). The possibility to eliminate oxalate byproduction by a knock-out strategy of the corresponding *oahA* gene has been demonstrated (24).

Citric acid can be accumulated in extremely high amounts, but this accumulation is only observed under a variety of rather strictly controlled nutrient conditions. In fact, during growth of *A. niger* in standard media for the cultivation of fungi, little if any citric acid is accumulated. The conditions required for optimum yields vary with the type of fermentation (see below), and are most critical in the submerged fermentation process (Table 1). However, the genetic basis of most of these conditions is unknown.

A notable exception is the effect of strong aeration. Dissolved oxygen tensions higher than those required for vegetative growth of *A. niger* stimulate citric acid fermentation, and sudden interruptions in the air supply cause an irreversible impairment of citric acid production without any harmful effect on mycelial growth (for review see [1]). The biochemical basis of this observation is the oxygen-dependent induction of an alternative respiratory pathway, whose presence is essential for citric acid accumulation, and which depends on a continuously maintained high oxygen tension. The gene encoding this alternative oxidase has been cloned but not yet functionally studied (25). Weiss and colleagues (26–28) detected that the assembly of the proton-pumping NADH:ubiquinone oxidoreductase is impaired during citric acid accumulation (27). This could be the reason for the importance of the activity of the alternative pathway, as a disruption of the gene encoding the NADH-binding subunit of complex I in a low-producing strain of *A. niger* increased its catabolic overflow. Interestingly however, this strain excreted much less citrate than its parent (28), emphasizing the fact that citric acid accumulation is not a monocausal process, and citrate accumulation in high amounts depends on a delicate balance of several factors whose biochemistry is not yet fully understood.

## B. Gluconic Acid

Upon incubation at high pH and with an excess of glucose, *A. niger* secretes a glucose oxidase that oxidizes glucose to D-glucono-δ-lactone and, because of the concomitant secretion of a lactonase, to gluconic acid. Gluconic acid is also formed by some bacteria; however, *A. niger* has become the main fungal source for industrial gluconate production which today can be estimated as 60,000 tons per year.

**Table 1** External Factors Influencing Citric Acid Production, Metabolic Basis, and Possible Genetic Solutions

| Factor | Optimal condition | Biochemical basis | Possible solutions |
|---|---|---|---|
| Carbon source type | Fat metabolizable | High glucolytic flux rate | Amplification of glycolitic regulator genes? |
| Carbon source concentration | >10% (w/v) | Deregulation of glycolysis? | Enzyme mutagenesis? |
| pH | <3.0 | Citrate transport, oxalic and gluconic acid by production | Transporter mutagenesis |
| | | | Respective gene disruption |
| Trace metals ($Mn^{2+}$) | Below $10^{-7}$ M | DNA, protein and lipid biosynthesis | NK |
| Dissolved oxygen tension | Very high | Citrate transport | Transporter mutagenesis |
| | | Non-ATP-yielding NADH reoxidation | Alternative oxidase expression |
| $NH_4^+$, phosphate concentration | Low | NK | Modulation of wide domain |
| | | | Regulatory genes? |

NK, not known; for details on data given, see (1–3).

The molecular genetic background of gluconic acid overproduction is very poorly understood. The gene encoding glucose oxidase of *A. niger* (*goxA*) has been cloned, and its amplification resulted in a two- to threefold increase in activity (29–31). It is most actively induced by high glucose concentrations and high aeration and at a pH above 4 (32). To protect itself against the arising hydrogen peroxide, *A. niger* also secretes multiple forms of catalases (33), among which one has been cloned and characterized (34). Swart et al. (35) described nine different complementation groups of glucose oxidase overproduction mutants. *GoxB*, *goxC* and *goxF* belong to linkage group II, *goxI* to linkage group III, *goxD* and *goxG* to linkage group V, *goxA* and *goxE* to linkage group VII, and the linkage of *goxH* is unknown. Their study also indicates that *goxA* overproduction is regulated by the carbon source and oxygen in an independent manner. Nothing is yet known about the gene encoding the lactonase.

## III. SECONDARY METABOLITES

### A. β-lactams

The discovery and development of the β-lactam antibiotics are among the most powerful and successful achievements of modern science and technology. Since Fleming's accidental discovery of the penicillin-producing mold, 70 years of steady progress has followed, and today the β-lactam group of compounds is the most successful example of natural product application and chemotherapy. Following on the heels of penicillin production by *P. chrysogenum* came the discoveries of cephalosporin formation by *Cephalosporium acremonium*, cephamycin, clavam, and carbapenem production by actinomycetes, and monocyclic β-lactam production by actinomycetes and unicellular bacteria. Each of these groups has yielded medically useful products and has contributed to the reduction of diseases throughout the world.

During the 1990s, major advances have been made on structural and regulatory biosynthetic genes and metabolic engineering of the pathways involved in fungal β-lactam formation. As a consequence, new semisynthetic compounds, especially those designed to combat resistance development, are being examined in the clinic, and unusual nonantibiotic activities of these compounds are being pursued (for review see [36–38]).

Commercial production of both antibiotics is exclusively achieved with strains developed by extensive strain improvement program involving mutagenesis, which over the years has resulted in the accumulation of these antibiotics in concentrations comparable to primary metabolites (30–40 g/l). Genetic approaches have subsequently been used to learn about the alterations in the strains during the mutagenesis programs and how the biosynthesis is controlled. Large scale genomic initiatives on *P. chrysogenum* or *C. acremonium* have not yet been reported. So far, commercial antibiotic production has not been reported to make use of recombinant strains.

## B.  Penicillin Biosynthesis

β-lactam antibiotics are formed after the condensation of three amino acids: L-cysteine, L-valine, and α-aminoadipic acid to yield the tripeptide δ-(L-alpha-aminoadipyl)-L-cysteinyl-D-valine as the first pathway-specific intermediate (Fig. 2). In *P. chrysogenum*, the prime industrial producer of penicillin, its further metabolism results in the formation of a thialozidine ring that forms the penicillin nucleus. *Aspergillus nidulans* also forms small amounts of penicillin and was used successfully as a model organism for the regulation of penicillin biosynthesis by several research groups. In fact, athough there are clear differences in some aspects of the regulation of penicillin biosynthesis by *A. nidulans* and *P. chrysogenum* (see below), work with the former has yielded several insights that would otherwise have only been difficult to obtain in *P. chrysogenum*. In addition, other *Penicillium* spp., such as *P. nalgiovense* isolates occurring in cured meat products, show penicillin production (39).

The *pcbAB, pcbC*, and *penDE* genes which encode the enzymes that catalyze the three steps of the penicillin biosynthesis, that is, ACV synthetase, isopenicillin N synthetase, and acyl-CoA:isopenicillin N acyltransferase, have been cloned and characterized from *P. chrysogenum* and *A. nidulans* (for review see [36]). The organization of the penicillin biosynthetic genes in *P. chrysogenum, P. notatum, P. nalgoviense*, and *A. nidulans* (*pcbAB, pcbC*, and *penDE*) is essentially the same (40): they are located in a cluster that occurs in chromosome I (10.4 Mb) of *P. chrysogenum* (Fig. 3), in chromosome II of *P. notatum* (9.6 Mb), and in chromosome VI (3.0 Mb) of *A. nidulans*. Interestingly, the mutation programs performed in industry have resulted in the amplification of these genes in higher producer strains: whereas only one copy is present in wild-type strains, between 5 and 50 copies of the cluster are present in industrial production strains, which are arranged in tandem repeats of 106.5- or 57.6-kb units, respectively (41, 42).

Interestingly, some penicillin-negative mutants of *P. chrysogenum* showed a deletion of 57.9 kb that corresponds exactly to the DNA fragment that is amplified in production strains (43). The borders of the deleted region showed the same junction point in all mutants so far tested, and the hexonucleotide repeat TTTACA, occurring in the junction region between the repeats, has been discussed as a hot spot for this first nonhomologous recombination (41). Thus, the tandem reiteration and deletion appear to arise by an initial chromosome misalignment and nonhomologous mitotic recombination between flanking sequences of the structural genes resulting in a duplication of the biosynthetic cluster (36, 41). Subsequent chromatid misalignment and homologous recombination

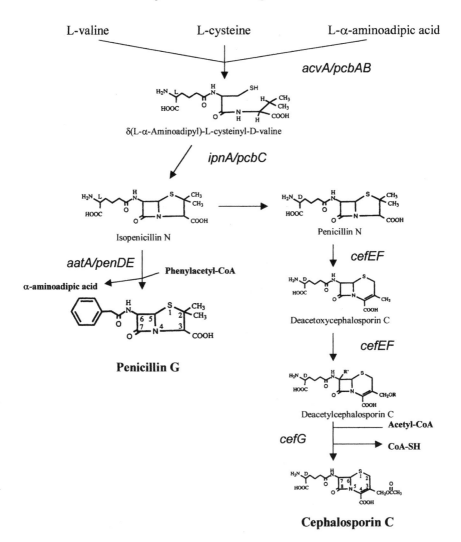

**Figure 2** Biosynthesis of penicillin and cephalosporin by *P. chrysogenum* and *A. chrysogenum*, respectively. Genes, whose products catalyze the respective step, are indicated in italics.

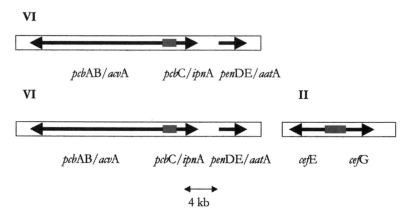

**Figure 3**   Genomic organization of the penicillin biosynthetic cluster in *P. chrysogenum* and *A chrysogenum*. The roman numbers indicate the chromosomal location of the genes shown.

could have led to the 50-fold amplification in the high producer strains. Thus the penicillin biosynthetic gene cluster is apparently located in an unstable genetic region, flanked by hot spots of recombination.

The biosynthesis of penicillin is subject to sophisticated genetic and metabolic regulation (44, 45). As the structural genes, *pcbC* and *pcbAB*, which encode two of the penicillin biosynthetic enzymes, are separated by a 1.16-kb intergenic region and transcribed divergently from one another, this dual promoter has received major attention. Reporter gene studies in *P. chrysogenum* and *A. nidulans* showed that the expression of each of these penicillin biosynthesis genes was found to be regulated by nitrogen repression, glucose repression, and growth stage control (46).

Glucose repression (carbon catabolite regulation) has long been supposed to be a major determinant of penicillin production. In *P. chrysogenum*, penicillin biosynthesis is repressed by glucose and other easily utilizable carbon sources, but not by lactose. This repression is enhanced by high phosphate concentrations, and results in the blockage of transcription of the *pcbAB, pcbC*, and *penDE* genes (47). Interestingly, neither in *P. chrysogenum* nor in *A. nidulans* does this effect of glucose involve the zinc finger DNA-binding protein CreA, which has been identified in several fungi as the main regulator in catabolite repression (48).

Nitrogen seems to control penicillin biosynthesis in *P. chrysogenum* but not in *A. nidulans* (36). Toward a molecular understanding of nitrogen catabolite repression of penicillin biosynthesis in *P. chrysogenum*, the *NRE* gene that encodes a Cys2-Cys2-type zinc finger protein with 60% identity to *A. nidulans areA* and 30% identity to *N. crassa nit-2* at the amino-acid level was cloned (49). A second GATA-binding protein, NreB, which shares several features with Dal80p/Uga43p and Gzf3p/Nil2p—both repressors in nitrogen metabolism in *S. cerevisia*—was also cloned from *P. chrysogenum* (50). Consistent with a role in the regulation of nitrogen metabolism, over-expression of *nreB* leads to repression of nitrate assimilatory genes. However, in vivo evidence for the involvement of *NRE* or *nreB* in penicillin biosynthesis has not yet been shown. Attempts to identify further nuclear proteins of *P. chrysogenum* that interact with the *pcbAB-pcbC* intergenic promoter region were initally done with mobility shift and DNA footprinting assays using crude and partially purified nuclear extracts. These studies detected binding of multiple DNA-binding proteins to different regions of this DNA segment. One abundant nuclear protein, detected by two groups, was found to bind to a single site in the intergenic promoter region at $-387$ to $-242$ relative to the *pcbAB* translational start codon and recognizing the sequence 5'-GCCAAGCC-3' or 5'-TGCCAAG-3', respectively (51, 52). The abundance of this DNA-binding protein was also related to ambient pH. As this nucleotide motif contains the consensus (5'-GGCARG-3') for binding of the pH regulator PacC—a Cys2-His2-zinc-finger transcription factor (53)—it is likely that both studies actually identified binding of PacC to the intergenic promoter. The *pacC* gene has originally been isolated from *A. nidulans* as an activator of transcription of the *pcbC* gene, and subsequently has been studied extensively (54). However, whereas in *A. nidulans* alkaline pH elevates *pcbC* transcript levels on either carbon source (strongly suggesting that *pcbC* is under direct *pacC* control), alkaline pH does not override the negative effects of a repressing carbon source in *P. chrysogenum* (47, 48). The corresponding *pacC* homologue of *P. chrysogenum* was isolated and characterized (55). It encodes a polypeptide with 64% identity to *A. nidulans* PacC, functionally complements an *A. nidulans pacC* null mutation, and contains three putative zinc fingers specifically recognizing a 5'-GCCARG-3' hexanucleotide. The gene also contains three PacC-binding sites in its 5'-upstream region, suggesting autoregulation.

Further DNA-binding protein complexes that regulate penicillin biosynthesis have been identified in *A. nidulans* and *P. chrysogenum*. Brakhage and colleagues have shown that the *A. nidulans* CCAAT motif-binding HapB-HapC-HapE complex is part of a PENR1 regulator of penicillin biosynthesis (56, 57). Despite the presence of similar CCAAT boxes in the *P. chrysogenum pcbAB-pcbC* intergenic region, the involvement of a PENR1 homologue has not been investigated in this fungus. Interestingly, mutations in the CCAAT box result in an increased (eightfold) transcription of *pcbAB* and a decreased expression of *pcbC*, indicating that this element is differently involved in the regulation of the two genes (56). The real contribution of PENR1 to the control of penicillin gene expression has not yet been revealed. The lack of PENR1 in a PENR1 mutant should, in theory, lead to higher transcript levels and higher penicillin production, as expression of the *pcbAB* gene in *A. nidulans* is known to be limiting for penicillin biosynthesis (58). However, expression from the *pcbAB* promoter was rather unaffected, and the penicillin titers even reduced (56). The authors explained these findings by the binding of a hypothetical repressor in the close vicinity of the CCAAT box. Experimental proof for this repressor, however, remains to be provided.

In *P. chrysogenum*, Martin and coworkers introduced a series of sequential deletions into the *pcbAB* promoter region in vivo, and identified three regions that produced a significant decrease in gene expression when deleted (59). Protein-DNA complexes were observed with two of these (boxes A and B). Uracil interference assays, point mutations, and both in vivo and in vitro deletion experiments revealed the sequence 5'-TAGTAA-3' to bind a transcriptional activator (PTA1), which is required for high-level expression of the *pcbAB* gene. Interestingly, this TAGTAA sequence resembles the target sequence of BAS2 (PHO2), a factor required for expression of several genes in yeasts. Last but not least, based on the identification of a regulatory protein of cephalosporin formation (see below), another novel transcription factor from *P. chrysogenum* has also been isolated, which shows significant sequence homology to human transcription factors of the regulatory factor X (RFX) family (60).

A proper description of the biosynthesis of fungal β-lactam antibiotics also requires detailed knowledge of the cell biology of the producing organisms, which involves a delineation of the compartmentalization of the biosynthetic pathways and of the consequential transport steps across both the cell-boundary plasma membrane and the organellar membranes. The localization of the enzymes involved in penicillin biosynthesis in *P. chrysogenum* has revealed a complicated pathway involving different intracellular locations (for review, see [61]): whereas the location of δ-(L-α-aminoadipyl)-L-cysteinyl-D-valine synthetase is still unresolved but may be associated with vacuolar membranes (62), and isopenicillin N-synthetase is cytosolic, the acyltransferase is confined to microbodies (63). Strains that express a mutant acyltransferase lacking the putative targeting signal for microbody proteins, located the mutant enzymes in vacuoles and in neighboring cytoplasm and did not produce penicillin (64). Toward an understanding of peroxisome formation and proliferation in *P. chrysogenum*, the *pex1* and *pex6* genes encoding proteins of the AAA-family of ATPases and involved in microsome biogenesis have been isolated (65).

The vacuole may play an ancillary role in the supply of precursor amino acids and in the storage of intermediates. This results in a significant number of transmembrane fluxes and pools of precursors, intermediates, and end and side products. Consequently, major research efforts have been directed toward both the cloning of the genes encoding the enzymes synthesizing the precursors and the transporters of some of the nutrients. Multiple copies of *lys1* encoding homocitrate synthase, the first enzyme of the lysine and α-aminoadipate synthesizing pathway, do not increase penicillin production (66), but disruption of the *lys2* gene, encoding α-aminoadipate reductase, results in a doubling of the flux from α-aminoadipate to penicillin (67). In industrial fermentations, *P. chrysogenum* uses sulfate as the source of sulfur for the biosynthesis of penicillin, and its assimilation seems to be limiting in some strains. Transport studies with *P. chrysogenum* plasma membranes fused with cytochrome c oxidase liposomes demonstrate that sulfate uptake is driven by the transmembrane pH gradient and not by the transmembrane electrical potential, using a symport of two protons with one sulfate anion (68). Two genes, *sutA* and *sutB*, whose encoded products belong to the SulP superfamily of sulfate permeases, were isolated (69). Expression of both *sutA* and *sutB* in *P. chrysogenum* is induced by growth under sulfur starvation conditions, with *sutA* being expressed to a much lower level than is *sutB*. Disruption of *sutB* resulted in a loss of sulfate uptake ability, indicating that *sutB* is the major sulfate permease involved in sulfate uptake by *P. chrysogenum*. In a high penicillin-yielding strain, *sutB* is effectively transcribed even in the presence of excess sulfate. This deregulation may facilitate the efficient incorporation of sulfur into cysteine and penicillin.

Uptake of phenylacetic acid, the side-chain precursor of benzylpenicillin, appears also to be a limiting factor in low-producing strains of *P. chrysogenum*, as 100% of the supplied phenylacetic acid was recovered in benzylpenicillin in fermentations with a high-yielding strain, whereas only 17% were incorporated by an early mutant of the Wisconsin series and the rest was metabolized. Accumulation of total phenylacetic acid-derived carbon in the cells was nonsaturable in both strains at high external concentrations of phenylacetic acid (250–3500 μM), and in the high-yielding strain at low phenylacetic acid concentrations (2.8–100 μM), indicating that phenylacetic acid enters the cells by simple diffusion (70). Thus, experiments to inhibit precurser metabolism were developed. In *A. nidulans*, phenylacetate is catabolized by homogentisate to fumarate and acetoacetate (71). Mutational evidence strongly suggested that phenylacetate is converted to homogentisate through two sequential hydroxylating reactions in positions 2 and 5 of the aromatic ring. A *phacA* gene, which encodes a cytochrome P450 protein and whose transcription is strongly induced by phenylacetate, was isolated and disrupted (72). The respective strains did not grow on phenylacetate but grew on 2-hydroxy- or 2,5-dihydroxyphenylacetate. Microsomal extracts of the disrupted strain were deficient in the NADPH-dependent conversion of phenylacetate to 2-hydroxyphenylacetate. Thus, the *phacA* gene product catalyzes the o'-hydroxylation of phenylacetate, the first step of *A. nidulans* phenylacetate catabolism. *PhacA* disruption increases penicillin production three- to fivefold, indicating that interference with catabolism favors antibiotic biosynthesis (72).

A further limitation in the incorporation of the precursor into penicillin was identified at the level of side-chain precurser activation. Thus, Minambres et al. (73) cloned the *pcl* gene encoding phenylacetyl-CoA ligase from *Pseudomonas putida. Penicillium chrysogenum* strains transformed with a construction bearing this gene exhibited a significant increase (between 1.8- and 2.2-fold higher) in the quantities of benzylpenicillin accumulated in the broths.

Also, the secretion of penicillin may constitute a limiting step in its production. Recently, the ABC transporter D encoding gene *atrD* of *A. nidulans* has been shown to be functionally involved in β-lactam formation (74). Export of antibiotics by ABC transporters would not be without precedent, as it has also been demonstrated in *Streptomyces* spp. (75, 76). However, the evidence available for *P. chrysogenum* so far does not rule out a requirement of *atrD* for one of the various other precursor or intermediate transport steps mentioned above.

In addition, Hicks et al. (77) recently showed that *A. nidulans* mutants bearing a constitutively activated (*fadA*-encoded) $G_\alpha$-protein overproduced penicillin, thus providing a first hint as to the signal-transducing pathway involved in β-lactam biosynthesis.

The progress in knowledge of the genetics of the individual steps of penicillin biosynthesis has enabled first approaches toward the production of recombinant antibiotics (for review also see [2]): toward production of the cephalosporin intermediates, 7-aminocephalosporanic acid (7-ACA) or 7-amino deacetoxy-cephalosporanic acid (7-ADCA), the *Streptomyces clavuligerus* expandase gene or the *A. chrysogenum* expandase-hydroxylase gene, with and without the acetyltransferase gene, were expressed in a penicillin production strain of *P. chrysogenum* (78). Growth of these transformants in media containing adipic acid as the side-chain precursor resulted in efficient production of cephalosporins having an adipyl side chain, proving that adipyl-6-APA is a substrate for either enzyme in vivo. Strains expressing expandase produced adipyl-7-ADCA, whereas strains expressing expandase-hydroxylase produced both adipyl-7-ADCA and adipyl-7-ADAC (aminodeacetylcephalosporanic acid). Strains expressing expandase-hydroxylase and acetyltransferase produced adipyl-7-ADCA, adipyl-7-ADAC, and adipyl-7-ACA. The adipyl side chain of these cephalosporins was easily removed with a *Pseudomonas*-derived amidase to yield the cephalosporin intermediates.

## C. Cephalosporin Biosynthesis

Cephalosporin C is the second major β-lactam and is produced by *Acremonium chrysogenum* (formerly named *Cephalosporium acremonium*). The first two steps in the biosynthetic pathway are virtually similar to those involved in penicillin biosynthesis by *P. chrysogenum*, but further biosynthesis and modification of the dihydrothiazine ring involves two different enzymes, that is, a deacetoxycephalosporin C synthetase/hydroxylase and a deacetylcephalosporin C acetyltransferase (Fig. 2). These genes are separated in two clusters (Fig. 3). Cluster I (*pcbAB-pcbC*) encodes the first two enzymes (α-aminoadipyl-cysteinyl-valine synthetase and isopenicillin N synthase) and is located in chromosome VII (4.6 Mb), and cluster II includes the *cefEF* and *cefG* genes (encoding deacetoxycephalosporin C synthetase/hydroxylase and a deacetylcephalosporin C acetyltransferase) and is located on chromosome I (2.2 Mb) (79). In contrast to *P. chrysogenum*, both clusters are present in a single copy in the *A. chrysogenum* genome, in wild type as well as in high cephalosporin-producing strains. It is unknown, at this time, if the *cefD* gene encoding isopenicillin epimerase is linked to any of these two clusters. The *cefG* gene, encoding acetyl coenzyme A:deacetylcephalosporin C acetyltransferase may be a rate-limiting step in cephalosporin C production. Sensitivity analysis of the biosynthesis of cephalosporin C in *C. acremonium* based on in vitro kinetic data indicated that, as in *A. nidulans* penicillin formation, δ-(L-α-aminoadipyl) L-cysteinyl D-valine (ACV) synthetase is the prime rate-limiting enzyme, but that increasing ACV synthetase enhances the production rate to a level that expandase/hydroxylase becomes rate limiting (80). Analysis of transformants containing additional copies of *cefG* showed a direct correlation between *cefG* copy number, *cefG* message levels, and cephalosporin C titers (81).

A molecular karyotype of an industrial strain of *A. chrysogenum* revealed eight chromosome bands ranging from approximately 1,700 kb up to more than 4000 kb. The total genomic content for this strain was estimated as at least 22,500 kb. The *pcbAB-pcbC* cluster resides on chromosome VI while the *cefEF* gene cluster is located on chromosome II(82). Interestingly, the β-isopropylmalate dehydrogenase gene (*leu2*), which is involved in the biosynthesis of the precurser amino acid valine, was also found to be linked to *pcbC* gene on chromosome VI. In contrast to *P. chrysogenum* penicillin production, no significant chromosome changes were detected in industrial cephalosporin C-producing mutants (83).

As in *P. chrysogenum*, the *A. chrysogenum pcbAB* and *pcbC* genes are orientated in opposite directions on the chromosomal DNA, and separated by a 1.2-kb putative promoter sequence. The *pcbC* promoter is at least five times stronger than the *pcbAB* promoter (84). Among the factors known to influence the biosynthesis of cephalosporin C by *A. chrysogenum* are exogenous methione addition, carbon catabolite repression, and nitrogen metabolite repression (85). Methionine stimulates by increasing the levels of transcripts of four genes (*pcbAB, pcbC, cefEF* and, to a slight extent, *cefG*) of the biosynthetic pathway (86). However, methionine also exerts a pleiotropic effect that coordinately regulates cephalosporin biosynthesis and differentiation, as it also stimulates mycelial fragmentation, whose relationship to the induction of the cephalosporin synthases has not yet been elucidated (87).

With respect to carbon catabolite regulation in *A. chrysogenum*, the *cre1* gene has been isolated and characterized (88). Interestingly, and in contrast to *creA/cre1* gene expression in *Trichoderma reesei* and *Aspergillus nidulans*, the *cre1* transcript level from an *A. chrysogenum* wild-type strain is increased in the presence of glucose. Remarkably, this glucose-dependent transcriptional up-regulation does not take place in an industrial cephalosporin producer strain of *A. chrysogenum*, suggesting that the deregulation of *cre1* is connected with the increased production rate in this strain (89).

The findings of nitrogen regulation of cephalosporin C formation correlate with the finding of 15 GATA motifs in the intergenic *pcbAB-pcbC* region of *A. chrysogenum*. However, the respective DNA-binding proteins have not yet been identified or studied.

Toward an identification of specific transactivators of cephalosporin biosynthesis, Schmitt and Kück (60) cloned the gene encoding a protein from *A. chrysogenum*, which binds to a nucleotide motif located 418 nt's upstream of the translational start in the *pcbC* promoter. The deduced polypeptide shows significant sequence homology to human transcription factors of the regulatory factor X (RFX) family and recognizes an imperfect palindrome typical of this family of transcription factors.

As for recombinant approaches towards increasing the yield of cephalosporin C, strategies applied were (a) increasing the dosage of the biosynthetic genes *cefEF* (deacetoxycephalosporin C expandase/hydroxylase) and *cefG* (deacetylcephalosporin C acetyltransferase) (81) or (b) enhancing the oxygen uptake by expressing a bacterial (*Vitreoscilla* sp.) oxygen-binding heme protein (90). The conversion of deacetylcephalosporin C to cephalosporin C is inefficient in most *A. chrysogenum* strains, because, as explained above, the *cefG* gene is expressed comparatively poorly. Expression of the *cefG* gene from the promoters of (a) the glyceraldehyde-3-phosphate dehydrogenase (*gpdA*) gene of *Aspergillus nidulans*, (b) the glucoamylase (*glaA*) gene of *A. niger*, (c) the glutamate dehydrogenase (*gdhA*) and (d) the isopenicillin N synthase (*pcbC*) genes of *P. chrysogenum*, led to very high steady-state levels of *cefG* transcript and to increased deacetylcephalosporin-C acetyltransferase protein concentration and enzyme activity in the transformants (91). As a result of a more efficient acetylation of deacetylcephalosporin C, cephalosporin production was increased two- to threefold. In addition, new biosynthetic capacities such as the production of 7-amino-cephalosporanic acid (7-ACA) or penicillin G have been achieved through the expression of the foreign genes *dao1* (encoding *Fusarium solani* D-amino acid oxidase) coupled with cephalosporin acylase or *P. chrysogenum penDE*(acyl-CoA:6-APA acyltransferase) respectively (92). Although the amounts have not yet been commercially significant, they represent a first microbial production of 7-ACA, thereby demonstrating the feasibility of introducing new biosynthetic capabilities into industrial microorganisms. Most recently, an alternative approach to production of 7-aminodeacetoxycephalosporanic acid was published, which was based on a replacement of the *A. chrysogenum cefEF* gene by the *Streptomyces clavuligerus cefE* gene, which resulted in considerably less byproduction of other cephalosporin intermediates (92a).

## D.  Cyclosporin

Cyclosporin A is a potent and clinically important immunosuppressive drug (SandimmunR). It is produced by the fungus *Tolypocladium niveum*. Cyclosporin A was introduced into clinical use in the late 1970s to reduce graft rejection after organ transplantation. This drug, a cyclic undecapeptide metabolite of the fungus *T. inflatum*, interferes with lymphokine biosynthesis; hence, its immunosuppressive activity. Recently, it has become clear that cyclosporins are also inhibitors of HIV-1 replication. Immunosuppressive and antiviral activities are distinct functions of the cyclosporins, but both functions require an interaction of the drug with cyclophilins (93).

Chemically, cyclosporins are cyclic molecules consisting of a high portion of methylated amino acids (Fig. 4). Molecular work on the biosynthesis of cyclosporin was almost restricted to the companies manufacturing it. The cloning of the corresponding cyclosporin synthetase was reported by Weber et al. (94). It contains an open reading frame of 45.8 kb, which encodes a peptide with a calculated M(r) of 1,689,243, which is the largest genomic ORF described so far. The predicted gene product contains 11 amino-acid-activating domains that are very similar to one another and to the domains of other peptide synthetases. Seven of these domains harbor N-methyltransferase functions. Mutants of *T. niveum* with disrupted versions of *simA* were engineered, which resulted in loss of the ability

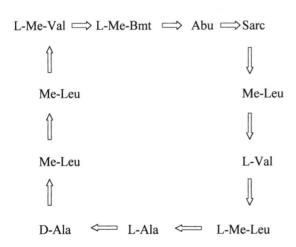

**Figure 4**   Chemical structure of cyclosporin. *MeBtm*, 4-(2-butenyl)-4-N-dimethyl-threonine; *Sar*, sarcosine; *MeLeu*, methylleucine; *MeVal*, methylvaline; *Abu*, α-amino-butyric acid; *Ala*, alanine; *Val*, valine. The *arrows* indicate the direction of the $-O=C \rightarrow N-H-$ peptide bond.

to produce cyclosporins, thereby proving the involvement of this synthetase in cyclosporin production (95).

To characterize production strains, a dispersed repetitive DNA element (cyclosporin production associated [CPA] element) was isolated, which exhibits similarities to a repeated sequence from Zea diploperennis (96). It appears to be strain specific, since it is absent from other related strains or fungi. Its copy number was estimated to be between 20 and 30 per haploid genome.

A 4,097 bp long hAT transposon named *restless* was identified in *T. inflatum*, which carries 20-bp inverted repeats, and an 8-bp target site duplication carries and encodes a long open reading frame interrupted by a single intron (97), which is the first identification of this type of class II transposon family (represented by the maize activator Ac), in filamentous fungi. The derived mRNA is subject to alternative splicing, resulting in the formation of two transcripts that encode polypeptides with significant homology to transposases from the hAT transposon family. All seven chromosomes carry a total of 15 copies of this 4.1-kb transposon in *T. inflatum*.

No molecular data are available as to the genetic regulation of cyclosporin A formation. However, *restless* was used for successful gene tagging (98) and *T. inflatum* mutants with a defect in nitrogen metabolism were isolated. One mutant carried a copy of the Restless element in a gene, which encoded a protein with high similarity to the *N. crassa* Nit4 regulator of nitrogen metabolism (*Tolypocladium* nitrogen regulator 1 [*tnirl*]).

Molecular investigations were also undertaken to learn how *T. niveum* renders itself resistant against cyclosporin. A 12-kDa peptidyl-prolyl-*cis/trans*-isomerase, which was found to be highly sensitive to the immunosuppressant FK-506 but not to cyclosporin, and another 17-kDa peptidyl-prolyl *cis-trans* was isomerase, which was inhibited by cyclosporin A in the nanomolar range, had been purified and investigated (99, 100). The gene of the latter cyclophilin was isolated and shown to belong to the cyclophilin A family members of eucaryotic and procaryotic origin (101).

## E. Gibberellins

Gibberellins (GAs) are a large family of isoprenoid plant hormones, some of which are bioactive growth regulators, controlling seed germination, stem elongation, and flowering. They represent the classic example of fungal plant growth regulators that are increasingly used in agriculture and horticulture. Particularly, the rice pathogen *Gibberella fujikuroi* (mating population C) is able to produce large amounts of gibberellic acid (Fig. 5). The economic importance of these plant hormones has led to an extensive study of the biochemistry and regulation of gibberellin biosynthesis. The main steps of the biosynthetic pathway have long

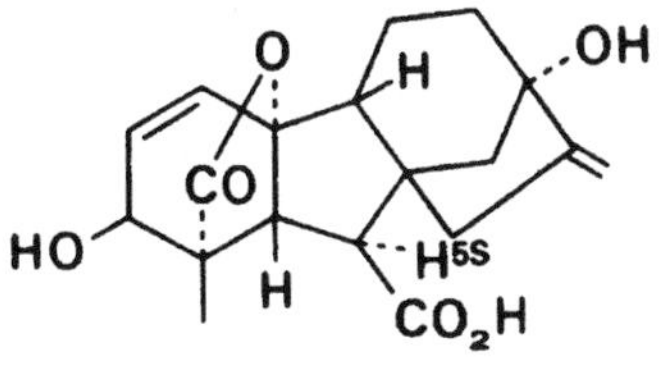

**Figure 5** Chemical structure of gibberellic acid.

been established from the identification of intermediates in wild-type *G. fujikuroi* and mutant strains (for review see [102, 103]).

Only some of the genes encoding the various steps of gibberellin biosynthesis have been cloned and characterized. Tudzynski et al. (104) performed differential screening of a *G. fujikuroi* cDNA library to clone and identify such genes. The deduced amino acid sequences of two (identical) clones, which were closely linked to each other, contained the conserved heme-binding motif of cytochrome P450 monooxygenases (FXXGXXXCXG). Further chromosome walking further revealed a putative geranylgeranyl diphosphate synthase gene, the copalyl diphosphate synthase gene, and a third P450 monooxygenase gene to be located next to these genes. Thus, the genes involved in the biosynthesis of gibberellins are closely linked in a gene cluster in *G. fujikuroi*. In contrast, the *ggs* gene from *G. fujikuroi*, which encodes a 418-aa geranylgeranyl diphosphate synthase—a key enzyme in isoprenoid and gibberellin precurser biosynthesis—is not linked to the farnesyl diphosphate synthase gene, indicating that the genes of the isoprenoid pathway are not clustered in the fungal genome (105).

Some of the pathway-specific genes have so far been cloned and characterized, for example the gene encoding ent-kaurene synthase (*Gcps/ks*), which converts geranylgeranyl diphosphate to copalyl diphosphate and ent-kaurene and whose sequence similarity suggests that it is also a bifunctional diterpene cyclase (106); and the gene coding for copalyl diphosphate synthase, which represents the first specific step of the gibberellin pathway as it branches off from the general isoprenoid (biosynthetic) pathway at geranylgeranyl disphosphate. Expression of the latter gene is strongly enhanced under conditions optimized for gibberellin biosynthesis and is reduced in the presence of high amounts of ammonium (106a).

Knowledge of the regulation of gibberellin biosynthesis is important for fermentation processes. Earlier reports described the effects of light, growth rate, inoculum size, and carbon and ammonium sources acting as regulators of gibberellic acid biosynthesis (102). In the latter case, ammonium inhibits the activity and represses the de nova synthesis of specific gibberellin-producing enzymes. Molecular evidence for the involvement of an *areA* homologue (*areA*-GF) has recently been reported: knock-out mutants were unable to use nitrogen sources other than ammonium and glutamine, and gave significantly reduced gibberellin production yields (107).

Besides nitrogen control, the biosynthesis of gibberellins is suppressed by glucose. This glucose effect can be overcome by the addition of mevalonic acid, suggesting that the key enzyme of the isoprenoid pathway (HMGCoA reductase) may be the target of carbon catabolite repression (102). However, the expression of the HMG-CoA reductase-encoding gene of *G. fujikuroi* is not influenced by glucose. Therefore, the glucose effect is obviously acting at a different step of gibberellin biosynthesis. Toward a molecular understanding of this repression, the *creA* gene of *G. fujikuroi* was recently isolated and characterized (108).

Little is known about the role of compartmentation and transport steps in gibberellin biosynthesis. However, evidence for the involvement of physically separated precursor pools has been reported (109).

## F. Ergot Alkaloids

The ergot peptide alkaloids represent a family of important metabolites produced by the ergot fungus *Claviceps purpurea*. They consist of an unusual cyclic arrangement of three amino acids having both lactam and lactone structures (Fig. 6). The tetracyclic D-lysergic acid is attached to the amino terminus of such cyclol structures through an amide bond. It forms the building block of ergot peptides, which are important therapeutic agents. Their synthesis originates from dimethylallyl pyrophosphate and tryptophan.

The first pathway-specific step of ergot alkaloid biosynthesis in the fungus *C. purpurea*, is catalyzed by the prenyltransferase, 4-(γ,γ-dimethylallyl)-tryptophan synthase. The corresponding gene, *dmaW*, isolated from strain ATCC

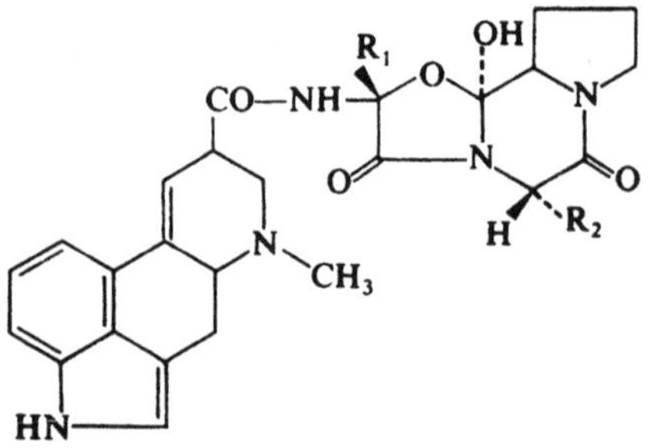

**Figure 6** Chemical structure of ergot alkaloids: ergotamine R1: —CH$_3$, R2: —CH$_2$—CH$_2$—CH$_2$—CH$_3$.

26245, encodes a 51 kDa polypeptide characterized by a putative prenyl diphosphate binding motif (110). In contrast, another gene (*cpd1*), also encoding the dimethylallyltryptophan synthase, was cloned from a strain of *C. purpurea* that produces alkaloids in axenic culture (111). The derived gene product shows only 70% similarity to the corresponding gene previously isolated from *Claviceps* strain ATCC 26245, which is likely to be an isolate of *C. fusiformis*. Analysis of the 3'-flanking region of *cpd1* revealed a second, closely linked ergot alkaloid biosynthetic gene named *cpps1*, which codes for a 356-kDa polypeptide showing significant similarity to fungal modular peptide synthetases. This protein contains three amino acid-activating modules, activating the three amino acids of the peptide portion of ergot peptide alkaloids during D-lysergyl peptide assembly. Further, chromosome walking revealed the presence of additional genes upstream of *cpd1*, which are probably also involved in ergot alkaloid biosynthesis: *cpox1* probably codes for a FAD-dependent oxidoreductase (which could represent the chanoclavine cyclase), and a second putative oxidoreductase gene, *cpox2*, is closely linked to it in inverse orientation. All four genes are expressed under conditions of peptide alkaloid biosynthesis.

Toward identification of other genes that are expressed during alkaloid synthesis in an axenic culture of *Claviceps* sp., a differential cDNA-screening was performed (112), resulting in the cloning of 10 genes for which alkaloid-synthesis-correlated expression was confirmed. They encoded dimethylallyltryptophan-synthase, the initial enzyme of the specific alkaloid pathway, and another gene with homology to the *Neurospora crassa ccg1* gene, coding for a clock-regulated putative general stress protein. In addition, seven cDNA clones, derived from the same gene, which is highly expressed under these conditions, contained typical hydrophobin domains and long stretches of asparagine/glycine repeats and thus probably represent a cell-wall constituent.

## G.  Polyketides

Polyketides, the ubiquitous products of secondary metabolism in microorganisms, are made by a process resembling fatty acid biosynthesis that allows the suppression of reduction or dehydration reactions at specific biosynthetic steps, giving rise to a wide range of often medically useful products. Fungal polyketides of major significance in biotechnology include griseofulvin and lovastatin. Unfortunately, both products have not received the same attention as the formation of toxic polyketides, such as aflatoxins by *Aspergillus parasiticus* and *A. flavus* or some other toxins by plant pathogenic fungi. Consequently, a bulk of information has accumulated on these subjects and the reader is referred to the respective chapters for further information.

Bingle et al. (113) suggested that fungal polyketide synthase gene sequences might be divided into two subclasses, designated WA-type and MSAS-

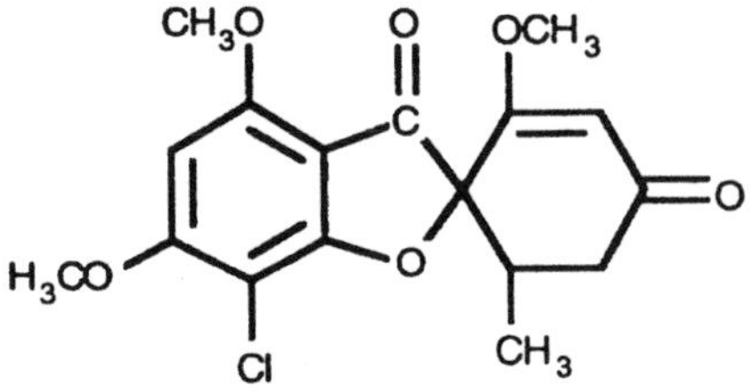

**Figure 7** Chemical structure of griseofulvin.

type. Respective polymerase chain reaction (PCR) primers amplified one or more PCR products from the genomes of a range of ascomycetous *Deuteromycetes*, and Southern blot analysis confirmed that the products obtained with each pair of primers emanated from distinct genomic loci. Sequence analysis showed that genes of the WA-type were closely related to genes involved in pigment and aflatoxin biosynthetic pathways, whereas the genes of the MSAS-type are closely related to 6-methylsalicylic acid synthase (MSAS)-encoding genes.

Griseofulvin (Fig. 7) is produced by various species of *Penicillium*, including *P. griseofulvu, P. janczewski*, and *P. patulum*, and is active against dermatophytic fungi of different species in the genera *Microsporum, Trychophyton*, and *Epidermophyton*. Because of its capacity to concentrate in the keratinous layer of the epidermis and its relatively low toxicity in man, griseofulvin has been extensively used in the therapy of dermatophytoses by oral administration. No details on the molecular genetics of griseofulvin formation have been published, but at least three different polyketide synthase genes were detected in the fungus (113).

Lovastatin, an HMG-CoA reductase inhibitor produced by the fungus *Aspergillus terreus*, is composed of two polyketide chains. One is a nonaketide that undergoes cyclization to a hexahydronaphthalene ring system and the other is a simple diketide, 2-methylbutyrate. The synthesis of the main nonaketide-derived skeleton was found to require a lovastatin nonaketide synthase (LNKS), plus at least one additional protein that interacts with LNKS and is necessary for the correct processing of the growing polyketide chain and production of the intermediate dihydromonacolin L. The noniterative lovastatin diketide synthase enzyme specifies formation of 2-methylbutyrate and interacts closely with an additional transesterase responsible for assembling lovastatin from this polyketide and monacolin (114).

Hendrickson et al. (115) isolated an *A. terreus* mutant that could not synthesize the nonaketide portion of lovastatin and was missing an approximately 250 kDa polypeptide normally present under conditions of lovastatin production. They also identified other mutants that produced lovastatin intermediates without the methylbutyryl sidechain and were missing a polypeptide of approximately 220 kDa. Cloning the gene encoding the former enzyme revealed that the poly- peptide contains catalytic domains typical of vertebrate fatty acid and fungal PKSs, plus two additional domains not previously seen in PKSs: a centrally lo- cated methyltransferase domain and a peptide synthetase elongation domain at the carboxyl terminus. The results provided evidence that the nonaketide and diketide portions of lovastatin are synthesized by separate large multifunctional PKSs.

## IV. CONCLUSIONS

It is evident from this chapter that the molecular genetics of industrial metabolite production has considerably advanced in the past years and has reached a stage where targeted manipulation of the producer organisms is possible and because of solid molecular genetic information, reasonable. However, whether these tech- niques will be introduced into the commercial production process will largely be subject to nonscientific decisions. There is little doubt that at the moment, organic acid production will stay "non-recombinant" because of the difficulties of intro- ducing the product into the food market. The situation looks different, however, when secondary metabolites are concerned, as there is a much better public accep- tance of recombinantly produced pharmaceutical compounds. Examples of such recombinant improvements have been shown in the chapters dealing with penicil- lin and cephalosporin biosynthesis. Yet production processes for the β-lactam antibiotics are already close to their theoretical yield, and thus the employment of recombinant strains, although providing a possible benefit to the rate of produc- tion, will continue to be the exception (as for byproduct elimination).

On the other hand, the experience collected, particularly on *P. chrysogenum* and *A. chrysogenum*, will undoubtedly be useful for opening new avenues for screening of novel metabolite producing fungi and their subsequent improve- ment. Rolf Prade's database of publicly available EST-sequences (http:// bioinfo.okstate.edu/pipeonline) of several (mostly plant pathogenic) filamentous fungi already contains a number of genes involved in secondary metabolite syn- thesis, which can be used to design probes for screening of other fungi of interest. Thus, even though many of the fungi currently investigated for metabolite pro- duction may not be subjects of EST—or of genomic sequencing themselves in the near future—the current genomic initiatives on other filamentous fungi will help to improve their efficacy and product spectrum. Also, the wealth of informa- tion now obtained on how the respective peptide or polyketide synthases work, how their reactions can be manipulated, and how the flux of their precursers can be channeled and regulated, offers the way for the recombinant production of new, tailor-made secondary metabolites.

## REFERENCES

1. CP Kubicek. Citric acid production. In: TW Nagodawithana, G. Reed, eds. Nutritional Requirements of Commercially Important Microorganisms. Milwaukee, WI: Esteekay Associates, Inc., 1998, pp. 236–257.
2. MF Wolschek, CP Kubicek. Biochemistry of citric acid accumulation by *Aspergillus niger*. In: B Kristiansen, M Mattey, J Linden, eds. Citric Acid Biotechnology. London: Taylor and Francis, pp. 11–32.
3. GJG Ruijter, CP Kubicek, J Visser. Production of organic acids by fungi. The Mycota, K Esser PA Lermke, eds., Vol. VI. in press, 2001.
4. GJG Ruijter, H Panneman, J Visser. Overexpression of phosphofructokinase and pyruvate kinase in citric acid producing *Aspergillus niger*. Biochim Biophys Acta 1334:317–326, 1997.
5. GJG Ruijter, H Panneman, D Xu, J Visser. Properties of *Aspergillus niger* citrate synthase and effects of citA overexpression on citric acid production. FEMS Microbiol Letts 184:35–40, 2000.
6. NV Torres, EO Voit, C González-Alcón. Optimisation of nonlinear biotechnological processes with linear programming: Application to citric acid production by *Aspergillus niger*. Biotechnol Bioeng 49:247–258, 1996.
7. HV Baker. *GCR1* of *Saccharomyces cerevisiae* encodes a DNA binding protein whose binding is abolished by mutations in the CTTCC sequence motif. Proc Natl Acad Sci USA 88:9443–9447, 1991.
8. H Uemura, Y Jigami. Role of GCR2 in transcriptional regulation of yeast glycolytic genes. Mol Cell Biol 12:3834–3842, 1992.
9. PK Brindle, JP Holland, CE Willet, MA Innis, MJ Holland. Multiple factors bind the upstream activation sites of the yeast enolase genes *ENO1* and *ENO2*: ABF1 protein, like repressor activator protein RAP1, binds *cis*-acting sequences which modulate repression or activation of transcription. Mol Cell Biol 10:4872–4895, 1990.
10. A Chambers, C Stanway, JSH Tsang, Y Henry, Al Kingsman, SM Kingsman. ARS binding factor 1 binds adjacent to RAP1 at the UASs of the yeast glycolytic genes *PGK* and *PYKI*. Nucleic Acids Res 18:5393–5399, 1990.
11. MA Huie, EW Scott, CM Drazinic, MC Lopez, IK Hornstra, TP Yang, HV Baker. Characterization of the DNA-binding activity of GCR1: *in vivo* Evidence for two GCR1-binding sites in the upstream activating sequences of *TPI* of *Saccharomyces cerevisiae*. Mol Cell Biol 12:2690–2700, 1992.
12. JB McNeil, P Dykshoorn, JN Huy, S Small. The DNA-binding protein RAP1 is required for efficient transcriptional activation of the yeast *PYK* glycolytic gene. Curr Genet 18:405–412, 1990.
13. PJ Punt, MA Dingemanse, A Kuyvenkofen, RDM Soede, PH Pouwels, CAMJJ van den Hondel. Functional elements in the promoter region of the *Aspergillus nidulans gpdA* gene encoding glyceraldehyde-3-phosphate dehydrogenase. Gene 93:101–109, 1990.
14. PJ Punt, C Kramer, A Kuyvenkofen, PH Pouwels, CAMJJ van den Hondel. An upstream activating sequence from the *Aspergillus nidulans gpdA* gene. Gene 120:62–73, 1992.

15.  JM Clements, CF Roberts. Transcription and processing signals in the 3-phospho-glycerate kinase (PGK) gene from *Aspergillus nidulans*. Gene 44:97–105, 1986.
16.  SJ Streatfield, S Toews, CF Roberts. Functional analysis of the expression of the 3′-phosphoglycerate kinase *pgk* gene in *Aspergillus nidulans*. Mol Gen Genet 233: 231–241, 1992.
17.  FG Falkner, R Mocikat, HG Zachau. Sequences closely related to an immunoglobulin gene promoter/enhancer element occur also upstream of other eukaryotic and of prokaryotic genes. Nucleic Acids Res 14:8819–8827, 1986.
18.  NV Torres. Modelling approach to control of carbohydrate metabolism during citric acid accumulation by *Aspergillus niger*:I. Model definition and stability of the steady-state. Biotechnol Bioeng 44:104–111, 1994.
19.  NV Torres. Modelling approach to control of carbohydrate metabolism during citric acid accumulation by *Aspergillus niger*: II. Sensitivity analysis. Biotechnol Bioeng 44:112–11, 1994.
20.  H Panneman, GJG Ruijter, HC van den Broeck, J Visser. Cloning and biochemical characterisation of *Aspergillus niger* hexokinace. The enzyme is strongly inhibited by physiological concentrations of trehalose 6-phosphate. Eur J Biochem 258:223–232, 1998.
21.  I Arisan-Atac, MF Wolschek, CP Kubicek Trehalose-6-phosphate synthase A affects citrate accumulation by *Aspergillus niger* under conditions of high glycolytic flux. FEMS Microbiol Lett 140:77–83, 1996.
22.  MF Wolschek, CP Kubicek. The filamentous fungus *Aspergillus niger* contains two "differentially regulated" trehalose-6-phosphate synthase encoding genes, *tpsA* and *tpsB*. J Biol Chem 272:2729–2735, 1997.
23.  CP Kubicek, G Schreferl-Kunar, W Wöhrer, M Röhr. Evidence for a cytoplasmic pathway of oxalate biosynthesis in *Aspergillus niger*. Appl Environ Microbiol 54: 633–637, 1988.
24.  H Pedersen, B Christensen, C Hjort, J Nielsen. Construction and characterization of an oxalic acid nonproducing strain of *Aspergillus niger*. Metab Eng 2:34–41, 2000.
25.  K Kirimura, M Yoda, S Usami. Cloning and expression of the cDNA encoding an alternative oxidase gene from *Aspergillus niger* WU-2223L. Curr Genet 34: 472–477, 1999.
26.  J Wallrath, J Schmidt, H Weiss. Concomitant loss of respiratory chain NADH: ubiquinone reductase (complex I) and citric acid accumulation in *Aspergillus niger*. Appl Microbiol Biotechnol 36:76–81, 1991.
27.  M Schmidt, J Wallrath, A Dörmer, H Weiss. Disturbed assembly of the respiratory chain NADH:ubiquinone reductase (complex I) in citric acid accumulating *Aspergillus niger* strain B 60. Appl Microbiol Biotechnol 36:667–672, 1992.
28.  C Prömper, R Schneider, H Weiss. The role of the proton-pumping and alternative respiratory chain NADH:ubiquinone oxidoreductases in overflow catabolism of *Aspergillus niger*. Eur J Biochem 216:223–230, 1993.
29.  KR Frederick, J Tung, RS Emerick, FR Masiarz, SH Chamberlain, A Vasavada, S Rosenberg, S Chakraborty, LM Schopfer. Glucose oxidase from *Aspergillus niger*. Cloning, gene sequence, secretion from *Saccharomyces cerevisiae* and kinetic analysis of a yeast-derived enzyme. J Biol Chem 265: 3793–3802, 1990.

30. H Whittington, S Kerry-Williams, K Bidgood, N Dodsworth, JF Peberdy, M Dobson, E Hinchliffe, DJ Balance. Expression of the *Aspergillus niger* glucose oxidase gene in *A. niger, A. nidulans* and *Saccharomyces cerevisiae*. Curr Genet 18:531–536, 1990.

31. CFB Witteveen, PJI van de Vondervoort, K Swart, J Visser. Glucose oxidase overproducing and negative mutants of *Aspergillus niger*. Appl Microbiol Biotechnol 33:683–686, 1990.

32. CFB Witteveen, PJI van de Vondervoort, HC van den Broeck, FAC van Engelenburg, LH de Graaff, MHBC Hillebrand, PJ Schaap, J Visser. Induction of glucose oxidase, catalase, and lactonase in *Aspergillus niger*. Cuff Genet 24:408–416, 1993.

33. CFB Witteveen, M Veenhuis, J Visser. Localization of glucose oxidase and catalase activities in *Aspergillus niger*. Appl Envir Microbiol 58:1190-1194, 1992.

34. T Fowler, MW Rey, P Vaha-Vahe, SD Power, RM Berka The catR gene encoding a catalase from *Aspergillus niger*: Primary structure and elevated expression through increased copy number and use of a strong promoter. Mol Microbiol 5:989–998, 1993.

35. K Swart, PJI van de Vondervoort, CFB Witteveen, J Visser. Genetic localization of a series of genes affecting glucose oxidase levels in *Aspergillus niger*. Curr Genet 18:435–439, 1990.

36. MA Penalva, RT Rowlands, G Turner. The optimization of penicillin biosynthesis in fungi. Trends Biotechnol 16:483–489, 1998.

37. J Nielsen. The role of metabolic engineering in the production of secondary metabolites. Curr Opin Microbiol 1:330–336, 1998.

38. JF Martin. New aspects of genes and enzymes for β-lactam antibiotic biosynthesis. Appl Microbiol Biotechnol 50:1–15, 1998.

39. F Laich, F Fierro, RE Cardoza, JF Martin. Organization of the gene cluster for biosynthesis of penicillin in *Penicillium nalgiovense* and antibiotic production in cured dry sausages. Appl Environ Microbiol 65:1236–1240, 1999.

40. F Fierro, S Gutierrez, B Diez, JF Martin. Resolution of four large chromosomes in penicillin-producing filamentous fungi: The penicillin gene cluster is located on chromosome II (9.6 Mb) in *Penicillium notatum* and chromosome I (10.4 Mb) in *Penicillium chrysogenum*. Mol Gen Genet 241:573–578, 1993.

41. F Fierro, JL Barredo, B Diez, S Gutierrez, FJ Fernandez. JF Martin. The penicillin gene cluster is amplified in tandem repeats linked by conserved hexanucleotide sequences. Proc Natl Acad Sci USA 92:6200–6204, 1995.

42. RW Newbert, B Parton, P Greaves, J Harper, G Turner. Analysis of a commercially improved *Penicillium chrysogenum* strain series: Involvement of recombinogenic regions in amplification and deletion of the penicillin biosynthetic gene cluster. J Ind Microbiol Biotechnol 19:18–27, 1997.

43. F Fierro, E Montenegro, S Gutierrez, JF Martin. Mutants blocked in penicillin biosynthesis show a deletion of the entire penicillin gene cluster at a specific site within a conserved hexanucleotide sequence. Appl Microbiol Biotechnol 44:597–604, 1996.

44. AA Brakhage. Molecular regulation of β-lactam biosynthesis in filamentous fungi. Microbiol Mol Biol Rev 62:547–585, 1998.

45. JF Martin. Molecular control of expression of penicillin biosynthesis genes in fungi: Regulatory proteins interact with a bidirectional promoter region. J Bacteriol 182:2355–2362, 2000.

46. B Feng, E Friedlin, GA Marzluf. A reporter gene analysis of penicillin biosynthesis gene expression in *Penicillium chrysogenum* and its regulation by nitrogen and glucose catabolite repression. Appl Environ Microbiol 60:4432–4439,1994.

47. S Gutierrez, AT Marcos, J Casqueiro, K Kosalkova, FJ Fernandez, J Velasco, JF Martin. Transcription of the *pcbAB, pcbC* and *penDE* genes of *Penicillium chrysogenum* AS-P78 is repressed by glucose and the repression is not reversed by alkaline pH. Microbiol. UK 145:317–324, 1999.

48. EA Espeso, J Tilburn, HN Arst Jr, MA Penalva. pH Regulation is a major determinant in expression of a fungal biosynthetic gene. EMBO J 12:3947–3956, 1993.

49. H Haas, B Bauer, B Redl, G Stöffler, GA Marzluf. Molecular cloning and analysis of *nre*, the major nitrogen regulatory gene of *Penicillium chrysogenum*. Curr Genet 27:150–158, 1995.

50. H Haas, K Angermayr, I Zadra, G Stöffler. Overexpression of *nreB*, a new GATA factor-encoding gene of *Penicillium chrysogenum*, leads to repression of the nitrate assimilatory gene cluster. J Biol Chem 272:22576–22582, 1997.

51. YW Chu, D Renno, G Saunders. Detection of a protein which binds specifically to the upstream region of the *pcbAB* gene in *Penicillium chrysogenum*. Curr Genet 28:184–189, 1995.

52. B Feng, E Friedlin, GA Marzluf. Nuclear DNA-binding proteins which recognize the intergenic control region of penicillin biosynthetic genes. Curr Genet 27:351–358, 1995.

53. J Tilburn, S Sarkar, DA Widdick, EA Espeso, M Orejas, J Mungroo, MA Penalva, HN Arst The *Aspergillus nidulans* PacC zinc finger transcription factor mediates regulation of both acidic- and alkaline expressed genes by ambient pH. EMBO J 14:779–790, 1995.

54. SH Denison. pH Regulation of gene expression in fungi. Fungal Genet Biol 29: 61–71, 2000.

55. T Suarez, MA Penalva. Characterization of a *Penicillium chrysogenum* gene encoding a PacC transcription factor and its binding sites in the divergent *pcbAB-pcbC* promoter of the penicillin biosynthetic cluster. Mol Microbiol 20:529–540, 1996.

56. KT Bergh, O Litzka, AA Brakhage. Identification of a major cis-acting DNA element controlling the bidirectionally transcribed penicillin biosynthesis genes *acvA* (*pcbAB*) and *ipnA* (*pcbC*) of Aspergillus nidulans. J Bacteriol 178:3908–3916, 1996.

57. O Litzka, TK Bergh AA Brakhage. The *Aspergillus nidulans* penicillin-bio-synthesis gene *aat* (*penDE*) is controlled by a CCAAT-containing DNA element Eur J Biochem 238:675–682, 1996.

58. J Kennedy, G Turner. δ-(α-aminoadipyl)-L-cysteinyl-D-valine synthetase is a rate limiting enzyme for penicillin production in *Aspergillus nidulans*. Mol Gen Genet 253:404-411, 1996.

59. K Kosalkova, AT Marcos, F Fierro, V Hernando-Rico, S Gutierrez, JF Martin. A novel heptameric sequence (TTAGTAA) is the binding site for a protein required

for high level expression of *pcbAB*, the first gene of the penicillin biosynthesis in *Penicillium chrysogenum*. J Biol Chem 275:2423–2430, 2000.

60. EK Schmitt, U Kück. The fungal CPCR1 protein, which binds specifically to beta-lactam biosynthesis genes, is related to human regulatory factor X transcription factors. J Biol Chem 275:9348–9357, 2000.

61. M van de Kamp, AJ Driessen, WN Konings. Compartmentalization and transport in beta-lactam antibiotic biosynthesis by filamentous fungi. Antonie Van Leeuwenhoek 75:41–78, 1999.

62. T Lendenfeld, D Ghali, M Wolschek, EM Kubicek-Pranz, CP Kubicek. Subcellular compartmentation of penicillin biosynthesis in *Penicillium chrysogenum*. J Biol Chem 268:665–671, 1993.

63. WH Muller, TP van der Krift AJ Krouwer HA Wosten, LH van der Voort, EB Smaal, AJ Verkleij. Localization of the pathway of the penicillin biosynthesis in *Penicillium chrysogenum*. EMBO J 10:489–495, 1991.

64. WH Muller, RA Bovenberg, MH Groothuis, F Kattevilder, EB Smaal, LH van der Voort, AJ Verkleij. Involvement of microbodies in penicillin biosynthesis. Biochim Biophys Acta 1116:210–213, 1992.

65. JAKW Kiel, RE Hilbrands, RAL Bovenberg, M Veenhuis. Isolation of *Penicillium chrysogenum PEX1* and *PEX6* encoding AAA proteins involved in peroxisome biogenesis. Appl Microbiol Biotechnol 54:238–242, 2000.

66. O Banuelos, J Casquiero, S Gutierrez, JF Martin. Overexpression of the *lys1* gene in *Penicillium chrysogenum*: Homocitrate synthase levels, alpha-aminoadipic acid pool and penicillin production. Appl Microbiol Biotechnol 54: 69–77, 2000.

67. J Casqueiro, S Gutierrez, O Banuelos, MJ Hijarrubia, JF Martin. Gene targeting in *Penicillium chrysogenum*: Disruption of the *lys2* gene leads to penicillin overproduction. J Bacteriol 181:1181–1188, 1999.

68. DJ Hillenga, HJ Versantvoort, AJ Driessen, WN Konings. Sulfate transport in *Penicillium chrysogenum* plasma membranes. J Bacteriol 178:3953–3956, 1996.

69. M van De Kamp, TA Schuurs, A Vos, TR van der Lende, WN Konings, AJ Driessen. Sulfur regulation of the sulfate transporter genes *sutA* and *sutB* in *Penicillium chrysogenum*. Appl Environ Microbiol 66:4536–4538, 2000.

70. SH Eriksen, B Jensen, I Schneider, S Kaasgaard, J Olsen. Utilization of side-chain precursors for penicillin biosynthesis in a high-producing strain of *Penicillium chrysogenum*. Appl Microbiol Biotechnol 40:883–887, 1994.

71. JM Fernandez-Canon, MA Penalva. Molecular characterization of a gene encoding a homogentisate dioxygenase from *Aspergillus nidulans* and identification of its human and plant homologues. J Biol Chem 270:21199–21205, 1995.

72. JM Mingot, MA Penalva, JM Fernandez-Canon Disruption of *phacA*, an *Aspergillus nidulans* gene encoding a novel cytochrome P450 monooxygenase catalyzing phenylacetate 2-hydroxylation, results in penicillin overproduction. J Biol Chem 274:14545–14550.

73. B Minambres, H Martinez-Blanco, ER Olivera, B Garcia, B Diez, JL Barredo, MA Moreno, C Schleissner, F Salto, JM Luengo. Molecular cloning and expression in different microbes of the DNA encoding *Pseudomonas putida* U phenylacetyl-CoA ligase. Use of this gene to improve the rate of benzylpenicillin biosynthesis in *Penicillium chrysogenum* J Biol Chem 271:33531–33538, 1996.

74. AC Andrade, JGM van Nistelrooy, RB Peery, PL Skatrud, MA de Waard. The role of ABC transporters from *Aspergillus nidulans* in protection against cytotoxic agents and in antibiotic production. Mol Gen Genet 263:966-977, 2000.

75. PG Guilfoile, CR Hutchinson. A bacterial analog of the *mdr* gene of mammalian tumor cells is present in *Streptomyces peucetius*, the producer of daunorubicin and doxcorubicin. Proc Natl Acad Sci USA 88:8553–8557, 1991.

76. E Fernandez, F Lombo, C Mendez, JA Salas. The ABC transporter is essential for resistance to the antitumor agent mithramycin in the producer *Streptomyces argillaceus*. Mol Gen Genet 251:692-698, 1996.

77. A Tag, J Hicks, G Garifullina, C Ake, TD Phillips, M Beremand, N Keller. G - protein signalling mediates differential production of toxic secondary metabolites. Mol Microbiol 38:658–665, 2000.

78. L Crawford, AM Stepan, PC McAda, JA Rambosek, MJ Conder, VA Vinci, CD Reeves. Production of cephalosporin intermediates by feeding adipic acid to recombinant *Penicillium chrysogenum* strains expressing ring expansion activity. Biotechnology 13:58–62, 1995.

79. S Gutierrez, F Fierro, J Casqueiro, JF Martin. Gene organization and plasticity of the beta-lactam genes in different filamentous fungi. Antonie Van Leeuwenhoek 75:81–94, 1999.

80. LH Malmberg, WS Hu. Identification of rate-limiting steps in cephalosporin C biosynthesis in *Cephalosporium acremonium*: A theoretical analysis. Appl Microbiol Biotechnol 38:122–128, 1992.

81. L Mathison, C Soliday, T Stepan, T Aldrich, J Rambosek. Cloning, characterization, and use in strain improvement of the *Cephalosporium acremonium* gene *cefG* encoding acetyl transferase. Curr Genet 23:33–41, 1993.

82. PL Skatrud, SW Queener. An electrophoretic molecular karyotype for an industrial strain of *Cephalosporium acremonium* Gene 78:331–338, 1989.

83. AW Smith, K Collis, M Ramsden, HM Fox, JF Peberdy. Chromosome rearrangements in improved cephalosporin C-producing strains of *Acremonium chrysogenum*. Curr Genet 19:235–237, 1991.

84. S Menne, M Walz, U Kück. Expression studies with the bidirectional *pcbAB-pcbC* promoter region from *Acremonium chrysogenum* using reporter gene fusions. Appl Microbiol Biotechnol 42:57–66, 1994.

85. JF Martin, S Gutierrez, FJ Fernandez, J Velasco, F Fierro, AT Marcos, K Kosalkova. Expression of genes and processing of enzymes for the biosynthesis of penicillins and cephalosporins. Antonie van Leeuwenhoek 65:227–243, 1999.

86. J Velasco, S Gutierrez, FJ Fernandez, AT Marcos, C Arenos, JF Martin. Exogenous methionine increases levels of mRNAs transcribed from *pcbAB, pcbC*, and *cefEF* genes, encoding enzymes of the cephalosporin biosynthetic pathway, in *Acremonium chrysogenum*. J Bacteriol 176:985–991, 1994.

87. AL Demain, J Zhang. Cephalosporin C production by *Cephalosporium acremonium*: The methionine story. Crit Rev Biotechnol 18:283–294, 1998.

88. K Jekosch, U Kück Glucose dependent transcriptional expression of the *cre1* gene in *Acremonium chrysogenum* strains showing different levels of cephalosporin C production. Curr Genet 37:388–395, 2000.

89. K Jekosch, U Kück. Loss of glucose repression in an *Acremonium chrysogenum*

β-lactam producer strain and its restoration by multiple copies of the *cre1* gene. Appl Microbiol Biotechnol 54:556–563, 2000.

90.  JA DeModena, S Gutierrez, J Velasco, FJ Fernandez, RA Fachini, JL Galazzo, DE Hughes, JF Martin. The production of cephalosporin C by *Acremonium chrysogenum* is improved by the intracellular expression of a bacterial hemoglobin. Biotechnology 11:926–929, 1993.

91.  S Gutierrez, J Velasco, AT Marcos, FJ Fernandez, F Fierro, JL Barredo, B Diez, JF Martin. Expression of the *cefG* gene is limiting for cephalosporin biosynthesis in *Acremonium chrysogenum*. Appl Microbiol Biotechnol 48: 606–614, 1997.

92.  T Isogai, M Fukagawa, I Aramori, M Iwami, H Kojo, T Ono, Y Ueda, M Kohsaka, H Imanaka. Construction of a 7-aminocephalosporanic acid (7ACA) biosynthetic operon and direct production of 7ACA in *Acremonium chrysogenum*. Biotechnology 9:188–191, 1991.

92a. J Velasco, J Luis Adrio, M Angel Moreno, B Diez, G Soler, JL Barredo. Environmentally safe production of 7-aminodeacetoxycephalosporanic acid (7-ADCA) using recombinant strains of *Acremonium chrysogenum*. Nat Biotechnol 18:857–861, 2000.

93.  M Thali. Cyclosporins: Immunosuppressive drugs with anti-HIV-1 activity. Mol Med Today 1:287–291, 1995.

94.  G Weber, K Schörgendorfer, E Schneider-Scherzer, E Leitner. The peptide synthetase catalyzing cyclosporine production in *Tolypocladium niveum* is encoded by a giant 45.8-kilobase open reading frame. Curr Genet 26:120–125, 1994.

95.  G Weber, E Leitner Disruption of the cyclosporin synthetase gene of *Tolypocladium niveum*. Curr Genet 26:461–467, 1994.

96.  F Kempken, C Schreiner, K Schörgendorfer, U Kück. A unique repeated DNA sequence in the cyclosporin-producing strain of *Tolypocladium inflatum* (ATCC 34921). Exp Mycol 19:305–313, 1995.

97.  F Kempken, U Kück. *Restless*, an active Ac-like transposon from the fungus *Tolypocladium inflatum*: Structure, expression, and alternative RNA splicing. Mol Cell Biol 16:6563–6572, 1996.

98.  F Kempken U Kück. Tagging of a nitrogen pathway-specific regulator gene in *Tolypocladium inflatum* by the transposon Restless. Mol Gen Genet 263: 302–308, 2000.

99.  C Lee, K Hoffmann, R Zocher. FK-506 binding protein from *Tolypocladium inflatum*: Resistance of FKBP/FK-506 complex against proteolysis Biochem Biophys Res Commun 182:1282–1287, 1992.

100. R Zocher, U Keller, C Lee, K Hoffmann. A seventeen kilodaltons peptidylprolyl cis-trans isomerase of the cyclosporin-producer *Tolypocladium inflatum* is sensitive to cyclosporin A. J Antibiot 45:265–268, 1992.

101. T Hornbogen, R Zocher. Cloning and sequencing of a cyclophilin gene from the cyclosporin producer *Tolypocladium niveum*. Biochem Mol Biol Int 36: 169–176, 1995.

102. B Bruckner. Regulation of gibberellin formation by the fungus *Gibberella fujikuroi*. Ciba Found Symp 171:129–137, 1992.

103. B Tudzynski. Biosynthesis of gibberellins in *Gibberella fujikuroi*: Biomolecular aspects. Appl Microbiol Biotechnol 52:298–310, 1999.

104. B Tudzynski, K Holter. Gibberellin biosynthetic pathway in *Gibberella fujikuroi*: Evidence for a gene cluster. Fungal Genet Biol 25:157–170, 1998.
105. K Mende, V Homann, B Tudzynski. The geranylgeranyl diphosphate synthase gene of *Gibberella fujikuroi*: Isolation and expression. Mol Gen Genet 255: 96–105, 1997.
106. T Toyomasu, H Kawaide, A Ishizaki, S Shinoda, M Otsuka, W Mitsuhashi, T Sassa. Cloning of a full-length cDNA encoding ent-kaurene synthase from *Gibberella fujikuroi*: Functional analysis of a bifunctional diterpene cyclase. Biosci Biotechnol Biochem 64:660–664, 2000.
106a. B Tudzynski, H Kawaide, Y Kamiya. Gibberellin biosynthesis in *Gibberella fujikuroi*: Cloning and characterization of the copalyl diphosphate synthase gene. Curr Genet 34:234–240, 1998.
107. B Tudzynski, V Homann, B Feng, GA Marzluf. Isolation, characterization and disruption of the *area* nitrogen regulatory gene of *Gibberella fujikuroi*. Mol Gen Genet 261:106–114, 1999.
108. B Tudzynski, S Liu, JM Kelly. Carbon catabolite repression in plant pathogenic fungi: Isolation and characterization of the *Gibberella fujikuroi* and *Botrytis cinerea creA* genes. FEMS Microbiol Lett 184:9–15, 2000.
109. CE Domenech, W Giordano, J Avalos, E Cerda-Olmedo. Separate compartments for the production of sterols, carotenoids and gibberellins in *Gibberella fujikuroi*. Eur J Biochem 239:720–725, 1996.
110. HP Tsai, H Wang, JC Gebler, CD Poulter, CL Schardl. The *Claviceps purpurea* gene encoding dimethylallyltryptophan synthase, the committed step for ergot alkaloid biosynthesis. Biochem Biophys Res Commun 216:119–125, 1995.
111. P Tudzynski, K Holter, T Correia, C Arntz, N Grammel, U Keller. Evidence for an ergot alkaloid gene cluster in *Claviceps purpurea*. Mol Gen Genet 261: 133–141, 1999.
112. C Arntz, P Tudzynski. Identification of genes induced in alkaloid-producing cultures of *Claviceps* sp. Curr Genet 31:357–360, 1997.
113. LE Bingle, TJ Simpson, CM Lazarus. Ketasynthase domain probes identify two subclasses of fungal polyketide synthase genes. Fungal Genet Biol 26: 209–223, 1999.
114. J Kennedy, K Auclair, SG Kendrew, C Park, JC Vederas, CR Hutchinson. Modulation of polyketide synthase activity by accessory proteins during lovastatin biosynthesis. Science 284:1368–1372, 1999.
115. L Hendrickson, CR Davis, C Roach, DK Nguyen, T Aldrich, PC McAda, CD Reeves. Lovastatin biosynthesis in *Aspergillus terreus*: Characterization of blocked mutants, enzyme activities and a multifunctional polyketide synthase gene. Chem Biol 6:429–439, 1999.

# 5

# Acquiring a New Viewpoint: Tools for Functional Genomics in the Filamentous Fungi

**Todd M. DeZwaan, Keith D. Allen, and Matthew M. Tanzer**
*Paradigm Genetics, Inc., Research Triangle Park, North Carolina, U.S.A.*

**Kiichi Adachi**
*Mitsui Global Strategic Studies Institute, Tokyo, Japan*

**Lakshman Ramamurthy**
*GlaxoSmithKline, Inc., Research Triangle Park, North Carolina, U.S.A.*

**Sanjoy Mahanty**
*Inspire Pharmaceuticals, Inc., Research Triangle Park, North Carolina, U.S.A.*

**Lisbeth Hamer**
*North Carolina State University, Raleigh, North Carolina, U.S.A.*

## I. ADVANCING HUMAN HEALTH AND AGRICULTURE WITH FILAMENTOUS FUNGAL FUNCTIONAL GENOMICS

Fungi are deep-rooted in human culture. The fungal kingdom includes some of the earliest cultivated organisms for food production and some of the most damaging pathogens to impact society (1). The production of beer, wine, bread, and many cheeses requires fungal metabolism. Fungi have also critically influenced the development of modern medicine. The antibiotic penicillin and the cholesterol lowering agent lovastatin from the filamentous fungi *Penicillium notatum* and

*Aspergillus terreus*, respectively, are examples of medically significant fungal metabolites. Fungal metabolites can also be potent toxins. For example, the food and feed molds *Aspergillus flavus* and *A. parasiticus* produce the carcinogenic agent aflatoxin (2), and the corn pathogen *Fusarium moniliforme* produces fumonisin, a toxin that causes serious, and often fatal, health problems in humans and animals (3). Many fungi are phytopathogens and cause severe yield loss of major agricultural crops such as wheat, rice and barley (4). Fungi are also the agents of an increasing number of opportunistic human infections (5). By understanding the mechanisms of such processes as metabolite production and host infection, we can enhance the biological output of fungi and aid the development of antifungal agents for agriculture and medicine. A way to gain this understanding is to determine the function of genes required for fungal metabolism and host interaction. Because many genes are involved in both of these processes, it is logical to undertake a functional genomic approach to investigate gene function on a large scale.

In 1996, the complete genome sequence of the budding yeast *Saccharomyces cerevisiae* was reported (6) and a number of bioinformatic and biological tools were developed for large-scale gene function analysis. Budding yeast is unicellular and uninucleate, and has a nonrepetitive genome size of 12,057 kilobases (7,8). *Saccharomyces cerevisiae* is very tractable to genetic manipulation. It is possible to produce diploids, mate strains of different mating type, transform cells, and easily create targeted gene disruptions in yeast (9). In contrast, the biological complexity of filamentous fungi prohibits the implementation of many of the molecular and functional genomic tools used in yeast. Compared to yeast, the filamentous fungi generally have larger genomes, are much less easy to manipulate genetically, and have more complex lifecycles (10–12). Filamentous fungi are multicellular, often multinuclear, and display a wide variety of cells and morphologies, ranging from infection structures to conidia-producing stalks (12). Many spore forms may exist for a single filamentous fungal species and some species depend on multiple hosts for the completion of their life cycle (4). Because of the large number of filamentous fungal genes without homologs in other organisms, bioinformatics tools must be adapted to interpret genome sequence data. The difficulties of transformation and targeted integration typical in filamentous fungi call for developments in gene alteration methodologies for high-throughput functional genomics. New strategies for functional analysis are required because filamentous fungi have a greater phenotypic repertoire than yeast. In this chapter, we discuss the computational and experimental methods currently available for functional analysis of filamentous fungal genomes. We also describe relevant methods in other organisms that may be applicable to filamentous fungi. Finally, future directions and challenges for functional genomics in filamentous fungi are discussed.

## II. GENOME ANNOTATION FOR FILAMENTOUS FUNGAL GENE DISCOVERY

Bioinformatic annotation of genome sequence data is an important preliminary step for functional genomics. Genome annotation can lead to gene discovery, the identification of novel gene activities, and can shed light on the many differences and similarities among fungal species. Sequence information is accumulating from human pathogens (*Aspergillus fumigatus, Candida albicans, Cryptococcus neoformans*, and *Pneumocystis carinii*), crop pathogens (*Botrytis cinerea, Fusarium sporotrichioides, Magnaporthe grisea, Mycosphaerella graminicola, Phytophthora infestans*, and *Phytophthora sojae*), toxigenic fungi (*A. flavus* and *A. parasiticus*), a biodegrading fungus (*Phanerochaete chrysosporium*), and model fungi (*S. cerevisiae, Aspergillus nidulans, Neurospora crassa*, and *Schizosaccharomyces pombe*), as shown in Table 1. These sequencing projects represent species from the two main fungal phyla, Ascomycetes and Basidiomycetes, as well as the stramenophile Oomycetes. A possible plan for gene function discovery starting with a complete or nearly complete genome sequence is shown in Figure 1. In this example, a fungal genome is annotated using a series of similarity search methods and comparative genomics. Gene prediction programs are used to identify novel genes. Mutant generation and functional analysis follows. Novel genes are a resource for functional discovery, and annotated genes are altered to produce a predicted functional outcome and to validate the in silico function assignment.

### A. Similarity Searching

Genome annotation of filamentous fungi or any other organism begins with the identification of all genes that have significant similarity to proteins in public databases. If computational resources allow, this is best done stepwise, using search tools with progressively increasing sensitivity and computational expense (Fig. 1). The majority of computationally derived gene annotations come from BLAST similarity searching. BLAST is a rapid heuristic search algorithm that is one of the most frequently used tools in bioinformatics (13). Any contiguous sequence elements (contigs), or portion thereof, that have a BLAST similarity alignment with proteins in the public database above a preassigned cutoff can be considered annotated and removed from further analysis. A typical cutoff for significant alignments is 35% identity over a stretch of at least 100 amino acids. Sequences with an alignment score below this cutoff are again searched against the public protein databases with a Smith-Waterman (SW) dynamic programming implementation (14), and sequence alignments above a predefined cutoff are selected as annotated. This step can require substantial computing time unless a

**Table 1** Fungal Genome Sequencing Projects[a]

| Phylum | Species | Genome sequencing status | EST sequences[b] | URL | Reference |
|---|---|---|---|---|---|
| Ascomycetes | Aspergillus fumigatus | BAC[c] end sequencing initiated | — | http://www.sanger.ac.uk/Projects/A_fumigatus/ | — |
| | Aspergillus flavus | — | 1,253 | http://www.genome.ou.edu/fungal.html | (135) |
| | Aspergillus nidulans | Three cosmids completed | 12,485 | http://www.genome.ou.edu/fungal.html | (135) |
| | Aspergillus parasiticus | Two cosmids completed | — | http://www.genome.ou.edu/fungal.html | (135) |
| | Botrytis cinerea | — | 6,558 | hhtp://www.genoscope.cns.fr/externe/English/Projects/Resultats/rapport.html | — |
| | Candida albicans | ~10X coverage | — | http://sequence-www.stanford.edu/group/candida/index.html | — |
| | Fusarium sporotrichoides | — | 7,495 | http://www.genome.ou.edu/fsporo.html | (136) |
| | Magnaporthe grisea | Chromosome 7 in progress | 1,607 | http://www.cals.ncsu.edu:8050/fungal_genomics/USDA-2000-Summary.htm; http://www.ncbi.nlm.nih.gov:80/htbin-post/Taxonomy/wgetorg?id=148305 | (137) |
| | Mycosphaerella graminicola | — | 1,158 | http://www.ncbi.nlm.nih.gov:80/htbin-post/Taxonomy/wgetorg?id=54734 | — |

| | | | | | |
|---|---|---|---|---|---|
| | Neurospora crassa | >10X coverage | 20,172 | http://www.genome.wi.mit.edu/ annotation/fungi/neurospora/; http://www.genome.ou.edu/ fungal.html | (135) |
| | Pneumocystis carinii | In progress | 3,896 | http://biology.uky.edu/Pc/ | — |
| | Saccharomyces cerevisiae | Completed | — | http://genome-www.stanford.edu/ Saccharomyces/ | (6) |
| | Schizosaccharo-myces pombe | >90% completed | — | http://www.sanger.ac.uk/Projects/ S_pombe/seqproj.shtml | — |
| Basidiomycetes | Cryptococcus neo-formans | ~2.8X cover-age(strain JEC21) | 2,333 (strain H99) 1,575 (strain JEC21) | http://www.genome.ou.edu/ cneo.html | (138) |
| | Phanerochaete chrysosporium | In progress | — | http://www.er.doe.gov/production/ ober/EPR/mig_cont.html | — |
| Oomycetes[d] | Phytophthora in-festans | — | 1,000 | http://www.ncgr.org/pgi/reports/ Pinfestans-hits.html | — |
| | Phytophthora sojae | In progress | 2,976 | http://www.ncgr.org/pgi/reports/ Psojae-hits.html; http://www.ncbi.nlm.nih.gov:80/htbin-post/Taxonomy/ wgetorg?id=67593 | (139) |

[a] For a complete listing of genome sequencing projects in all organisms visit GOLD™: Genomes Online Database at http://wit.integratedgenomics.com/ GOLD/.

[b] Refers to the number of high quality sequences produced in each project. Includes redundant sequences and sequences from both ends of cloned cDNA inserts

[c] BAC, Bacterial artificial chromosome.

[d] Stramenophiles, not true fungi.

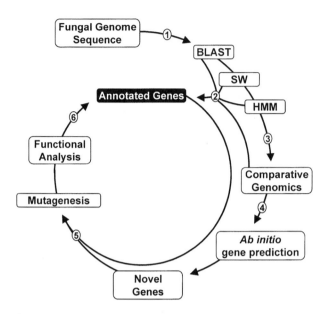

**Figure 1** Fungal gene function discovery. Fungal genome annotation includes computational and experimental methods. Computational analysis begins with a series of similarity search tools that have incremental analytic capacity (1). Annotated genes are accumulated at each step (2). Comparative genomics provides additional gene annotation (3). Ab initio gene prediction tools are used to identify novel genes that do not have significant similarity to genes in the database (4). Mutagenesis followed by analysis of the mutant phenotype (functional analysis) is performed to experimentally determine the gene function. Novel genes are mutated to discover new gene functions, and genes that are annotated computationally are mutated to confirm the in silico functional prediction and to uncover any additional, previously unidentified gene functions (5). Experimentally determined gene functions are an additional resource for gene annotation (6).

hardware-accelerated SW version is available. For example, the TimeLogic De-Cypher® system accelerates calculations in SW dynamic programming algorithms with field programmable gate arrays (FPGAs). Field programmable gate arrays are specially designed modifiable hardware circuits that can be rewired and reprogrammed during their operation (15). Generally, SW will increase the number of annotated sequences by about 10%. Sequences remaining that are not annotated can be searched against a profile hidden markov model (HMM) database, such as Pfam (16), using a specially designed search tool, such as HMMer (17). Hidden markov model searching is more sensitive that any pairwise sequence comparison algorithm because it incorporates all of the information contained in a multiple sequence alignment into the search. A limitation is that, although the Pfam database is steadily growing, it only represents a fraction of all proteins.

## B.  Comparative Genomics

Typically, only about 50% of the predicted genes in a new genome can be meaningfully annotated by performing similarity searches of public protein databases. This value may be lower for filamentous fungi. When the *N. crassa* genome was searched against public protein databases, less than 40% of the genome could be annotated (18). For further annotation, a comparative genomics approach can be taken (Fig. 1). Comparative genomics is useful for identifying conserved novel genes and for predicting similarities and differences in global biological processes. The databases used for genome-level comparisons consist of complete or extensive genome sequence datasets or nonredundant unigene assemblies of EST sequence datasets. Unigene assemblies are normally used instead of EST collections to maximize the contig length and reduce the overall size of the dataset. Comparison of a genome with itself is useful for identifying genes related by duplication within the genome that may have divergent functions (paralogs). Comparison with another species' genome reveals genes that are ancestrally related and may have a common function across species lines (orthologs). Genome comparisons can also be exploited to identify conserved regulatory regions.

BLASTN is often used for DNA comparisons of a genome with a unigene assembly or the genome of a related species. The utility of BLASTN is limited, however, because it is unable to handle large insertions that will arise at intronic regions when genome and EST sequences are compared. BLASTN will present individual exon matches as separate pieces of the overall hit and manual sorting or some form of computational postprocessing of the aligned pieces will be required for assembly of the complete alignment. Another limitation of BLASTN is its requirement for a contiguous word match. The default state of the National Center for Biotechnology Information (NCBI) BLASTN program uses a minimum word size of 11. This can result in significant matches being missed if there is substantial divergence resulting from wobble at the third base position of many codons. The program can be modified to run with a word size as low as seven, but the performance degradation is substantial. Wobble aware bulk aligner (WABA) overcomes these drawbacks of BLASTN because it handles introns and wobble mismatches without compromising performance (19). Wobble aware bulk alignment was used to compare 8 million base pairs (bps) of the nematode *Caenorhabditis briggsae* genome with the entire 97 million bp genome of the related species *C. elegans*. A large number of uncharacterized open reading frames (ORFs) and regulatory regions were identified. These two species share about 80% sequence identity within coding regions and 30% identity in noncoding regions. This method should be valuable for comparison of fungal species that are related at a similar evolutionary distance to *C. elegans* and *C. briggsae*.

To compare the genomes of divergent organisms, TBLASTX can be used. This algorithm compares DNA sequences at the protein level by translating both the query and subject sequences in all six reading frames. Each DNA-DNA comparison requires 36 independent protein-protein comparisons in order for all pairwise protein comparisons to be performed. Comparison of DNA sequences with TBLASTX is often informative but it is computationally demanding because of the number of alignments required. Aligning genome sequences and proteins from divergent species can reveal highly conserved gene functions and metabolic pathways. When the yeast genome was compared to the genome of *C. elegans*, the number of proteins responsible for fundamental biological processes was similar in both organisms (20). Analogous comparative genomic studies of divergent fungal species may identify conserved fungal-specific processes that can be used for the development of broad-spectrum antifungal compounds with little or no cross-reactivity or toxicity to humans. Toxicity to humans is a shortcoming of the current generation of pharmaceutical antifungal agents (21).

## C.  Gene Prediction

Sequence similarity searching is a useful discovery tool, but it may not identify novel genes because they often lack significant similarity to known genes. Coding regions of the genome have features that can facilitate gene identification independent of similarity alignments. Coding regions have a different base composition than noncoding regions because of their protein translation information content. Often associated with coding regions are highly conserved nucleotide elements that are critical for gene function. This includes transcription and translation start sites, intron donor, acceptor and lariat branch sites, and polyadenylation sites. Gene prediction programs have been developed to identify gene-coding regions based on these features. For example, the Pombe program has been used to identify genes in *S. pombe* based on multiple coding and coding-associated features, including exon base composition, GC ratios, intron donor and acceptor sites, and lariat branch sites (22). Gene prediction programs that are based on coding region features can be used to identify novel genes because they do not depend on similarity to known genes (Fig. 1).

Pombe and most other gene prediction programs are ab initio tools that require a species-specific data training set consisting of reliably annotated genes for which the entire gene structure is known (i.e., the exact transcriptional and translational start sites, all intron-exon boundaries, and the polyadenylation sites). Experimentally determined gene structures can be incorporated into the training set. Similarity searching can also reveal entire gene structures that can be used for constructing a training set. This is particularly true when genomic and EST datasets from the same organism are compared, and ab initio gene prediction is facilitated by parallel genome and EST sequencing efforts, such as those in *M. grisea, N. crassa, C. neoformans, P. carinii,* and *P. sojae* (Table 1). GeneWise is a similarity search tool that is useful for determining gene structure when comparing DNA sequence with a protein database (23). GeneWise uses an HMM that takes into account frame shifts in the DNA data and models introns, returning a gene structure prediction whose reliability is proportional to the strength of the protein match. In general, ab initio gene prediction programs perform well at finding genes, but they do best at identifying internal exons, whereas initial and terminal exons and short exons are more likely to be missed. The accuracy of gene prediction improves when ab initio tools are used in conjunction with similarity-based gene structure determination.

## III. GENERAL APPROACHES FOR FILAMENTOUS FUNGAL GENE ALTERATION

Early filamentous fungal genetic studies involved random mutagenesis with either radiation or chemicals, and assignment of gene function was based on phenotypic analysis of the mutants. These studies resulted in detailed chromosome maps that facilitated later molecular genetic and genomic studies of *N. crassa* and *A. nidulans* (24,25). Ultimately gene function was only correlated with a map location, not a specific gene sequence. With the development of filamentous fungal transformation systems, shown in Table 2, it became possible to clone genes by complementing mutant phenotypes and thereby assign functions to specific genes. Subsequent to the initial transformation of *N. crassa* (26), multiple procedures were developed for transformation of numerous filamentous fungal species. Transformation procedures differ in the type of fungal material that is used, and each has certain advantages that may influence their utility in different species (Table 2). With the advent of fungal transformation, the classic approach of random mutagenesis and cloning by complementation has largely been superseded by transformation-dependent approaches, such as insertional mutagenesis, antisense RNA-mediated gene inhibition and repeat-induced mutagenesis. These approaches are essential to the development of functional genomics in filamentous fungi.

### A. Insertional Mutagenesis

Insertional mutagenesis has two advantages over classic mutagenesis. Insertion mutants can be selected directly on the appropriate medium because the inserted DNA typically carries an auxotrophic or drug-resistance marker. This type of

**Table 2** Transformation Methods for Filamentous Fungi

| Method | Fungal material transformed[b] | Advantage | Key reference(s) |
|---|---|---|---|
| PEG[a]/CaCl2 | Spheroplasts and protoplasts[b] | Easy and inexpensive to implement | (51) |
| Electroporation | Protoplasts, conidia, and germinating conidia | Protoplasting can be omitted; good trans-formation efficiency | (140,141) |
| Microprojectile bombardment | Conidia, mycelia and fungal colonies | Protoplasting can be omitted; applicable to obligate fungi | (142,143) |
| Agrobacterium tumefaciens TDNA transfer | Protoplasts, conidia, and germinating conidia | Easy and inexpensive to implement; proto-plasting can be omitted; good transfor-mation efficiency | (144,145) |

[a] PEG, Polyethylene glycol.
[b] Spheroplasts are formed by partial removal of the cell wall from fungal hyphae or spores with cell wall-digesting enzymes. Protoplasts are formed by complete or nearly complete removal of the cell wall.

selection is not possible with chemical or radiation mutagenesis. Insertional mutagenesis facilitates gene recovery. The inserted DNA and adjacent genomic region can be recovered from mutants in the form of plasmids for sequencing. Insertional mutagenesis is currently the basis of numerous functional genomic strategies in fungi and other organisms (27). Insertional mutagenesis strategies include restriction enzyme-mediated integration (REMI), transposon-mediated insertion and targeted integration via homologous recombination.

## 1. REMI

DNA introduced into some fungi can integrate in the genome and produce mutations. Simply transforming a fungus with a piece of foreign DNA carrying a selectable marker can produce insertions at unspecified sites, and REMI is often used to increase the frequency of insertion (28). REMI is based on the cotransformation of a linear piece of foreign DNA with the restriction enzyme used for linearization. The enzyme cleaves at genomic restriction sites, and the foreign DNA is integrated into these sites by the host DNA repair machinery. The result is improved transformation efficiency without large-scale genome rearrangement (29). REMI is a versatile tool for fungal gene modification. Many fungal species have been subjected to REMI mutagenesis with great success (28). Reporter genes have been introduced into the plant pathogens *U. maydis* and *Pyrenophora teres* by REMI to trap infection-specific promoters and enhancers (28). REMI has also been used for promoter tagging to activate gene transcription in *Aspergillus niger* (30). Despite its utility as a forward genetic tool in fungi, REMI is a highly nonstochastic process and is not sufficient to achieve genomic saturation (28,31).

## 2. Transposon-Mediated Insertion

Transposons are mobile DNA elements that were discovered by Barbara McClintock in her studies of maize phenotypes (32). Type I transposons, or retroelements, are mobilized to DNA by reverse transcription of an RNA intermediate. In contrast, type II transposons are mobilized from DNA to DNA through the cut-and-paste action of a transposon-encoded transposase. Transposons are widespread, and transposon-based approaches are used to create gene insertions in many bacterial and eukaryotic species (33). Four transposon-based approaches that have applications to fungi are shown in Figure 2.

Transposon insertions can be produced in vivo by either endogenous or heterologous transposable elements. Examples of endogenous elements that are commonly used for mutagenesis are the type I Ty1 element in *S. cerevisiae* and numerous type II Tn elements in bacteria. Heterologous transposons are used in organisms lacking a well-developed endogenous transposon system. For example, the maize transposon system *Enhancer/Suppressor-mutator* (En/ Spm) has

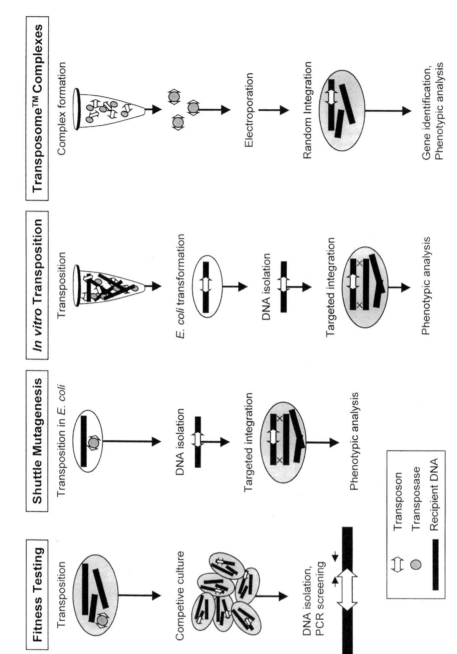

**Figure 2**

recently been used for large-scale transposon mutagenesis of the model plant *Arabidopsis thaliana* (34,35). Type I and type II transposons are present in the filamentous fungi (36), and active in vivo transposition has been observed in numerous species including *N. crassa, Tolypocladium inflatum, Fusarium oxysporum,* and *M. grisea* (37–40). Progress has been made in developing filamentous fungal transposons as gene alteration tools. The impala transposon of *F. oxysporum* was recently engineered to confer hygromycin resistance and could be mobilized both endogenously and heterologously in the closely related species *F. moniliforme* (41). Control of transposon mobility is critical when using in vivo transposition for mutagenesis. The presence of active transposase can result in remobilization of transposon insertions, and unregulated transposition can ultimately lead to a loss of viability arising from the accumulation of harmful mutations (42). Typically, control of transposon mobility is achieved with two-component systems in which the transposase is supplied in trans. The transposase can be removed via counterselection or outcrossing. Control of the yeast Ty1 element is typically achieved via transient expression from a galactose-inducible promoter (43).

A population of mutants can be constructed by in vivo transposon mutagenesis that can be used for assessing gene function en masse. In one such approach, poor competitors for nutrient resources are identified in a population of mutants by their reduced fitness in co-culture (fitness testing) (Fig. 2). In genetic footprinting, fitness testing of a population of yeast mutants is done by PCR screening with gene-specific and transposon-specific primers. The relative abundance of a particular gene insertion in a competitive culture is determined under a variety of growth conditions (44). Genetic footprinting has been used to functionally

---

**Figure 2** Transposon-based mutagenesis approaches with applications to fungi. Fitness testing: An organism is subjected to in vivo transposon mutagenesis and a population of independent mutants is cultured. Competition for nutrients occurs among the mutants. DNA isolation followed by PCR screening is used to determine the mutant profile before and after the competitive culture step to identify mutants with reduced fitness. Shuttle mutagenesis: *E. coli* serves as a surrogate host for in vivo transposition. The recipient DNA is typically a plasmid or another type of autonomous replicon. The mutated DNA is isolated and used for targeted integration in the organism being studied. Phenotypic analysis is performed to determine gene function. In vitro transposition: This method is similar to shuttle mutagenesis, except transposition takes place in a cell-free system instead of a surrogate host. Plasmids, cosmids, or another type of autonomous replicon may be used as the recipient DNA. The mutated DNA is transformed into *E. coli* and prepared for targeted integration followed by phenotypic analysis of the mutant. Transposome™ complexes: Transposon-transposase complexes are generated in a cell-free system and electroporated into the organism of choice where transposition occurs. Mutant recovery for gene identification and phenotypic analysis follows.

analyze 261 (97%) of the predicted ORFs on yeast chromosome V (45). Fitness testing was performed under seven selection conditions to assess, for example, auxotrophy, respiration, temperature sensitivity, and salt tolerance. Many previously known mutant phenotypes were confirmed, and new defects were associated with about 30% of the mutants. Significantly, specific growth defects were associated with 11 out of the approximately 100 mutants representing novel genes on chromosome V.

Advantages of this method include rapid genome-scale determination of gene function, as it is performed on a mixed population of cells, and applicability to any free-living fungus that is amenable to insertional mutagenesis. A disadvantage is that subsequent recovery of gene mutations is not possible because poor competitors are eliminated from the population.

Signature tagged mutagenesis (STM) is a transposon-based fitness testing approach that is used in pathogen systems to identify genes that are essential for virulence (46). Multiple unique signature tags are cloned into transposons that are used for in vivo transposon mutagenesis. Pools of mutants are used as inoculum, and the signature tag profiles of the surviving pathogen populations from the host and from the inoculum are compared. Loss of a signature tag in the host indicates disruption of an essential virulence function. In contrast to genetic footprinting, gene mutations are recovered from STM because they are present in the inoculum. STM has recently been applied to two human fungal pathogens but not using transposons. *Aspergillus fumigatus* was mutagenized by REMI and restriction enzyme-independent integration, and subjected to STM screening. Two mutants were identified that demonstrated significantly reduced competition in mice (47). One mutant had an insertion in a novel gene and the other an insertion in PabaA, a $p$-aminobenzoic acid synthase gene. When *Candida glabrata* was subjected to STM screening, 14 independent insertions were identified in the promoter of a cell-wall protein (EPA1) that is required for adhesion to epithelial cells (48).

Shuttle mutagenesis is an in vivo transposon mutagenesis approach that is well suited for functional genomic analysis. A genomic library from an organism of interest can be mutagenized in Escherichia coli with a bacterial transposon, and the mutated genes can be shuttled into the organism of interest for targeted integration (Fig. 2) (49). The altered gene alleles are maintained as plasmids in *E. coli*, facilitating subsequent recovery of desired alleles. Shuttle mutagenesis has recently been used for large-scale analysis of the yeast genome (50). In this study, a multifunctional transposon was used that enabled gene expression analysis, protein localization, and mutant phenotyping. This study identified more than 300 small ORFs that were not identified by the yeast genome sequencing project, including ORFs that were in antisense orientation within annotated ORFs. Although shuttle mutagenesis could be used for functional analysis in filamentous fungi, the low frequency of homologous recombination in these organisms (51) would necessitate more extensive screening of transformants to identify targeted integration events than is required in yeast.

It is possible to mobilize transposons in cell-free systems that contain the recipient DNA, the transposon, and the appropriate enzymes (e.g., transposase or RNA integrase) (Fig 2) (52–55). Such in vitro transposition (IVT) systems were first used as a means for directed sequencing of large fragments of DNA (52). A few IVT systems result in a nearly random pattern of transposon insertion (56). In vitro transposition has been used as a tool for gene discovery in a number of microbial systems. For example, IVT was used to mutagenize a plasmid library from the industrial microorganism *Streptomyces coelicolor* (57). The insertion alleles were introduced back into the genome by targeted integration, and the *S. coelicolor* mutants were screened for changes in phenotype. The transposon insertions were characterized by comparison of the flanking DNA sequence with the *S. coelicolor* genome sequence, and genes for pigmentation, antibiotic sensitivity, hyphal morphology, and sporulation were identified. In vitro transposition-based sequencing and gene disruption technologies have been integrated for functional genomic analysis of filamentous fungi in a method called TAGKO™ (transposon arrayed gene knock out) (56). Filamentous fungal genomic libraries are subjected to IVT and sequenced from the insertion site for gene discovery. After detailed bioinformatic analysis, genes are selected and the same IVT construct used for sequencing is the basis for targeted gene disruption in the fungus. Cosmid libraries are used for TAGKO because the larger regions of homology flanking the targeted gene improve the efficiency of homologous recombination. TAGKO is effective for skim sequencing of filamentous fungal genomes and fungal gene disruption.

In vitro construction of Transposome™ complexes consisting of a transposon and the corresponding transposase can be used to create transposon insertions in organisms lacking an in vivo transposition system (58) (Fig. 2). The complex of transposase and transposon is created in the early steps of IVT, but fails to mobilize in vitro because no catalytic cations are provided. Transposomes are introduced into the host by electroporation and, because of the intracellular cation concentration, mobilization into the host genome follows. Transposome technology has been efficiently applied to three bacterial species and to budding yeast, and may be useful in filamentous fungi without a known in vivo transposon system (58).

## 3.  Site-Selected Mutagenesis

Random insertional mutagenesis is not efficient for generating targeted mutations. However, insertion mutant collections can be used as a resource for identifying mutations in genes of interest when extensive genome sequence is available. Site-selected mutagenesis is a reverse genetic method in which an organism of known genomic sequence is subjected to insertional mutagenesis to generate a population of mutant strains. Gene-specific primers and primers specific to the inserted DNA are used to identify mutants with insertions in genes of interest (59). This method has been used for functional genomic analysis of *A. thaliana* (34,35,60). *Arabidopsis thaliana* is diploid and mutants are maintained as heterozygotes and crossed to homozygosity for phenotype analysis. As fungal genome sequences accumulate, site-selected mutagenesis could be used for large-scale gene function analysis. Ascomycete fungi are often haploid, and appropriate cultivation will be necessary for propagation of conditional lethal mutants. Lethal mutants will not be isolated in haploid fungi by this method.

### 4. Targeted Integration

In *S. cerevisiae*, genome-wide targeted gene deletions have been constructed (61,62). Budding yeast is amenable to this approach because efficient targeted disruption can be performed with PCR-derived constructs that contain a total of only about 100 base pairs of sequence identity (i.e., 50 base pairs on each flank of a selectable marker) (63,64). Gene deletions are introduced in a diploid strain so a wild-type copy of the gene remains to compensate for any deleterious effects of the mutation. To examine the phenotypic consequences of the deletion, the heterozygous diploid mutant is resolved by meiotic spore formation to produce the haploid mutant. As with genetic footprinting, genome-scale collections of deletion mutants can be used to evaluate gene function en masse. This is done through the use of bar codes—unique sequence tags that are introduced during construction of the deletion cassette that can be used for fitness testing of multiple deletion strains in a competitive culture. To identify strains with reduced fitness, the bar codes are PCR-amplified from competitive cultures and hybridized to arrays containing their complement (62). This type of PCR-based gene disruption strategy may be suited to other fungal species with efficient homologous recombination such as *Ashbya gosypii* and the human pathogen *C. albicans* (65,66)

### B.  RNA Inhibition

Gene expression can be inhibited at the level of the messenger RNA (mRNA) using in vivo antisense RNA expression. Antisense is a poorly understood process but it may be mediated by double stranded RNA-directed RNA decay or interference with mRNA processing, transport or translation (67). Genes that are critical for growth of *C. albicans* have been identified through the use of antisense RNA inhibition (68). Complementary DNA (cDNA) fragments were directionally cloned in the antisense orientation under control of an inducible promoter. Eighty-six of 2,000 *C. albicans* transformants exhibited impaired growth on inducing medium, each expressing a different antisense cDNA fragment. The 86 different cDNA fragments corresponded to genes involved in a variety of cellular processes and included genes known to be essential in other organisms and 33 orphan genes (i.e., novel genes of unknown function). Antisense inhibition has been applied to the Ascomycete filamentous fungi *N. crassa, A. nidulans, Aspergillus niger, A. oryzae,* and *Penicillium chrysogenum* (69–73). However, antisense inhibition was not successful for constructing uridine auxotrophs in the Basidiomycete *Ustilago maydis,* suggesting that it may not be useful in all fungal species or effective at all gene loci (74).

## C.  Repeat-Induced Mutagenesis

Repeat-induced point mutation (RIP) and MIP (methylation induced premeiotically) are two methods of repeat-induced mutagenesis that are used for creating targeted gene mutations. Repeat-induced point mutation is a process by which duplicated sequences are subject to modification by point mutation and methylation (75). In *N. crassa,* RIP occurs during the sexual cycle between cell fusion and nuclear fusion and requires pairing of homologous sequences present in the haploid nuclei of premeiotic hyphae (76,77). Although the exact mechanism of RIP is unknown, it appears that it results from cytosine methylation and G:C to A:T mutations (75). Repeated-induced point mutation was discovered in *N. crassa* and has been exploited for mutagenesis in this fungus. It is not known how widespread RIP is in other fungal species, but an RIP-like process is active in *Podospora anserina,* indicating that it is not exclusive to *N. crassa* (78). Methylation-induced premeiotically is a process, similar to RIP, that occurs in *Ascobolus immersus* (79). As with RIP, MIP is caused by repeats and leads to gene inactivation. However, whereas RIP is irreversible and involves both cytosine methylation and G:C to A:T mutations, MIP is reversible and only appears to involve cytosine methylation (79,80). Repeat-induced point mutation and MIP are both efficient processes for gene disruption and may become useful functional genomic tools. This is particularly true in the case of RIP, as the *N. crassa* genome was recently completed (Table 1).

## D.  Evaluating the Stringency of Gene Mutations

Fungal mutants are the basis for gene function determination. As such, it is critical to know the extent to which gene function is disrupted with a particular gene alteration method. Null mutations are the most stringent gene alterations and result in a complete loss of gene function. If targeted deletion is used to remove the entire gene-coding region, the resultant mutation is null. In all other gene disruption strategies discussed above, gene alteration can yield null mutations, as well as less stringent partial loss of function mutations and even silent mutations. In untargeted insertional mutagenesis approaches such as REMI and transposon mutagenesis, the translation reading frame downstream from the insertion site is frame-shifted and the location of the insertion in the ORF may be an indicator of the stringency of a gene mutation. Insertions in the 3' region of an ORF are more likely to result in a less stringent partial loss of function mutation because the 5' region of the ORF is retained in the proper reading frame. In contrast, insertions in the 5' region of an ORF are more likely to be null mutations because more of the gene is frame-shifted. In insertional approaches, examining multiple mutant alleles for each gene will increase the probability that a null phenotype has been assessed and will also provide an internal control because multiple mutant alleles of the same gene should have consistent phenotypes. In genetic footprinting and STM, it may be possible to infer how stringent mutant alleles are by examining the kinetics of their disappearance in competitive culture. Insertions that severely compromise fitness rapidly disappear from the competitive culture and may be null mutations. For all mutagenesis approaches, the only way to conclusively determine that a null mutation is formed is by measuring gene expression at the mRNA or protein level. Analyzing the gene expression of every mutant is impractical for high-throughput analysis but, to evaluate how robust a method is for making null mutations, gene expression analysis on a random sampling of mutants should be performed.

## IV.  METHODS FOR STUDYING ESSENTIAL FILAMENTOUS FUNGAL GENES

Identifying essential genes in Ascomycete fungi can be challenging because many of them are haploid during the majority of their lifecycle and, in many species, no sexual stage has been identified. In diploid organisms, mutants with alterations in one copy of an essential gene can be propagated as heterozygotes because a second wild-type copy of the gene is present. When two such heterozygotes are mated, a viable:nonviable segregation ratio of 3:1 in the diploid progeny is indicative of a recessive lethal gene mutation. Alternatively, essential genes can be identified in diploids by resolving the heterozygous mutant into the wild-type and mutant haploids by meiotic sporulation, as described for the targeted gene deletion projects in yeast (61,62). A viable:nonviable segregation ration of 2:2 in the haploid progeny is indicative of a recessive lethal mutation. Of the 6,237 genes in the yeast genome, approximately 15% are essential for viability (81). A similar percentage of essential genes is expected from filamentous fungi. Traditionally, essential genes in filamentous fungi are identified in temperature-sensitive mutants after a shift to the restrictive temperature. A limitation of this strategy is that cloning by complementation must follow to recover the essential gene. In addition, many essential genes cannot be rendered thermolabile, and there may be species-specific constraints on mutagenesis (82).

## A. Identification of Essential Genes

Identification of essential fungal genes has been facilitated by strategies specifically designed for this purpose. In heterokaryotic gene disruption, an intrinsic feature of filamentous fungal biology is exploited to identify essential genes (83). Ascomycete filamentous fungi have multinucleate vegetative hyphae, and it is possible to disrupt essential genes in one nucleus and maintain the mutant nuclei in the presence of wild-type nuclei. Such fungi are heterokaryotic because they have more than one type of nucleus. Heterokaryotic gene disruption has been applied successfully in *A. nidulans* where a number of essential genes were characterized, including bimA, a conserved component of the anaphase-promoting complex, and cnaA, a calmodulin-dependent protein phosphatase catalytic subunit (84,85). Upon conidiation of *A. nidulans* heterokaryons, uninucleate spores are produced and spores with an essential gene mutation fail to propagate. This allows for limited cytological analysis of the mutants by examining germinated conidia; however, the cytology is complicated because both wild-type and mutant spores are present during sample preparation.

To identify essential genes in the opportunistic fungal pathogen *C. albicans*, a novel targeted integration method has been developed (86). *Candida albicans* is diploid but does not undergo meiotic division, so essential genes cannot be identified by meiotic sporulation. A DNA construct called UAU1 is used to identify essential *C. albicans* genes. UAU1 is composed of an ARG4 gene that confers arginine prototrophy (Arg$^+$), flanked by overlapping URA3 deletion derivatives that confer uracil prototrophy (Ura+) if they recombine in the region of overlap to form an intact URA3 gene. Gene deletion constructs can be prepared in which regions of identity to the gene of interest flank UAU1. When this construct is transformed into *C. albicans* and homologous recombination occurs at one of the diploid copies of the gene, the strain becomes Arg$^+$. When the Arg$^+$ transformant is cultured, two recombination events can occur. First, the gene deletion can become homozygous through either gene conversion or mitotic recombination. Second, the overlapping URA3 deletion derivatives in UAU1 can recombine and the strain becomes Arg$^+$ Ura$^+$. Finally, if the homozygous deletion occurs in an essential gene, a triplication of the chromosome carrying the targeted gene can occur. When Arg+ Ura+ transformants are screened by PCR, transformants that are recovered as triplication-bearing segregants but not as homozygous deletants are inferred to have a mutation in an essential gene. Gene targeting with UAU1 was used in *C. albicans* to cross-validate three essential *S. cerevisiae* genes. Two of these were also essential in *C. albicans*, but the third gene, CDC25

(Cdc25p is an activator of Ras proteins), was not (86). These results emphasize the need for experimental cross-validation of predicted orthologs and the limitations of in silico functional assignment.

Genomic analysis and mapping by in vitro transposition (GAMBIT) is used to identify essential genes in organisms of known sequence (87). Open reading frames from a genomic sequence are PCR-amplified and cloned and IVT is used to create multiple insertion alleles of each ORF. The mutant alleles are introduced back into the gene locus by homologous recombination and pools of mutants are subjected to fitness testing. The competitive culture used for fitness testing is screened by PCR with gene-specific and transposon-specific primers, and if an insertion confers a lethal phenotype it is not possible to recover the insertion allele from the mutant pool. Genonomic analysis and mapping by in vitro transposition depends on efficient homologous recombination and a complete or nearly complete genome sequence for its implementation. The low frequency of homologous recombination and the lack of available sequence data currently limit the application of GAMBIT to most filamentous fungi.

## B.  Functional Analysis of Essential Genes

Construction of mutations in essential genes is an important discovery tool, but little or no subsequent functional information can be determined, as these mutations result in lethality. Functional information about essential genes can be acquired by producing a partial loss of essential gene function. Sheltered disruption is a variation of heterokaryotic gene disruption that has been developed in *N. crassa* to produce a partial loss of function of essential genes (88). A heterokaryotic strain is constructed from two starting strains, one carrying a gene that confers resistance to benomyl (ben$^R$) and the other a gene that confers resistance to p-fluorophenylalanine (fpa$^R$). Either the ben$^R$ or the fpa$^R$ nucleus carries essential gene mutations that are introduced into this heterokaryon. The mutant allele can be forced to predominate in the heterokaryon by growth on medium containing either benomyl or p-fluorophenylalanine, depending on which nucleus the mutation occurred in. This was demonstrated in studies of the *N. crassa* Mom-22 gene that encodes an essential component of the mitochondrial protein import complex. When Mom-22 was disrupted in the fpa$^R$ nucleus of a ben$^R$/fpa$^R$ heterokaryon, the disrupted allele predominated when the heterokaryon was grown on p-fluorophenylalanine, whereas the wild-type allele was depleted. Detailed analysis of the Mom-22-depleted heterokaryon revealed roles for Mom-22 in the import of many mitochondrial preproteins, assembly of a normal mitochondrial protein import complex, and establishing or maintaining normal mitochondrial morphology.

Conditional mutations that yield a uniform arrest phenotype under restrictive conditions can be used to determine the function of essential genes. TnAra-Out is a method that has been used to construct conditional mutations in the prokaryotic pathogen *Vibrio cholerae* (89). A transposon carrying an arabinose-inducible promoter is fused to genes by in vivo transposition. If the insertion occurs upstream from an essential gene, the mutant will be dependent on the presence of arabinose in the culture medium for survival. This technique is directly applicable to the analysis of essential genes in yeast and filamentous fungi because inducible promoters have been well characterized in these organisms. For example, the GAL4 promoter of *S. cerevisiae* and the alcA promoter of *A. nidulans* have been extensively used for the inducible expression of endogenous and heterologous proteins (90,91).

Conditional gene targeting is used in mammalian cell culture models to study essential genes (92). This involves the Cre/loxP site-specific recombination system in which the Cre recombinase catalyzes recombination between two loxP sites (93). loxP Sites are introduced into two or more noncoding regions of an essential gene, and the altered gene is used to replace the wild-type allele by targeted gene replacement. When the Cre recombinase is introduced into the mutant mammalian cell line by a second transformation, recombination between the loxP sites occurs, creating a gene disruption. A variation of conditional gene targeting will most likely work in fungi, as the Cre/loxP system has been successfully implemented in yeast, and the existence of inducible fungal promoters would facilitate conditional expression of the Cre recombinase (94). Cre/loxP has been used in yeast to reduce transposon insertions to smaller epitope tag insertions that can be used for cytological studies of the protein encoded by the disrupted gene (95).

Comparative genomic analysis followed by the construction of conditional mutations has been used to study novel essential genes in bacteria. These genes were identified in *E. coli* by genome sequence comparison with *Mycoplasma genitalium* (96). It is postulated that *M. genitalium* contains the minimal set of genes required for survival because it only encodes 517 genes, the smallest gene complement of any free-living organism (97). More than 50% of the *M. genitalium* genes are essential under laboratory conditions, so genes that are conserved in *M. genitalium* may be required for survival in *E. coli*. Twenty-six *E. coli* genes of unknown function were found to have orthologs in *M. genitalium* (96). When the corresponding disruption mutants were constructed, six of these failed to produce null alleles and were characterized as putative essential genes. The genes were confirmed to be essential after construction of conditional gene mutants. Importantly, these essential *E. coli* genes were cross-validated in *Bacillus subtilis* and *S. cerevisiae*. *Bacillus subtilis* contains orthologs of all six of the *E. coli* genes but only five are essential. *Saccharomyces cerevisiae* contains orthologs of five of the six *E. coli* genes, but only one is essential. As fungal genome sequences accumulate, comparative genomics can be used to identify putative essential genes, and these can be validated and functionally analyzed by constructing conditional mutations.

## V. HIGH-THROUGHPUT APPROACHES FOR FILAMENTOUS FUNGAL GENE FUNCTION DISCOVERY

Analysis of mutant phenotypes can lead to the discovery of novel gene functions and confirmation of in silico gene function predictions (Fig. 1). Phenotypic analysis of filamentous fungi has a long history. Beadle and Tatum developed their "one gene-one enzyme hypothesis" from their studies of *N. crassa* amino acid and vitamin metabolism (98). In the intervening years, it has become relatively straightforward to determine how a limited number of fungal gene mutations affect a specific phenotype. The challenges for functional genomics are to examine large numbers of fungal mutants under many different growth conditions and compile the functional data in a readily accessible format. An example of large-scale gene function analysis is illustrated in Figure 3. Fungal mutants are subjected to a series of simultaneous comprehensive analyses to acquire a comprehensive understanding of gene function. The data are stored in a readily accessible manner in a gene function database.

### A. Growth Profiling

The function of fungal genes can be determined by analysis of mutants with the gene disrupted. A large number of conditions must be examined for each mutant because fungi adapt to the environment with respect to temperature, pH, osmotic balance, and nutrient availability, and adaptation plays an important role in fungal growth and development (99). Analysis of the effects of environmental conditions in fungal mutants can be used to categorize genes required for adaptive changes in growth rate or morphological structures under those conditions. This is illustrated by many plant pathogenic fungi that respond to environmental cues present on the host plant surface for infection structure differentiation (100).

High-throughput growth profiling strategies have been developed for yeast. European network for functional analysis of yeast genes (EUROFAN) has developed a phenotype analysis platform to characterize orphan genes that includes tests for lethality, temperature sensitivity, growth on fermentable and nonfermentable carbon sources, and auxotrophy (8). A petri plate system has been designed to test the responses to 21 different conditions, including thermotolerance, osmotolerance, ethanol sensitivity, and sensitivity to chemicals affecting protein phosphorylation, and protein and nucleic acid synthesis (101). Also, a microtiter plate-based chemotyping system has been designed to simultaneously analyze the growth of yeast mutants when exposed to approximately 300 inhibitory compounds, including antifungal agents, metal ions, and a diverse array of chemical inhibitors (102). Finally, a phenotype macroarray has been used to simultaneously analyze the growth of 576 mutant yeast strains under 20 different test conditions, including various growth inhibiting compounds and temperatures (50).

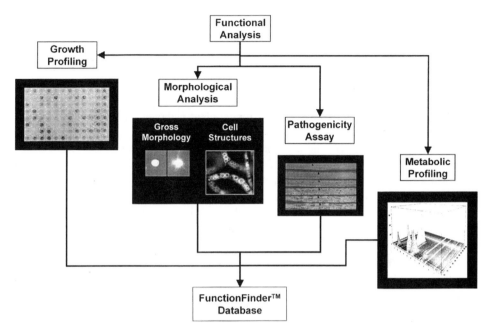

**Figure 3** Large-scale compilation of gene function data. The methodologies depicted are used in the GeneFunction Factory™ at Paradigm Genetics, Inc. for analysis of filamentous fungal mutants. Growth profiling reveals alterations in the use of specific nutrients such as amino acids and carbon sources. Nutrient utilization assays are performed using Phenotype Microarrays™ containing a tetrazolium dye that is used as an indirect measure of growth. Darker wells correlate with more robust growth. Morphological analysis reveals alterations in the growth habit and cellular organization of a mutant. Pathogenicity assays are an example of analytical tools that can be developed to examine species-specific phenomena. Other examples might include assays of circadian rhythmicity or mating. Metabolic profiling is used to identify alterations in the metabolite composition of a mutant. Fungal mutants are examined through the use of multiple methodologies to gain a complete description of gene function. All the data derived from these analyses are stored in the FunctionFinder™ bioinformatics system that has datamining and automated query capabilities to facilitate extraction of gene function information.

The methods used for growth profiling in yeast may not be suitable for the filamentous fungi. Under most conditions, filamentous growth precludes the production of dense colony arrays on a single plate, which complicates the transfer of material by replica plating. Determination of filamentous fungal growth in the presence of a large number of nutrient sources is possible using methods designed for clinical identification of bacterial and yeast species. Biolog, Inc.

offers a filamentous fungus species identification plate based on 95 carbon source utilization tests (FF) and a series of plates (Phenotype Microarrays™ [PM]) for the analysis of growth on various carbon, nitrogen, phosphate, and sulfate sources, and various metabolites designed for bacteria and fungi (Fig. 3). The Biolog PM plates include a tetrazolium dye color indicator that is a measure of respiratory activity (103). The Biolog system has recently been used to analyze nitrogen utilization in two filamentous fungal plant pathogens, *M. grisea* and *M. graminicola*, with mutations in hydroxyphenylpyruvate dehydrogenase (HPD4), a gene required for tyrosine and phenylalanine catabolism. Large-scale growth analysis of these mutants was performed on the Biolog nitrogen plate, which includes tests of each of the 20 amino acids as the sole nitrogen source. Growth is impaired when phenylalanine or tyrosine is used as the sole nitrogen source, possibly because of the accumulation of a toxic intermediate (56). The data are consistent with the role of HPD4 in phenylalanine and tyrosine catabolism. Nutrient analysis in this format could greatly increase the speed of acquisition and depth of knowledge about metabolic gene function. However, multicellular filamentous fungi offer a challenge to produce uniform and reproducible growth.

## B.  Morphological Analysis

Fungi undergo a variety of morphological changes during their life cycle, and microscopy has become an increasingly powerful tool for studying fungal mutants (Fig. 3). The cell morphology of a mutant can be an indicator of gene function. For instance, *A. nidulans* bimC mutants fail to separate their spindle poles and divide their nucleus (104). BimC encodes a kinesin-related protein that acts on overlapping spindle microtubules to establish and elongate a bipolar mitotic spindle (105). Data banks containing mutant phenotypes can be used to identify putative modes of action of inhibitory compounds. As an example, in a recent phenotype-based screen for inhibitors of mitosis in mammalian cells, the compound monastrol was found to inhibit mitosis by preventing the separation of spindle poles (106). Monastrol appears to specifically disrupt the action of the kinesin-related protein Eg5, the mammalian ortholog of *A. nidulans* BimC. Digital image analysis is becoming a valuable tool for quantification of morphological features. Detailed analysis of *Puccinia mentha* (mint rusts) urediniospores revealed several measurable parameters that distinguished peppermint-and spearmint-specific races (107). Digital image analysis has also been applied to mycelial systems for the study of foraging characteristics and responses to nutrient availability, both in industrial and natural systems (108, 109).

Advances in high-throughput imaging technology facilitate genome-scale analysis of fungal mutants. High content screening (HCS) is the automated extraction of multicolor information derived from specific fluorescent reagents incorporated into cells (110). Cell structures can be labeled with fluorescent reagents in fixed and living cells. Fluorescent labels can be used in living cells as environmental indicators for such parameters as pH and ion concentrations, and physiological indicators of biochemical changes, such as enzyme activity and cytoskeletal flux. The ArrayScan™ system from Cellomics, Inc. can be used for high-throughput microtitre plate-based HCS. The ArrayScan system can assay up to four different fluorescent label systems and derive morphometric information regarding cell and organelle shape, size, and distribution. It compiles the HCS data in graphical format that enables visualization at a range of magnifications and includes images from each fluorescent channel as well as the overlay of all channels. High content screening has currently only been applied to mammalian cells using ArrayScan but the technology should be adaptable to fungi.

## C. Gene Expression Profiling

Gene expression profiling (GEP) of the genes in a genome can lead to an understanding of the timing and coordination of gene activation, conditions that promote or repress gene function, and the interplay among genes and pathways. The uniform nature of mRNA and its ability to hybridize to its cognate gene make it very amenable to high-throughput analysis. Several techniques can be used to analyze mRNA levels, each requiring multiple molecular biological steps and sequencing to identify the expressed genes (111–115). DNA/RNA hybridization techniques use hybridization arrays of clones or oligonucleotide sequences representing all or most of the expressed genes in a genome. Yeast genome arrays are available, as is a nearly complete primer set for PCR amplification of the ORFs. These yeast expression arrays have been used to examine the diauxic shift, modes of action of drugs, and mitogen-activated protein kinase (MAPK) signaling (116–119). Additional fungal genome sequences will enable GEP studies of the transcription and synchronized expression of, for example, species-specific genes, fungal-specific genes, or broadly conserved genes. These studies may reveal regulatory differences and similarities between fungal species and between fungi and other eukaryotes.

## D. Proteomics

Proteomics is the study of protein function on a genomic scale (120). Proteomics picks up where GEP stops and has the advantage of being closer to the actual cellular function of a gene. Proteomic studies that examine differential protein expression, localization, protein-protein interactions and complex formation, and posttranslational modifications have been described (120–124). Proteomics methods applied to yeast may also find application to the filamentous fungi. Examples include shuttle mutagenesis for the introduction of reporter genes for protein localization studies (50), genome-scale GST-tagging for affinity purification and identification of biochemical functions (125), and the yeast two-hybrid system for large-scale detection of protein-protein interactions and construction of detailed protein interaction maps (126). Genome-scale two-hybrid analysis has been a particularly useful gene function discovery tool. It has led to the identification of two-hybrid interactions between proteins of known and unknown function, unrecognized interactions between proteins involved in the same biological process, and the integration of individual biological events into cellular processes (127).

Mass spectrometry can be used to analyze large numbers of intact proteins and peptides (120). The samples are typically derived from crude protein mixtures, and by isotopically labeling proteins, accurate quantitative data can be obtained. Peptides derived from protein proteolysis can be identified with very high sensitivity. In particular, matrix assisted laser desorption ionization (MALDI) mass spectrometry can produce tryptic peptide mass maps that are sufficiently accurate to search databases of calculated tryptic peptide masses and identify the proteins of known sequence that are represented in these databases. A recent study of yeast spindle pole bodies (SPB) exemplifies the power of MALDI for functional genomics (128). A highly enriched SPB preparation was analyzed by MALDI, followed by database searching. A total of 43 proteins were identified, including five of eight spindle components known to localize near the SPB, seven of nine previously identified SPB components, and numerous proteins not known to localize to the SPB, or of unknown function. Protein localization studies of this latter class revealed seven new SPB components. Similar MALDI and database studies of organelles and cellular structures can be applied on a large-scale in filamentous fungi.

## E. Metabolic Profiling

Metabolic profiling is the large-scale display of gene function at the metabolite level (Fig. 3). This provides a direct link between an organism's genotype and its metabolic network and can identify gene functions that are related through regulation at the metabolic level. At present, metabolic profiling has only been performed with gas chromatography coupled to mass spectrometry (GC/MS). The sensitivity of mass spectrometry enables deconvolution of complex chromatographic peaks, allowing for shorter chromatographic run times and higher throughput (129). With this approach, hundreds of compounds can be analyzed from a single extract preparation (130). High-throughput GC/MS represents a paradigm shift from the traditional analytical chemistry approach to studying metabolites, which was highly focused and quantitative, and low-throughput (129). Compromising on such analytical parameters as the metabolite extraction efficiency, which is often below analytical standards but good enough for comparative profile analysis, facilitates high-throughput metabolic profiling. Metabolic

**Table 3** Selected Fungal Functional Genomic Databases[a]

| | Database | Fungi examined | Description | URL | Reference |
|---|---|---|---|---|---|
| Gene function databases | Bioknowledge™ Library | S. cerevisiae; S. pombe; C. albicans | Heavily annotated gene function information with accumulated knowledge from the scientific literature | http://www.proteome.com/index.html | (81) |
| | EUROFAN[b] | S. cerevisiae | Experimental gene function analysis of all novel ORFs | http://www.mips.biochem.mpg.de/proj/eurofan/ | — |
| | Triples | S. cerevisiae | Phenotypic, protein localization, and gene expression data from a transposon insertion mutant collection | http://ygac.med.yale.edu/triples/triples.htm | (146) |
| | MIPS[c] | S. cerevisiae; N. crassa | Annotated resource of genomic information that includes a functional categories catalogue | http://www.mips.biochem.mpg.de/proj/yeast/; http://www.mips.biochem.mpg.de/proj/neurospora/ | (147) |
| | PathCalling® Yeast Interaction Database | S. cerevisiae | A protein interaction database from two massive two-hybrid screens in yeast | http://portal.curagen.com/extpc/com.curagen.portal.servlet.PortalYeastList | (127) |

| | | | | |
|---|---|---|---|---|
| Saccharomyces Genome Deletion Project | S. cerevisiae | Information resource for the Stanford yeast genome deletion project with protocols and strain information | http://www-sequence.stanford.edu/group/yeast_deletion_project/deletions3.html | (67) |
| Biochemical pathway databases | | | | |
| KEGG[c] | S. cerevisiae; S. pombe | Pathways are depicted with their associated enzymes, substrates, and end products with links to sequence databases | http://www.genome.ad.jp/kegg/kegg2.html | (132) |
| WIT[c] | S. cerevisiae | Gene function and metabolic pathway database | http://wit.mcs.anl.gov/WIT2/ | — |
| Comparative genomic database | | | | |
| Genolevures | S. cerevisiae and 13 related yeast species | Comparison of hemiascomycetes using genomic random sequence tags for molecular evolution studies | http://natchaug.labri.u-bordeaux.fr/Genolevures/Genolevures.php3 | (133) |

[a] For a complete listing of functional genomic and molecular biology databases see Nucleic Acids Res 29(1), 2001.
[b] EUROFAN exists as a subset of the MIPS database
[c] MIPS, Munich Information Center for Protein Sequences; KEGG, Kyoto Encyclopedia of Genes and Genomes; WIT, What Is There?

profiling is in its infancy and has only been described in two studies of *A. thaliana* and of potato (*Solanum tuberosum*) (130, 131). In both studies, data-mining tools revealed metabolic profile relationships between plant lines that matched the outcome predicted by the genotypes. Metabolic profiling shows great promise for fungal functional genomics because it can be used to assess the primary and secondary metabolic capacity of an organism.

## F.  Fungal Gene Function Databases

Current fungal functional genomic databases are primarily focused on yeast, as shown in Table 3, but their structure and design are of general utility. Gene function databases, biochemical pathway databases, and comparative genomic databases have been developed that support functional genomics research efforts. Gene function databases typically include information from large-scale phenotypic analyses. The Bioknowledge™ library from Proteome, Inc. integrates both large-scale analyses and previously published functional information into a relational database (81). Biochemical pathway databases are used to determine how genes and molecules are interconnected. KEGG is a particularly useful biochemical pathway database that maintains catalogs of genes for organisms from genome sequencing projects, and a catalog of enzymes, substrates, products, and other compounds (132). Metabolic and regulatory pathways can be depicted from reference organisms in KEGG to reconstruct, for example, organism-specific pathways. The development of comparative genomic databases for fungal gene function analysis is in its infancy. One of the new databases, the Genolevures database, is based on a large-scale comparative analysis between the complete genome of *S. cerevisiae* and random sequence tags representing 0.2–0.4 genome equivalents from 13 other Hemiascomycetous yeast species (133). Functions of numerous Ascomycete-specific gene classes, including cell-wall biosynthesis, pheromone response, and regulation of sterol biosynthesis and amino acid metabolism, were among the findings of this study (134).

## VI.  GREAT EXPECTATIONS FOR FILAMENTOUS FUNGAL FUNCTIONAL GENOMICS

Many tools are in place for implementing fungal functional genomics. The completion of the genome sequence of multiple filamentous fungi is expected in the near future (Table 1). Annotation of these genomes is likely to reveal many novel fungal genes and provide unprecedented insight into fungal biology. Comparative studies of fungal genomes and metabolism can be used to identify fungal-specific essential genes and metabolic pathways, and species-specific gene functions. Numerous gene alteration strategies exist that can be tailored to a given filamentous fungal species or to meet a desired analytical need. As examples, fungi that have in vivo transposon systems could be used to develop genetic footprinting methods, and an in vivo transposon-based site-selected mutagenesis strategy may have utility for reverse genetic screens. High-throughput phenotyping methods have been developed for experimentally determining gene function. Growth profiling of genome-scale collections of fungal mutants may facilitate the dissection of primary metabolic and drug resistance pathways. Gene expression profiling may lead to the identification of inducible filamentous fungal promoters that can be developed into genetic systems for the conditional expression of essential genes. Yeast proteomic tools for protein localization and two-hybrid analysis can be developed for use in the filamentous fungi. Finally, metabolic profiling can be used for screening bioactive filamentous fungal secondary metabolites to identify genes required for toxin or antibiotic production.

Despite these successes, a number of important challenges exist. Detailed gene structures are needed for each sequenced fungal species to support ab initio gene prediction. Transformation-based gene alteration methods will need to be scaled up for high-throughput production of filamentous fungal mutants to be used for large-scale gene function analysis. This is likely to be difficult in many filamentous fungal species because of poor transformation efficiency, low frequency of homologous recombination, and slow growth. Identification and functional analysis of essential filamentous fungal genes may require the development of special methods in species that do not readily form heterokaryons or have a known inducible promoter. For complete phenotypic assessment, high-throughput methods will need to be developed for morphology profiling to determine the function of cell type-specific or stage-specific genes. As we have seen in this chapter, many creative strategies have been developed for gene function analysis in filamentous fungi. This legacy of creativity will serve the filamentous fungal community well in the development of new strategies for functional genomic analysis.

## ACKNOWLEDGMENTS

We acknowledge Patrick Hurban, Craig Liddell, Jeffrey Shuster, and Jeff Woessner for helpful discussions and critical reading of the manuscript. Thanks to Amy Skalchunes, Blaise Darveaux, Ryan Heiniger, and Clive Lo for preparing images used to construct Figure 3.

## REFERENCES

1.  GW Hudler. Magical Mushrooms, Mischievous Molds. Princeton, NJ: Princeton University Press, 1998, p248.

2. WO Ellis, JP Smith, BK Simpson, JH Oldham. Aflatoxins in food: Occurrence, biosynthesis, effects on organisms, detection, and methods of control. Crit Rev Food Sci Nutr. 30(4):403–439, 1991.

3. WF Marasas. Fumonisins: History, world-wide occurrence and impact. Adv Exp Med Biol 392:1–17, p1996.

4. GN Agrios. Plant Pathology. 4th ed. San Diego, CA: Academic Press, Inc., 1997, 635.

5. R Fussle. Diagnosis of fungal infections. Mycoses. 40(Suppl 2):13–15, 1997.

6. A Goffeau, BG Barrell, H Bussey, RW Davis, B Dujon, H Feldmann, F Galibert, JD Hoheisel, C Jacq, M Johnston, EJ Louis, HW Mewes, Y Murakami, P Philippsen, H Tettelin, SG Oliver. Life with 6000 genes. Science 274(5287):546, 563–567, 1996.

7. JR Pringle, LH Hartwell. The Saccharomyces cervisiae life cycle. In: JN Strathern, EW Jones, JR Broach, eds. The Molecular Biology of the Yeast Saccharomyces— Life Cycle and Inheritance. Cold Spring Harbor, NY: Cold Spring Harbor Laboratory Press, 1981, pp97–142.

8. A Goffeau. Four years of post-genomic life with 6000 yeast genes. FEBS Lett 480:37–41, 2000.

9. F Sherman. An introduction to the genetics and molecular biology of the yeast Saccharomyces cerevisiae. http://dbb.urmc.rochester.edu/labs/Sherman_f/yeast/Index.html, 1998.

10. DM Kupfer, CA Reece, SW Clifton, BA Roe, RA Prade. Multicellular ascomycetous fungal genomes contain more than 8000 genes. Fungal Genet Biol. 21:364–372, 1997.

11. PA Lemke. The Thom Award address. Industrial mycology and the new genetics. J Ind Microbiol 14(5):355–364, 1995.

12. CJ Alexopolous, CW Mims. Introductory Mycology. 3rd ed. New York: John Wiley And Sons, Inc., 1979, p632.

13. S Altschul, W Gish, W Miller, E Myers, D Lipman. Basic local alignment search tool. J Mol Biol 215:403–410, 1990.

14. TF Smith, MS Waterman. Identification of common molecular subsequences. J Mol Biol 147(1):195–197, 1981.

15. J Villasenor, WH Mangione-Smith. Configurable computing. Sci Amer 0697: 1997.

16. A Bateman, E Birney, R Durbin, SR Eddy, KL Howe, ELL Sonnhammer. The Pfam protein families database. Nucleic Acids Res 28(1):263–266, 2000.

17. S Eddy. HMMER 2.1.1 Profile hidden Markov models for biological sequence analysis. http://hmmer.wustl.edu/, 2001.

18. EL Braun, AL Halpern, MA Nelson, DO Natvig. Large-scale comparison of fungal sequence information: Mechanisms of innovation in Neurospora crassa and gene loss in Saccharomyces cerevisiae. Genome Res 10:416–430, 2000.

19. WJ Kent, AM Zahler. Conservation, regulation, synteny, and introns in a large-scale C. briggsae-C. elegans genomic alignment. Genome Res 10(8):1115–1125, 2000.

20. S Chervitz, L Aravind, G Sherlock, C Ball, E Koonin, S Dwight, M Harris, K Dolinski, S Mohr, T Smith, S Weng, J Cherry, D Botstein. Comparison of the

complete protein sets of worm and yeast: Orthology and divergence. Science. 282(5396):2022–2028, 1998.

21.  S Arikan, JH Rex. New agents for treatment of systemic fungal infections. Emerging Drugs. 5(2):135–160, 2000.

22.  MQ Zhang, T Chan. Pombe: A gene-finding and exon-intron structure prediction system for fission yeast. Yeast. 14(8):701–710, 1998.

23.  E Birney, R Durbin. Using GeneWise in the Drosophila annotation experiment. Genome Res 10(4):547–548, 2000.

24.  J Clutterbuck. The Aspergillus nidulans linkage map. http://www.fgsc.net/mirror/www.gla.ac.uk/Acad/IBLS/molgen/aspergillus/index.html, 2000.

25.  DD Perkins, A Radford, MS Sachs. The Neurospora Compendium: Chromosomal Loci. San Diego, CA: Academic Press, Inc., 2000, p254.

26.  ME Case, M Schweizer, SR Kushner, NH Giles. Efficient transformation of *Neurospora crassa* by utilizing hybrid plasmid DNA. Proc Natl Acad Sci USA 76(10): 5259–5263, 1979.

27.  PSR Coelho, A Kumar, M Snyder. Genome-wide mutant collections: toolboxes for functional genomics. Curr Opin Microbiol 3:309–315, 2000.

28.  FJ Maier, W Schafer. Mutagenesis via insertional–or restriction enzyme-mediated integration (REMI) as a tool to tag pathogenicity related genes in plant pathogenic fungi. Biol Chem 380:855–864, 1999.

29.  RH Scheistl, TD Petes. Integration of DNA fragments by illegitimate recombination in *Saccharomyces cerevisiae*. Proc Natl Acad Sci USA 88:7585–7589, 1991.

30.  JR Shuster, MB Connelley. Promoter-tagged restriction enzyme-mediated insertion (PT-REMI) mutagenesis in Aspergillus niger. Mol Gen Genet 262:27–34, 1999.

31.  JA Sweigard, AM Carroll, B Valent. Restriction enzyme-mediated integration in the rice blast fungus. In: Long Ashton Int. Symp, 15th. IACR, Septoria on cereals: A study in pathosystems. Long Ashton Res. Sta, Bristol, England, 1999.

32.  B McClintock. Carnegie Institute of Washington Yearbook 47:155–169, 1948.

33.  L Hamer, T DeZwaan, MV Montenegro-Chamorro, SA Frank, JE Hamer. Recent advances in large-scale transposon mutagenesis. Curr Opin Chem Biol 5:67–73, 2001.

34.  E Speulman, PLJ Metz, G van Arkel, B te Lentel Hekkert, WJ Stiekema, A Pereira. A two-component enhancer-inhibitor transposon mutagenesis system for functional analysis of the Arabidopsis genome. Plant Cell 11:1853–1866, 1999.

35.  AF Tissier, S Marillonnet, V Klimyuk, K Patel, MA Torres, G Murphy, JDG Jones. Multiple independent defective suppressor-mutator transposon insertions in Arabidopsis: A tool for functional genomics. Plant Cell 11:1841–1852, 1999.

36.  F Kempken. Transposons in filamentous fungi. http://homepage.ruhr-uni-bochum.de/Frank.Kempken/fkpilztnp.htm#fig1, 2000.

37.  EB Cambareri, J Helber, JA Kinsey. Tad1-1, an active LINE-like element of Neurospora crassa. Mol Gen Genet 242(6):658–665, 1994.

38.  F Kempken, U Kuck. restless, an active Ac-like transposon from the fungus Tolypocladium inflatum: Structure, expression, and alternative RNA splicing. Mol Cell Biol 16(11):6563–6572, 1996.

39.  E Gomez-Gomez, N Anaya, MI Roncero, C Hera. Folyt1, a new member of the

hAT family, is active in the genome of the plant pathogen Fusarium oxysporum. Fungal Genet Biol. 27(1):67–76, 1999.

40. H Nakayashiki, K Kiyotomi, Y Tosa, S Mayama. Transposition of the retrotransposon MAGGY in heterologous species of filamentous fungi. Genetics 153(2):693–703, 1999.

41. A Hua-Van, JA Pamphile, T Langin, MJ Daboussi. Transposition of autonomous and engineered impala transposons in Fusarium oxysporum and a related species. Mol Gen Genet 264(5):724–731, 2001.

42. WR Engels. P elements in Drosophila. http://www.wisc.edu/genestest/CATG/engels/Pelements/Pt.html, 1996.

43. JD Boeke, DJ Eichinger, G Natsoulis. Doubling Ty1 element copy number in Saccharomyces cerevisiae: host genome stability and phenotypic effects. Genetics 129(4):1043–1052, 1991.

44. V Smith, D Botstein, PO Brown. Genetic footprinting: A genomic strategy for determining a gene's function given its sequence. Proc Natl Acad Sci USA 92: 6479–6483, 1995.

45. V Smith, KN Chou, D Lashkari, D Botstein, PO Brown. Functional analysis of the genes of yeast chromosome V by genetic footprinting. Science 274:2069–2074, 1996.

46. M Hensel, J Shea, C Gleeson, M Jones, E Dalton, DW Holden. Simultaneous identification of bacterial virulence genes by negative selection. Science. 269:400–403, 1995.

47. JS Brown, A Aufauvre-Brown, J Brown, JM Jennings, HJ Arst, DW Holden. Signature-tagged and directed mutagenesis identify PABA synthetase as essential for Aspergillus fumigatus pathogenicity. Mol Microbiol 36:1371–1380, 2000.

48. BP Cormack, N Ghori, S Falkow. An adhesin of the yeast pathogen Candida glabrata mediating adherence to human epithelial cells Science 285:578–582, 1999.

49. HS Seifert, EY Chen, M So, F Heffron. Shuttle mutagenesis: A method of transposon mutagenesis for Saccharomyces cerevisiae. Proc Natl Acad Sci USA. 83:735–739, 1986.

50. P Ross-Macdonald, PSR Coelho, T Roemer, S Agarwal, A Kumar, R Jansen, K-H Cheung, A Sheehan, D Symoniatis, L Umansky, M Heidtman, FK Nelson, H Iwasaki, K Hager, M Gerstein, P Miller, GS Roeder, M Snyder. Large-scale analysis of the yeast genome by transposon tagging and gene disruption. Nature 402: 413–418, 1999.

51. JR Fincham. Transformation in fungi. Microbiol Rev 53(1):148–170, 1989.

52. S Devine, J Boeke. Efficient integration of artificial transposons into plasmid targets in vitro: A useful tool for DNA mapping, sequencing and genetic analysis. Nucleic Acids Res 22:3765–3772, 1994.

53. J Vos, I Baere, RH Plaster K. Transposase is the only nematode protein required for in vitro transposition of Tc1. Genes Dev 10:755–761, 1996.

54. D Lampe, M Churchill, HM Robertson. A purified mariner transposase is sufficient to mediate transposition in vitro. EMBO J 15:5470–5479, 1996.

55. I Goryshin, W Reznikoff. Tn5 in vitro transposition. J Biol Chem. 273:7367–7374, 1998.

56. L Hamer, K Adachi, MV Montenegro-Chamorro, MM Tanzer, SK Mahanty, BA

Darveaux, DJ Lampe, TM Slater, L Ramamurthy, TM DeZwaan, GH Nelson, JR Shuster, J Woessner, JE Hamer. Gene discovery and gene function assignment in filamentous fungi. Proc Natl Acad Sci USA 2001.

57. AM Gehring, JR Nodwell, SM Beverley, R Losick. Genomewide insertional mutagenesis in Streptomyces coelicolor reveals additional genes involved in morphological differentiation. PNAS 97:9642–9647, 2000.

58. IY Goryshin, J Jendrisak, LM Hoffman, R Meis, WS Reznikoff. Insertional transposon mutagenesis by electroporation of released Tn5 transposition complexes. Nat Biotechnol 18:97–100, 2000.

59. K Kaiser, SF Goodwin. "Site-selected" transposon mutagenesis of Drosophila. Proc Natl Acad Sci USA 87:1686–1690, 1990.

60. PJ Krysan, JC Young, MR Sussman. T-DNA as an insertional mutagen in Arabidopsis. Plant Cell 11(12):2283–2290, 1999.

61. B Dujon. European Functional Analysis Network (EUROFAN) and the functional analysis of the Saccharomyces cerevisiae genome. Electrophoresis 19:617–624, 1998.

62. EA Winzeler, DD Shoemaker, A Astromoff, H Liang, K Anderson, B Andre, R Bangham, R Benito, JD Boeke, H Bussey, AM Chu, C Connelly, K Davis, F Dietrich, SW Dow, M El Bakkoury, F Foury, SH Friend, E Genatelen, G Giaever, JH Hegemann, T Jones, M Laub, H Liao, N Liebundguth, DJ Lockhart, A Lucau-Danila, M Lussier, N M'Rabet, P Menard, M Mittmann, C Pai, C Rebischung, JL Revuelta, L Riles, CJ Roberts, P Ross-MacDonald, B Scherens, M Snyder, S Sookhai-Mahadeo, RK Storms, S Veronneau, M Voet, G Volckaert, TR Ward, R Wysocki, GS Yen, K Yu, K Zimmermann, P Philippsen, M Johnston, RW Davis. Functional characterization of the S. cerevisiae genome by gene deletion and parallel analysis. Science 285:901–906, 1999.

63. A Baudin, O Ozier-Kalogeropoulos, A Denouel, F Lacroute, C Cullin. A simple and efficient method for direct gene deletion in Saccharomyces cerevisiae. Nucleic Acids Res 21(14):3329–3330, 1993.

64. MC Lorenz, RS Muir, E Lim, J McElver, SC Weber, J Heitman. Gene disruption with PCR products in Saccharomyces cerevisiae. Gene 158(1):113–117, 1995.

65. RB Wilson, D Davis, AP Mitchell. Rapid hypothesis testing with Candida albicans through gene disruption with short homology regions. J Bacteriol. 181(6):1868-1874, 1999.

66. J Wendland, Y Ayad-Durieux, P Knechtle, C Rebischung, P Philippsen. PCR-based gene targeting in the filamentous fungus Ashbya gossypii. Gene 242(1–2): 381–391, 2000.

67. JE Bourque. Antisense strategies for genetic manipulations in plants. Plant Sci 105:125–149, 1995.

68. MD De Backer, B Nelissen, M Logghe, J Viaene, I Loonen, S Vandoninck, R de Hoogt, S Dewaele, FA Simons, P Verhasselt, G Vahoof, R Contreras, WHML Luyten. An antisense-based functional genomics approach for identification of genes critical for growth of Candida albicans. Nat Biotechnol 19:235–241, 2001.

69. S Tentler, J Palas, C Enderlin, J Campbell, C Taft, TK Miller, RL Wood, CP Selitrennikoff. Inhibition of Neurospora crassa growth by a glucan synthase-1 antisense construct. Curr Microbiol 34(5):303–308, 1997.

70. B Hoffmann, SK LaPaglia, E Kubler, M Andermann, SE Eckert, GH Braus. Developmental and metabolic regulation of the phosphoglucomutase-encoding gene, pgmB, of Aspergillus nidulans. Mol Gen Genet 262(6):1001–1011, 2000.
71. C Ngiam, DJ Jeenes, PJ Punt, CAMJJ van den Hondel, DB Archer. Characterization of a foldase, protein disulfide isomerase A, in the protein secretory pathway of Aspergillus niger. Appl Environ Microbiol 66(2):775–782, 2000.
72. N Kitamoto, S Yoshino, K Ohmiya, N Tsukagoshi. Sequence analysis, overexpression, and antisense inhibition of a β-xylosidase gene, xylA, from Aspergillus oryzae KBN616. Appl Environ Microbiol 65(1):20–24, 1999.
73. I Zadra, B Abt, W Parson, H Haas. xylP promoter-based expression system and its use for antisense downregulation of the Penicillium chrysogenum nitrogen regulator NRE. Appl Environ Microbiol 66(11):4810–4816, 2000.
74. JP Keon, JW Owen, JA Hargreaves. Lack of evidence for antisense suppression in the fungal plant pathogen Ustilago maydis. Antisense Nucleic Acid Drug Dev 9:101–104, 1999.
75. EU Selker. Premeiotic instability of repeated sequences in Neurospora crassa. Annu Rev Genet 24:579–613, 1990.
76. EU Selker, EB Cambareri, BC Jensen, KR Haack. Rearrangement of duplicated DNA in specialized cells of Neurospora. Cell 51(5):741–752, 1987.
77. JR Fincham, IF Connerton, E Notarianni, K Harrington. Premeiotic disruption of duplicated and triplicated copies of the Neurospora crassa am (glutamate dehydrogenase) gene. Curr Genet 15(5):327–334, 1989.
78. A Hamann, F Feller, HD Osiewacz. The degenerate DNA transposon Pat and repeat-induced point mutation (RIP) in Podospora anserina. Mol Gen Genet 263(6): 1061–1069, 2000.
79. J-L Rossignol, M Picard. Ascobolus immersus and Podospora anserina: Sex, recombination, silencing and death. In: JW Bennett, LL Lasure, eds. More Gene Manipulations in Fungi. San Diego, CA: Academic Press, Inc., 1991, pp 266–290.
80. C Goyon, G Faugeron. Targeted transformation of Ascobolus immersus and de novo methylation of the resulting duplicated DNA sequences. Mol Cell Biol 9(7): 2818–2827, 1989.
81. MC Costanzo, ME Crawford, JE Hirschman, JE Kranz, P Olsen, LS Robertson, MS Skrzypek, BR Braun, KL Hopkins, P Kondu, C Lengieza, JE Lew-Smith, M Tillberg, JI Garrels. YPD(TM), PombePD(TM) and WormPD(TM): Model organism volumes of the BioKnowledge(TM) library, an integrated resource for protein information. Nucleic Acids Res 29(1):75–79, 2001.
82. SD Harris, L Cheng, TA Pugh, JR Pringle. Molecular analysis of Saccharomyces cerevisiae chromosome I. On the number of genes and the identification of essential genes using temperature-sensitive-lethal mutations. J Mol Biol 225(1):53–65, 1992.
83. SA Osmani, DB Engle, JH Doonan, NR Morris. Spindle formation and chromatin condensation in cells blocked at interphase by mutation of a negative cell cycle control gene. Cell 52:241–251, 1988.
84. KL O'Donnell, AH Osmani, SA Osmani, NR Morris. bimA encodes a member of the tetratricopeptide repeat family of protiens and is required for the completion of mitosis in Aspergillus nidulans. J Cell Sci 99:711–719, 1991.

85. C Rasmussen, C Garen, S Brining, RL Kincaid, RL Means, AR Means. The calmodulin-dependent protein phosphatase catalytic subunit (calcineurin A) is an essential gene in Aspergillus nidulans. EMBO J 13(11):2545–2552, 1994.

86. E Enloe, A Diamond, AP Mitchell. A single-transformation gene function test in diploid Candida albicans. J Bacteriol 182(20):5730–5736, 2000.

87. B Akerley, E Rubin, A Camilli, D Lampe, H Robertson, JJ Mekalanos. Systematic identification of essential genes by in vitro mariner mutagenesis. Proc Natl Acad Sci USA 95:8927–8932, 1998.

88. FE Nargang, K-P Kunkele, A Mayer, RG Ritzel, W Neupert, R Lill. 'Sheltered disruption' of Neurospora crassa MOM22, an essential component of the mitochondrial protein import complex. EMBO J 14(6):1099–1108, 1995.

89. N Judson, JJ Mekalanos. TnAraOut, A transposon-based approach to identify and characterize essential bacterial genes. Nat Biotechnol 18:740–745, 2000.

90. B Felenbok. The ethanol utilization regulon of Aspergillus nidulans: the alsA-alcR system as a tool for the expression of recombinant proteins. J Biotechnol 17(1): 11–17, 1991.

91. T Fukasawa, Y Nogi. Molecular genetics of galactose metabolism in yeast. Biotechnology 13:1–18, 1989.

92. Y Xiao, DT Weaver. Conditional gene targeted deletion by Cre recombinase demonstrates the requirement for the double-strand break repair Mre1 1 protein in murine embryonic stem cells. Nucleic Acids Res 25(15):2985–2991, 1997.

93. N Sternberg, D Hamilton. Bacteriophage P1 site-specific recombination. I. Recombination between loxP sites. J Mol Biol 150(4):467–486, 1981.

94. B Sauer. Functional expression of the cre-lox site-specific recombination system in the yeast Saccharomyces cerevisiae. Mol Cell Biol 7(6):2087–2096, 1987.

95. P Ross-Macdonald, A Sheehan, GS Roeder, M Snyder. A multipurpose transposon system for analyzing protein production, localization, and function in Saccharomyces cerevisiae. Proc Natl Acad Sci USA 94:190–195, 1997.

96. F Arigoni, F Talabot, M Peitsch, MD Edgerton, E Meldrum, E Allet, R Fish, T Jamotte, ML Curchod, H Leferer. A genome-based approach for the identification of essential bacterial genes. Nat Biotechnol 16(9):851–856, 1998.

97. CA Hutchinson, SN Peterson, SR Gill, RT Cline, O White, CM Fraser, HO Smith, J Venter. Global transposon mutagenesis and a minimal Mycoplasma genome. Science 286:2165–2169, 1999.

98. GW Beadle. Biochemical genetics. Chem Revs 37:15–96, 1945.

99. ADM Rayner, ZR Watkins, JR Beeching. Self-integrion — an emerging concept from the fungal mycelium. In: NAR Gow, GD Robson, GM Gadd, eds. The Fungal Colony. Cambridge: Cambridge University Press, 1999, pp1–24.

100. PE Kolattukudy, LM Rogers, D Li, C-S Hwang, MA Flaishman. Surface signaling in pathogenesis. Proc Natl Acad Sci U S A. 92:4080–4087, 1995.

101. MM Bianchi, G Sartori, M Vandenbol, A Kaniak, D Uccelletti, C Mazzoni, J Di Rago, G Carignani, PP Slonimski, L Frontali. How to bring orphan genes into functional families. Yeast 15:513–526, 1999.

102. K Rieger, M El-Alama, G Stein, C Bradshaw, PP Slonimski, K Maundrell. Chemotyping of yeast mutants using robotics. Yeast 15:973–986, 1999.

103. PJ Mussenden, C Bucke, G Saunders, T Keshavarz. Use of INT to determine the

respiratory activity of immobilized and free Penicillium chrysogenum. J Chem Tech Biotechnol 60:39–44, 1994.

104. AP Enos, NR Morris. Mutation of a gene that encodes a kinesin-like protein that blocks nuclear division in A. nidulans. Cell 60(6):1019–1027, 1990.

105. AS Kashina, GC Rogers, JM Scholey. The bimC family of kinesins: Essential bipolar mitotic motors driving centrosome separation. Biochim Biophys Acta 1357(3):257–271, 1997.

106. TU Mayer, TM Kapoor, SJ Haggerty, RW King, SL Schreiber, TJ Mitchison. Small molecule inhibitor of mitotic spindle bipolarity identified in a phenotype-based screen. Science 286:971–974, 1999.

107. DA Johnson, TA Ball, WM Hess. Image analysis of urediniospores that infect Mentha. Mycologia 91:1016–1020, 1999.

108. DG O'Shea, PK Walsh. The effect of culture conditions of the morphology of the dimorphic yeast Kluyveromyces marxianus var. marxianus NRRLy2415: A study incorporating image analysis. Appl Microbiol Biotechnol 53:316–322, 2000.

109. DP Donnelly, L Boddy, MF Wilkins. Image analysis—a valuable tool for recording and analysing development of mycelial systems. Mycologist 13:120–125, 1999.

110. KA Giuliano, RL DeBiasio, RT Dunlay, A Gough, JM Volosky, J Zock, J Pavlakis, DL Taylor. High-content screening: A new approach to easing key bottlenecks in the drug discovery process. J Biomolec Screen 2(4):249–259, 1997.

111. K Okubo, N Hori, R Matoba, T Niiyama, A Fukushima, Y Kojima, K Matsubara. Large scale cDNA sequencing for analysis of quantitative and qualitative aspects of gene expression. Nat Genet 2(3):173–179, 1992.

112. P Liang, L Averboukh, K Keyomarsi, R Sager, AB Pardee. Differential display and cloning of messenger RNAs from human breast cancer versus mammary epithelial cells. Cancer Res 52(24):6966–6968, 1992.

113. VE Velculescu, L Zhang, B Vogelstein, KW Kinzler. Serial analysis of gene expression. Science 270:484–487, 1995.

114. DJ Duggan, M Bittner, Y Chen, P Meltzer, JM Trent. Expression profiling using cDNA microarrays. Nat Genet 21 supplement:10–14, 1999.

115. RJ Lipshutz, SPA Fodor, TR Gingeras, DJ Lockhart. High density synthetic oligonucleotide arrays. Nat Genet 21 supplement:20–24, 1999.

116. JL DeRisi, VR Iyer, PO Brown. Exploring the metabolic and genetic control of gene expression on a genomic scale. Science 278:680–686, 1997.

117. MJ Marton, JL DeRisi, HA Bennett, VR Iyer, MR Meyer, CJ Roberts, R Stoughton, J Burchard, D Slade, H Dai, DEJ Bassett, LH Hartwell, PO Brown, SH Friend. Drug target validation and identification of secondary drug target effects using DNA microarrays. Nat Med 4(11):1293–1301, 1998.

118. TR Hughes, MJ Marton, AR Jones, CJ Roberts, R Stoughton, CD Armour, HA Bennett, E Coffey, H Dai, YD He, MJ Kidd, AM King, MR Meyer, D Slade, PY Lum, SB Stepaniants, DD Shoemaker, D Gachotte, K Chakraburtty, J Simon, M Bard, SH Friend. Functional discovery via a compendium of expression profiles. Cell 102:109–126, 2000.

119. CJ Roberts, B Nelson, MJ Marton, R Stoughton, MR Meyer, HA Bennett, YD He, H Dai, WL Walker, TR Hughes, M Tyers, C Boone, SH Friend. Signaling

and circuitry of multiple MAPK pathways revealed by a matrix of global gene expression profiles. Science 287:873–880, 2000.

120. JS Andersen, M Mann. Functional genomics by mass spectrometry. FEBS Lett 480:25–31, 2000.

121. B Futcher, GI Latter, P Monardo, CS McLaughlin, JI Garrels. A sampling of the yeast proteome. Mol Cell Biol 19:7357–7368, 1999.

122. MJ Page, B Amess, RR Townsend, R Parekh, A Herath, L Brusten, MJ Zvelebil, RC Stein, MD Waterfield, SC Davies, MJ O'Hare. Proteomic definition of normal human luminal and myoepithelial breast cells purified from reduction mammoplasties. Proc Natl Acad Sci U S A 96:12589–12594, 1999.

123. M Larsson, S Graslund, L Yuan, E Brundell, M Uhlen, C Hoog, S Stahl. High-throughput protein expression of cDNA products as a tool for functional genomics. J Biotech 80:143–157, 2000.

124. A Pandey, M Mann. Proteomics to study genes and genomes. Nature 405:837–846, 2000.

125. MR Martzen, SM McCraith, SL Spinelli, FM Torres, S Fields, EJ Grayhack, EM Phizicky. A biochemical genomics approach for identifying genes by the activity of their products. Science 286:1153–1155, 1999.

126. P Legrain, L Selig. Genome-wide protein interaction maps using two-hybrid systems. FEBS Lett 480:32–36, 2000.

127. P Uetz, L Giot, G Cagney, TA Mansfield, RS Judson, JR Knight, D Lockshon, V Narayan, M Srinivasan, P Pochart, A Qureshi-Emili, Y Li, B Godwin, D Conover, T Kalbfleisch, G Vijayadamodar, M Yang, M Johnston, S Fields, JM Rothberg. A comprehensive analysis of protein-protein interactions in Saccharomyces cerevisiae. Nature 403:623–631, 2000.

128. PA Wigge, ON Jensen, S Holmes, S Soues, M Mann, JV Kilmartin. Analysis of the Saccharomyces spindle pole by matrix-assisted laser desorption/ionization (MALDI) mass spectrometry. J Cell Biol 141(4):967–977, 1998.

129. N Glassbrook, C Beecher, J Ryals. Metabolic profiling on the right path. Nat Biotechnol 18:1142–1143, 2000.

130. O Fiehn, J Kopka, P Dormann, T Altmann, RN Trethewey, L Willmitzer. Metabolite profiling for plant functional genomics. Nat Biotechnol 18:1157–1161, 2000.

131. U Roessner, A Luedemann, D Brust, O Fiehn, T Linke, L Willmitzer, AR Fernie. Metabolic profiling allows comprehensive phenotyping of genetically or environmentally modified plant systems. Plant Cell 13:11–29, 2001.

132. M Kanehisa, S Goto. KEGG: Kyoto encyclopedia of genes and genomes. Nucleic Acids Res 28:27–30, 2000.

133. F Tekaia, G Blandin, A Malpertuy, B Llorente, P Durrens, C Toffano-Nioche, O Ozier-Kalogeropoulos, E Bon, C Gaillardin, M Aigle, M Bolotin-Fukuhara, S Casaregola, J de Montigny, A Lepingle, C Neuveglise, S Potier, J-L Souciet, M Wesolowski-Louvel, B Dujon. Genomic exploration of the Hemiascomycetous yeasts: 3. methods and strategies used for sequence analysis and annotation. FEBS Lett 487:17–30, 2000.

134. A Malpertuy, F Tekaia, S Casaregola, M Aigle, F Artiguenave, G Blandin, M Bolotin-Fukuhara, E Bon, P Brottier, J de Montigny, P Durrens, C Gaillardin, A Lepingle, B Llorente, C Neuveglise, O Ozier-Kalogeropoulos, S Potier, W Saurin,

C Toffano-Nioche, M Wesolowski-Louvel, P Wincker, J Weissenbach, J-L Souciet, B Dujon. Genomic exploration of the Hemiascomycetous yeasts: 19. Ascomycete-specific genes. FEBS Lett 487:113–121, 2000.

135. BA Roe, D Kupfer, H Zhu, J Gray, S Clifton, R Prade, J Loros, J Dunlap, M Nelson. The Aspergillus nidulans and the Neurospora crassa sequencing project. www.genome.ou.edu/fungal, 2001.

136. B Roe, Q Ren, A Peterson, D Kupfer, H Lai, M Beremand, A Peplow, A Tag. The Fusarium sporotrichioides cDNA Sequencing Project. www.genome.ou.edu/fsporo, 2000.

137. H Zhu, BP Blackmon, M Sasinowski, RA Dean. Physical map and organization of chromosome 7 in the rice blast fungus, Magnaporthe grisea. Genome Res 9(8): 739–750, 1999.

138. B Roe, D Kupfer, J Lewis, S Yu, K Buchanan, D Dyer, J Murphy. The Cryptococcus neoformans cDNA Sequencing Project. www.genome.ou.edu/cneo, 2001.

139. D Qutob, PT Hraber, BW Sobral, M Gijzen. Comparative analysis of expressed sequences in Phytophthora sojae. Plant Physiol 123(1):243–254, 2000.

140. BN Chakraborty, NA Patterson, M Kapoor. An electroporation-based system for high-efficiency transformation of germinated conidia of filamentous fungi. Can J Microbiol 37(11):858–863, 1991.

141. S-C Wu, K-S Ham, AG Darvill, P Albersheim. Deletion of two endo-beta-1,4-xylanase genes reveals additional isozymes secreted by the rice blast fungus. Mol Plant Microbe Interact 10(8):700–708, 1997.

142. M Lorito, CK Hayes, A Di Pietro, GE Harman. Biolistic transformation of Trichoderma harzianum and Gliocladium virens using plasmid and genomic DNA. Curr Genet 24(4):349–356, 1993.

143. P Chaure, SJ Gurr, P Spanu. Stable transformation of erysiphe graminis, an obligate biotrophic pathogen of barley. Nat Biotechnol 18(2):205–207, 2000.

144. MJ de Groot, P Bundock, PJ Hooykaas, AG Beijersbergen. Agrobacterium tumefaciens-mediated transformation of filamentous fungi. Nat Biotechnol 16(9):839–842, 1998.

145. RO Abuodeh, MJ Orbach, MA Mandel, A Das, JN Galgiani. Genetic transformation of Coccidioides immitis facilitated by Agrobacterium tumefaciens. J Infect Dis 181(6):2106–2110, 2000.

146. A Kumar, K Cheung, P Ross-Macdonald, PSR Coelho, P Miller, M Snyder. TRIPLES: A database of gene function in Saccharomyces cerevisiae. Nucleic Acids Res 28:81–84, 2000.

147. HW Mewes, K Heumann, A Kaps, F Pfeiffer, S Stocker, D Frishman. MIPS: A database for genomes and protein sequences. Nucleic Acids Res 27:44–48, 1999.

# 6

# Functional and Comparative Genomics of Cyanobacteria

**Bradley L. Postier and Robert L. Burnap**
*Oklahoma State University, Stillwater, Oklahoma, U.S.A.*

With six phylogenetically diverse cyanobacterial genomes completely sequenced or nearing completion, and several others in progress, meaningful comparative genomics is becoming possible, and their role as model systems in understanding processes in higher plants, for example, is becoming increasingly useful. For years, cyanobacteria have been used as model systems to study homologous processes occurring in plants owing to their similarities to photosynthetic physiology and metabolism and because of their comparatively simple and easily manipulated genetic systems. With the release of these fully annotated cyanobacterial genomic sequences, this approach is poised to include advances in comparative genomics, functional genomics, and metabolic engineering. The use of genomic information provides a new level of analysis that ranges from the formulation of testable hypotheses on the structure and function of individual genes to the more extensive accounting of the genetic differences underlying species-specific niche specializations.

## I. INTRODUCTION

### A. Cyanobacterial Systems

Some of earth's oldest fossils derive from cyanobacteria-dominated mats that filled the shallow Archean seas more than 3.5 billion years ago (Schopf 1993). A significant factor in the ecological success of the cyanobacteria was likely to have been the evolutionary invention, presumably within the cyanobacteria or their evolutionary antecedents, of the photosynthetic $H_2O$-splitting mechanism. This enzymatic innovation enabled the use of $H_2O$ as a source of electrons to satisfy the demand for reductant that is required for the biosynthetic processes and thus, for organismal growth and proliferation. Before this innovation, reductant for assimilatory metabolism must have been obtained from other external sources such as the $Fe^{2+}$, $H_2S$, or abiotically generated reduced organic compounds postulated to have been accumulated through chemical and photochemical processes occurring early in earth's history. The consequence of the innovation of using $H_2O$ as a source of electrons was the gradual accumulation of oxygen in the early earth's atmosphere as a result of the release of molecular oxygen, $O_2$, which is the principal waste product of the light-driven $H_2O$-splitting reaction. The rise of oxygenic photosynthesis thereby transformed the earth's atmosphere from that of a reducing environment containing trace levels of geochemically generated oxygen to one in which oxygen became freely available. With the ability to tap into a virtually limitless supply of electrons ($H_2O$) to sustain anabolic metabolism, the prokaryotic oxygenic photosynthesis, perhaps due largely to the cyanobacteria, became the base of ecosystem productivity in the Archean world and, at the same time, set the stage for the advent of respiratory metabolism as the dominant mode of heterotrophic existence. To this day, cyanobacteria remain one of the largest and most diverse groups of eubacteria and, accordingly, occupy a variety of key positions within the biosphere. Furthermore, if one considers the likelihood that the chloroplasts of all extant and extinct plant and algal species are indeed the specialized descendents of endosymbiotic cyanobacteria or prochlorophytes of the Cambrian epoch, the centrality of the cyanobacterial lineage to past and current global primary productivity can be appreciated. From a practical perspective, the genetic affinities between cyanobacteria and chloroplasts, with the noted exception of the lack of enzymes to synthesize chlorophyl *b* in cyanobacteria, has rendered cyanobacteria as models of plantlike photosynthesis. As discussed below, this has provided a genetic resource for the elucidation of metabolic pathways that have proven difficult to clarify using traditional genetic and biochemical approaches in the more complex plant systems.

As prokaryotes, cyanobacteria do not contain bona fide membranous subcellular organelles. However, cyanobacteria generally exhibit a high degree of subcellular structure when viewed with an electron microscope. The ability of some species to differentiate into different cell types, such as the nitrogen-fixing heterocyst cell-type of the filamentous genera, extends the range of internal complexity (Whitton and Potts 2000). Cyanobacterial cells possess an extensive internal membrane system called thylakoids, observed in a variety of species-specific arrangements, often appearing in thin-section electron micrographs as concentric rings of appressed membrane pairs that are immediately beneath the cytoplasmic membrane and surround the central cytoplasm (Gantt 1994). The pairs of appressed membranes enclose the thylakoid lumen, analogous to the situation found in higher plant chloroplasts. As in higher plants, the thylakoid membranes contain the photosynthetic machinery, but unlike higher plants, the thylakoid membranes also contain an extensive set of multiprotein complexes associated with respiratory metabolism, including NADH dehydrogenase and cytochrome $c$ oxidase. The production of the thylakoid increases the total cellular membrane surface area several-fold and permits increased numbers of membrane-bound photosynthetic complexes to form. Recent evidence suggests that the thylakoids are formed, or at least expanded, by fusion of vesicles originating from the surrounding cytoplasmic membrane. On the cytoplasmic surface of the thylakoid membranes are numerous structures, called phycobilisomes, which are large pigment-protein complexes composed of water-soluble phycobiliproteins and serve as light-harvesting antennae for the photosynthetic reaction centers. Phycobilisomes are found in all cyanobacterial groups with the exception of the Prochlorophytes, which, as discussed below, have replaced the phycobilisome with intrinsic, membrane-bound, light-harvesting antennae.* Another photosynthetic structure visible by electron microscopy is the polyhedral body, that is, the carboxysome, which is the site of carbon fixation and, accordingly, is composed primarily of the major carbon-fixing enzyme, ribulose bisphosphate carboxylase-oxygenase (Kaplan et al 1994). Other subcellular inclusions visible in electron micrographs are cyanophycin granules composed of a copolymer of aspartate and arginine. Cyanophycin and phycocyanin are both used extensively as nitrogen reserve molecules, with the latter often comprising up to 50% of the total cellular protein content. In broad morphological terms, cyanobacteria occur either as unicellular or multicellular, filamentous forms. In addition to being major contributors to the fixation of carbon through photosynthesis, cyanobacteria occupy key positions in many ecosystems owing to the ability of some species to fix nitrogen. Examples of nitrogen fixation are found in both morphological forms, with the filamentous cyanobacteria differentiating specialized cell types (heterocysts) to isolate the $O_2$-sensitive N-fixing enzyme (nitrogenase) from the $O_2$-yielding photosynthetic reactions and the unicellular forms achieving the separation temporally by performing N-fixation at night or at periods of low light to avoid $O_2$ poisoning of the nitrogenase (Chen et al 1999)

---

* While Prochlorophytes do not assemble phycobilisomes, genomic sequence analyses indicate the presence of intact phycobiliprotein genes. Features such as this support the close taxonomic affinities between the phycobilisome-containing cyanobacteria and the Prochlorophytes deduced from rRNA sequence analyses. As discussed below, this contrasts with earlier expectations that the Prochlorophytes represent a separate prokaryotic group that split from the cyanobacterial line and subsequently produced representatives that gave rise to green plant and algal chloroplasts.

Although oxygenic photosynthesis is their primary source of energy, a number of cyanobacterial species are capable of using external carbon such as glucose, glycerol, and amino acids to supplement their endogenous photosynthetic and metabolic capabilities. In some cases, these capabilities have been used to enhance the genetic and biochemical analysis of these organisms. For example, the endogenous or selected capability of using glucose as an exogenous energy and carbon source has been used to rescue otherwise lethal mutations in the photosynthetic apparatus. The photosynthetic mechanisms involving the light-driven electron transport reactions and $CO_2$ fixation by the Calvin cycle are highly homologous to the corresponding processes in higher plants and algae and, accordingly, the corresponding genes reflect these deep similarities.

Although amino acid sequence identities between core photosynthetic proteins of higher plants and cyanobacteria are generally very high and in the range of 60% to 90% sequence identity, the more peripheral components of the mechanism are typically less conserved or even entirely different. The light-harvesting antennae of the reaction centers are a good example of this divergence. In the case of higher plants, virtually all light harvesting is carried out by the intrinsic light-harvesting chlorophyll protein complexes (LHCs), whereas in cyanobacteria and red algae this function is generally carried out by very large peripheral membrane protein complexes called phycobilisomes. On the other hand, another group of cyanobacteria, the Prochlorophytes, lack phycobilisomes and, like higher plants, rely on intrinsic membrane protein complexes that contain chlorophyl *b*. Initially, this similarity was thought to indicate that the Prochlorophytes descended from the same line that gave rise to modern green algal and higher plant chloroplasts and, conversely, that phycobilisome-containing cyanobacteria gave rise to the chloroplasts of red algae. However, closer analysis of the chlorophyll pigment-protein complexes revealed that the genes of the membrane-bound chlorophyll-protein complexes bore little resemblance to the higher plant complexes and were most closely related to a stress-induced protein called isiA (iron-stress induced) found in the phycobilisome-containing cyanobacteria (Tomitani et al 1999) These findings and consideration of the enzymatic simplicity of producing chlorophyl *b* from chlorophyll *a* has led to the conclusion that the appearance of chlorophyl *b* probably occurred independently in several species lines during the course of evolution and, perhaps, occurred after the endosymbiosis event that led to modern plant chloroplasts. This "postendosymbiosis" hypothesis for the evolution of the modern higher plant and green algal type of LHCs is consistent with phylogenetic analyses that find the phycobilisome-containing endosymbionts/ chloroplasts of the taxonomically enigmatic *Cyanophora paradoxa* as sharing the last common ancestor of both the cyanobacteria and chloroplasts. This has implications regarding the similarities and contrasts between the higher plants and cyanobacteria discussed below, as it is incorrect to assume that closer homologs to higher plant genes would be found in the Prochlorophytes rather than in other cyanobacteria.

## B.  Cyanobacterial Sequencing Efforts

According to the National Center for Biotechnology Information (NCBI) at the National Institutes of Health, more than 800 full genome sequences, including 172 Eukaryota, 112 bacterial, and 13 archaeal are complete, or "in process," and have been submitted to their database as of the summer of 2001. In this increasingly data-rich environment, the ability to efficiently analyze and find meaningful patterns to specific research problems is a critical task. One important consideration regarding the utility of the cyanobacterial sequences obtained is the phyletic diversity of genomes analyzed. Although the choice of genomes to be sequenced inevitably reflects consideration of its use as a model system in the scientific community, these choices are also reflecting fundamental scientific issues regarding the type of information that can be obtained from phyletic position. The choice of the first sequenced cyanobacterium (for example see Kaneko et al 1996), *Synechocystis* sp PCC 6803 depended, to a large measure, on its widespread, although not exclusive (Golden 1988), use as a cyanobacterial model system for the study of the oxygenic photosynthetic mechanism (Williams 1988). On the other hand, choices for the subsequent cyanobacterial genomes that have been sequenced and those in process also reflect additional evolutionary and physiological considerations. In the case of the six cyanobacterial genomes now available and discussed below, comparative information will undoubtedly be developed based on these additional considerations. Moreover, the utilization of diverse sequences will enable better comparisons within the currently sequenced cyanobacterial genomes, as well as with other prokaryotes and higher organisms such as the fully sequenced plant model *Arabidopsis thalliana* and the green algal model *Chlamydomonas reinhardtii*, which is being sequenced. It is apparent, for example, that the cyanobacterial genomes contain significant numbers of genes for two-component regulatory systems and ABC-type transporters (Kaneko and Tabata 1997) that may be meaningfully compared with better characterized genes in other bacteria to infer testable hypotheses regarding function (Mizuno et al 1996). At the same time, the recently sequenced *Arabidopsis* genome contains both two-component regulatory genes and ABC transporters, the latter of which occur in large numbers. This raises the question of the extent to which cyanobacterial models can be used to explore hypotheses regarding the function of genes such as these in making inferences regarding the function of the higher plant homologs.

Currently, six cyanobacterial genomic sequences are available at least in part: the freshwater unicellular species *Synechocystis* sp PCC 6803, mentioned above; the marine species *Synechococcus*; two ostensibly similar marine Prochlorophytes isolated from different depths; and two filamentous *Nostoc* species capable of differentiation and nitrogen fixation. As mentioned above, *Synechocystis* sp PCC 6803 has been extensively used for detailed molecular analysis of the enzymes and regulatory features underlying oxygenic photosynthesis. Important for this purpose, *Synechocystis* is capable of photoheterotrophic growth, which allows the construction and rescue of otherwise lethal mutations in the photosynthetic mechanism (Williams 1988). Additionally, *Synechocystis* is easily transformable with exogenous DNA and undergoes homologous double crossover recombination (Williams 1988). This has permitted a number of genetic strategies, such as replacement of wild-type alleles with copies that are mutated in vitro, allowing the study of structure-function relationships of target proteins (Golden 1988; Jansson et al 1987; Labarre et al 1989; Marraccini et al 1993; Poncelet et al 1998; Vermaas et al 1986; Williams 1988). When combined with isotopic labeling, this has allowed the detailed analysis of the function of specific parts of the photosynthetic reaction centers down to the level of individual atoms comprising catalytic amino acid sidechains (Breton et al 1999; Debus et al 2001; Diner et al 1998; Hienerwadel et al 1997; Noguchi et al 1999). Additionally *Synechocystis* is capable of growth under a variety of conditions, including high and low salt, light, and temperature (Bhaya et al 2000). As with other photoautotrophic strains, it must respond to the environment and adapt the photosynthetic and metabolic machinery to balance energy production with $CO_2$ fixation and other energetically demanding activities, such as the maintenance of transmembrane ion homeostasis. Considering the relative ease of protein and genetic manipulation in addition to its diverse mechanisms of adaptation identifies this strain as a valuable prokaryote for functional genomics.

Marine *Synechococcus* strains occupy a variety of important niche positions in oceanic and coastal marine environements and figure prominently in the marine food webs (Waterbury et al 1986). *Synechococcus* strains are well studied for their enigmatic mechanism of motility. They are capable of traveling in a gliding motion at up to 25 $mmsec^{-1}$ and use no microscopically obvious extracellular machinery for this activity. Most importantly it is taxic, not to light, but to trace metals and nutrients at concentrations $10^{-9}$ to $10^{-10}$ M in accordance with the paucity, patchiness, or both of nutrients often existing in its marine habitat. Compared with freshwater species, the obligately marine *Synechochoccal* strain requires high $Na^+$, $Cl^-$, $Mg^{2+}$, and $Ca^{2+}$ concentrations for growth, which, when considered together with the other features, makes it a model for nutrient sensing and chemotactic motility, as well as for the study of photosynthetic capacity in marine environments (Brahamsha, 1996; Brahamsha, 1999).

*Prochlorococcal* strains may contain the smallest genomes, around 2 Mb, of all known photoautotrophic organisms, but as they dominate the temperate and tropical oceans they are speculated to be the most abundant and contribute a significant fraction of the photosynthesis in the worlds oceans, making them an important factor in global carbon cycling and the Earth's atmospheric fluxes of carbon dioxide and oxygen (Chisholm et al 1988; Urbach et al 1998). In addition to chlorophyll *b*, mentioned above, they also contain an unusual combination of divinyl derivatives of chlorophyll *a*, α-carotene, zeaxanthin, and a type of phycoerythrin. Two ecotypes of *Prochlorophytes* have been sequenced. According to the Department of Energy's Joint Genome Institue (DOE JGI), one strain (MED4) is capable of growth at greater depths than the second (MIT9313). Although these two strains representing the two ecotypes have been sequenced, annotation of only MED4 is complete at the time of this writing. At first glance, though, the two genomes differ in several respects, including a 0.3 Mb range in size from 1.8 to 2.1 Mb and a highly contrasting G+C content.

*Nostoc punctiforme* and *Anabaena* are two representatives of the filamentous Nostocs, which are capable of differentiating into four cellular types. The vegetative state is most common and is responsible for asexual reproduction and general metabolism under normal conditions. Heterocysts are specialized cells capable of fixing dinitrogen. Akinetes are spore-like bodies that develop under growth-limiting conditions. Finally, hormogonia are short fragments of filaments capable of motility. Nostocs are found in freshwater and tropical, temperate, and polar terrestrial systems, and can also be found rarely in marine samples (Potts 2000).

## C.  Sequence Comparisons and Statistics

*Synechocystis* sp PCC6803 was the first cyanobacterium to have its complete genome sequenced. When the sequence was first completed, 3,168 potential proteins were identified, of which nearly 40% still are not linked to characterized proteins with specified biological activity, as seen in Table 1, which is a compilation of data from several database resources, including the DOE-JGI (*http://www.jgi.doe.gov/*), Integrated Genomics database "ERGO" (*http://wit. integratedgenomics.com/Igwit/*), and Cyanobase (*http://www.kazusa.or jp/ cyanol/*). Many of the genes coding for metabolic functions could be categorized, and *Synechocystis* has already been a valuable resource for the study of photosynthesis, respiration, and the biosynthesis of prosthetic groups and the like. In addition, 80 genes for two-component signal transduction have been identified. These have become interesting targets for knock-out mutagenesis and phenotypic analysis, as *Synechocystis* is known for its capacity to acclimate to changing environments, possibly using two-component regulatory systems (Suzuki et al 2001).

First comparisons of these five genomes reveal a number of distinguishing characteristics. *Prochlorococal* strains appear as good examples of what may prove to be among the smallest genomes of photoautotrophic organisms. These strains have developed a niche in which they are highly competitive and are

**Table 1**  Statistical Summaries and Comparisons of Five Cyanobacterial Genome Sequencing Efforts

| | Prochlorococcus marinus (MED4) | | Synechococcus marinus | | Synechocystis PCC6803 | | Anabaena | | Nostoc punctiforme | |
|---|---|---|---|---|---|---|---|---|---|---|
| | Total | % Total | Total | % Total | Total | % Total | Total | % Total | Total | % Total |
| Length, Mb | 1.67 | — | 2.75 | — | 3.57 | — | 7.20 | — | 9.22 | — |
| Coding sequence, Mb | 1.4 | 85.9 | 2.1 | 75.2 | 3.1 | 86.7 | 5.8 | 79.8 | 7.2 | 77.9 |
| G+C content | — | 30.90 | — | 58.46 | — | 47.20 | — | 41.20 | — | 41.50 |
| Total orfs | 1645 | — | 2447 | — | 3356 | — | 5826 | — | 7569 | — |
| Orfs with assigned function | 1128 | 68.6 | 1358 | 55.5 | 1978 | 58.9 | 3281 | 56.3 | 3158 | 41.7 |
| Hypothetical orfs | 70 | 4.3 | 46 | 1.9 | 1262 | 37.6 | 191 | 3.3 | 83 | 1.1 |
| Orfs with no assignment | 447 | 27.2 | 1043 | 42.6 | 116 | 3.5 | 2354 | 40.4 | 4328 | 57.2 |

Data for this table were compiled from three database resources: ERGO, Cyanobase, and the DOE JGI, as referenced in Table 2.

postulated to be the most abundant photosynthetic organisms on earth and contribute a large fraction of the ocean's primary productivity (Partensky et al 1999). However, they lack the ability to survive well outside of it, possibly because of their streamlined genome, which accompanies their exceedingly small cellular dimension ($<$ 1 µM). Notice the linear relationship between genome length and the number of open reading frames. The two *Nostoc* genomes are nearly five times greater in length and in number of open reading frames maintaining a similar coding sequence percentage. Each of these genomes is densely packed with coding information, averaging from one open reading frame per thousand bases in the finished *Prochloroccus* MED4 strain, to at most one open reading frame in 1,218 bases in *Nostoc punctiforme*. Interestingly, the two *Prochlorococcus* strains have widely different G+C contents (data not shown), with that of MIT9313 falling closer to the *Synechococcus* strain. *Synechocystis* has a larger proportion of hypothetical genes (genes with no functional assignment) than the other organisms. This may be because of the differences in orf prediction matrices used to analyze the genomes and their default functional assignment cutoff values, or the fact that when it was sequenced, fewer genome sequences were available for comparison. Also, it is likely that different definitions of "hypothetical" are used between the various databases. Cyanobase annotates all genes with slight to no apparent sequence similarity to previously characterized genes from Genebank as hypothetical genes. ERGO, however, defines them as unknown genes and assigns a hypothetical class to them if they are loosely associated with another characterized gene. Differences such as these also might contribute to the discrepancy between the total number of open reading frames assigned by each of the databases, as ERGO includes an additional 188 open reading frames. Interestingly, the two *Nostocs* have a larger supply of genes with no known homologs. This makes them excellent resources for novel genes and protein products. Other obvious findings include the development in the organisms with larger genomes of more functional divergence, such as cell differentiation, filamentation, and nitrogen fixation, all of which require additional genetic information.

## D. Comparative Genomics

Comparative genomics is literally the study of two or more genomes, comparing them either through nucleic acid or amino acid sequence similarity and dissimilarity or through mapping gene positions, analysis of gene function, and studies on evolution. A principal aim is to study all these aspects and develop an understanding of how to link genotype with phenotype (Clark 2000) and understand the evolution of entire ensembles of genes that may move en block across species barriers through "horizontal" gene transfer mechanisms. An example of this type of analysis with cyanobacteria, still in its early stages, was presented recently by the Chisholm group at the latest Cyanobacterial Workshop (Rocap 2001). This work requires the sequencing and analysis of complete genomes for the DOE JGI focusing on two *Prochlorococcus* genomes. Nearing completion of the sequencing and analysis of the second strain, direct comparison of gene amino acid sequence similarity enables the clustering of genes based simply on their presence or absence from each of the two organisms or from other cyanobacterial genomes such as *Synechocystis* PCC6803.

Figure 1 illustrates several examples of how genes are clustered in this manner. This analysis begins with the first completed sequenced genome depicted by example I. Once a second genome is complete, the total contents of both are compared through homology of each gene's amino acid sequence to its counterpart's genome and scored for hits based on a predetermined level of sequence similarity required to specify two genes as orthologs. Example II could represent the two genomes of *Synechocystis* PCC6803, and *Prochlorococcus* MED4, with the area of overlap representing the number of orthologous genes present. These two cyanobacteria are moderately related, and a large number of genes may show homology, as shown here with two overlapping circles. Some examples of orthologous genes might encode for the photosynthetic complexes or for common

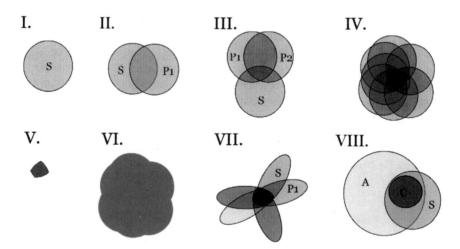

**Figure 1** Hypothetical Venn diagrams representing the extent of homologous genes from various whole genome sequences. Each circle represents a sequenced genome. The overlapping areas of two or more circles represents the amount of homology between those genomes. Parts I–VI represent the accumulating results of sequencing efforts on cyanobacterial genomes. Parts VII and VIII depict the relationships between cyanobacteria and other sequenced organisms such as other bacterial species or *Arabidopsis* and its chloroplast genomes. (A) *Arabidopsis* genome; (C) Chloroplast genome; (P1) *Prochlorococcus* genome 1; (P2) *Prochlorococcus* genome 2; (S) *Synechocystis* genome

metabolic pathways such as lipid biosynthesis or conserved regulatory genes. Clearly, this analysis depends on the sequence similarity cut-off values used to score pairs of genes as orthologous. Extending this analysis to include three genomes, now including the *Prochlorococcus* strain, MIT9313, is shown in hypothetical example III. The two strains of *Prochlorococcus* represented by the top two circles show much more extensive homology with each other than either show toward the strain *Synechocystis*. In addition, *Synechocystis* has many more genes showing no homology to either *Prochlorococcus* strain, even while there are still many examples of nonorthologous genes between the two *Prochlorococcal* strains, denoted here by the light green areas of each circle. As a result of increasing the stringency for scoring hits, the amount of overlap between *Synechocystis* and *Prochlorococcus* MED4 would be less than that depicted in example II. Patterns such as this might aid in phylogenetic analysis and, in principal, should tend to reflect phylogenetic trees based rRNA sequence analysis if the extent of horizontal gene transfer is minimal. However, this is probably not a valid assumption when considering prokaryotic systems. Possibly, more cases of horizontal gene transfer would become apparent as more genomes from cyanobacteria are sequenced and it becomes possible to develop a situation such as that represented by example IV, two important components of which are the focus of example V, which represents the core set of genes found in all cyanobacterial strains and example VI, which represents the total genetic variation of all cyanobacterial strains. As a result, example IV might also suggest which strains are more susceptible to horizontal gene transfer and which are incapable. Additionally, two distantly related cyanobacterial strains shown in green in example VII are depicted as they would relate to other organisms whose genomes have been sequenced. This demonstrates how they each have evolved by building on the successes of a core set of genes. This highly conserved class might be well represented by DNA replicase, ribosomal proteins, and tRNAs. In a more complex variation example VIII depicts the relationship between a cyanobacterium, *Arabidopsis*, and its chloroplast. It would be intriguing to find out how the set of highly conserved "core" genes compare to chloroplast and plant genomes. Unfortunately we must better define that core before those hypotheses can be made.

A common result from every sequencing project thus far has been the assignment, depending on the stringency criteria applied, of roughly 25% of each genome with no discernable sequence homology to gene sequences already deposited in the databases.* These genes are typically characterized as hypothetical,

---

* The following argument was first brought to the attention of the authors during a presentation by Prof. Don Bryant of the Pennsylvania State University. The analysis of the proteases is based on a talk given by Prof. W. Pearson and appearing in the World Wide Web at http://www.people. virginia.edu/~wrp/talk95/prot_talk12-95_17.html

unknown, or unique. On the surface it would appear as if every genome carries a large portion of novel and, therefore, uncharacterized genes. However, it is possible that some amino acid sequences have diverged to the extent that statistically significant sequence homologies are no longer detectable and yet still retain a high degree of tertiary structure homology. In short, a protein's structural conservation might be enough to maintain function even in the face of radical sequence divergence and evolution. As a result, many of the genes with no sequence homology may actually have functional homology to known and well-characterized genes. For example, three members of the serine protease superfamily are clearly homologous at the tertiary structure level and in terms of the catalytically active residues of the active site. Nevertheless, two of the structurally homologous proteases, bovine chymotrypsin and *Staphylococcus griseus* trypsin, share strong sequence similarity (expectations values smaller than $\sim 10^{-20}$), whereas the third related sequence, *S. griseus* protease A, has a very low sequence similarity score compared to the others (expectation values greater that 10). Given these considerations, it is possible that a large proportion of the hypothetical genes from related species are truly orthologous and are still functionally related despite the loss of coherent similarity at the primary amino acid sequence level. As a result, it will be impossible to determine evolutionary links or protein function through sequence analysis for many of these genes.

## E. Gene Analysis Tools: Beyond the Blast

There is already a great variety of bioinformatic tools available on the Internet that one can use to identify biologically important genes or proteins. Table 2 is

**Table 2**  Abbreviated List of Web-Based Genomic Information Services

| Provider/service | Web link |
|---|---|
| National Center for Biotechnology Information | http://www.ncbi.nlm.nih.gov |
| The Institute for Genomic Research | http://www.tigr.org |
| Kyoto Encyclopedia of Genes and Genomes | http://www.tokyo-center.genome.ad.jp/kegg/kegg.html |
| DOE Joint Genome Institute | http://www.jgi.doe.gov/ |
| Pedant | http://pedant.gsf.de/ |
| Mips | http://mips.gsf.de/ |
| Integrated Genomics | http://wit.integratedgenomics.com/Igwit/ |
| SDSC Biology Workbench | http://workbench.sdsc.edu/ |
| Cyanobase (KDRI) | http://www.kazusa.or.jp/cyano/ |

a partial list of sites used frequently for just such analysis. The availability of these databases and their powerful techniques are designed for use without an extensive background in writing computational algorithms. This allows the biologist to investigate without the need for programming skills. Tools are currently available for sequence alignments, hydropathy profiles, and domain analysis. Also available are protein motif and secondary structure assignment algorithms.

Many prokaryotic organisms possess an operonic structure for transcriptional expression of genes linked to the same metabolic pathway, thus enabling the regulation of entire pathways with one promoter. Even where polycistronic mRNAs are not produced, functionally connected genes sometimes remain physically connected at the same chromosomal region, perhaps shuffled in order. As a result, the gene order or proximity of functionally related genes for a particular pathway may be preserved. Therefore, the existence of patterns of gene clustering and gene linkage alignments allows the generation of hypotheses on unknown gene function on the basis of physical association with better defined genes. Importantly, there are multigenomic bioinformatic approaches exploiting this type of conserved gene clustering. The motivation for these tools is to provide clues to the function of unknown genes, even if they themselves are not clustered with functionally related genes in the species in which they are found. Integrated Genomics Inc. offers a database called ERGO, available at their website referenced in Table 2. After selecting a gene of interest in the database, it is possible to use a query tool that provides analysis of regions of conserved gene order and clustering among related species. This clustering tool allows ready access to information regarding the clustering of orthologous genes around the query gene. These so-called "pinned regions" may be conserved operons found across more than one species and thus consist of stretches of genomic sequence from two or more organisms containing more than one orthologous gene when the corresponding stretches of DNA from the different species are compared. In some cases, gene order is preserved; in other cases, the gene order is rearranged, yet the clustering remains. In still other cases, one or more of the genes present in the operon of one species has been lost from the corresponding region in the other species. These typically are found in closely related organisms, but in the case of several ribosomal protein complex genes, are common to many of the prokaryotes.

Researchers in many areas of plant biology are using the cyanobacterial databases to search for homologous proteins with specific enzymatic activities. This is particularly important for proteins involved in chloroplast structure and function; for example, chloroplast envelope proteins, protein translocation systems, phytochromes, vitamin synthesis, and ion uptake. For example, Shintani and Della Penna describe the introduction of the *Synechocystis* gene γ-tocophorol methyltransferase (γ-TMT), important in the metabolic processing of α-tocopherol (vitamin E), into *Arabidopsis* seeds increasing vitamin E production (Shintani and DellaPenna, 1998). Here they have used the sequence from a cloned carrot gene called p-hydroxyphenylpyruvate dioxygenase (HPPDase) known to be involved in the metabolic pathway of vitamin E synthesis. From the sequence of the carrot HPPDase gene, they identified the *Synechocystis* ortholog slr0090. Through gene clustering, they were able to identify a candidate open reading frame slr0089 for γ-TMT, which was immediately upstream of the ortholog to carrot HPPDase. This gene (γ-TMT) was capable of increasing vitamin E production once cloned and expressed in *Arabidopsis* seeds. Currently, only a few examples exist for using cyanobacterial genes for metabolic enhancement of other organisms. However, given the diversity of genes in the five sets of genomic sequences from cyanobacterial strains, efforts in these areas will undoubtedly continue for some time.

## F.  Functional Genomics in a Cyanobacterium

Recent advances in technology have given rise to techniques that take advantage of the availability of full genome sequences and provide vast amounts of information in a short period. With the development of the Internet-based databases, the wealth of genomic sequences, and the diverse set of bioinformatic tools available, it is now possible to make a large number of hypotheses concerning the functions of the multitude of genes revealed through sequencing efforts. Assignment of gene function, referred to as "annotation," is a predictive tool and the annotations are essentially a set of hypotheses. This is fundamentally different from the initial stages of molecular database formation. In the early stages, individual gene sequences were typically deposited by one research group that had isolated the gene after extensive characterization (e.g., enzymatic activities) or on the basis of the physical characterization of mutants for which detailed phenotypic information had been developed. Consequently, the reliability of the deposited gene annotation had the strength of the associated biochemical and genetic characterizations preceding it. The genomic sequence era now inverts this process and at the same time builds upon the earlier and concurrent biochemical and genetic characterizations to generate the hypothetical gene function assignments. Although bioinformatic approaches are excellent at predicting biological function and generating hypotheses, it is obviously important to test these hypotheses. In the case of *Synechocystis* sp PCC 6803, one such strategy combines the use of various technologies including gene expression arrays, PCR-mediated knock-out strategies, and high-throughput screening of mutants for functional analysis of the *Synechocystis* PCC6803 genome as is demonstrated below. The strategy diagrammed in Figure 2 exemplifies the types of technologies necessary to perform molecular biological analysis on a genomic scale. An explanation with illustrations of the techniques used and some of the information gathered would help develop the concepts, which we wish to explore here.

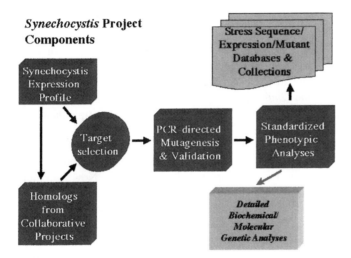

**Figure 2** Functional genomics strategy for *Synechocystis* PCC6803.

## 1. Application of Functional Genomics to a Physiological Problem: Salt Stress in Cyanobacteria

Salt stress in cyanobacteria remains poorly understood at the molecular level, although classic mutational approaches have recently identified a set of structural genes important for coping with this stress. *Synechocystis* sp PCC6803 represents an excellent model for osmoregulation, as it tolerates up to 1.2 M NaCl. From a physiological point of view, salt stress adaptation in cyanobacteria is a rich phenomenon with a characteristic sequence of events that ensue immediately after an initiation of salt shock and proceeding over the course of 24 to 48 hours as the cells adapt to the new regimen. The physiological adjustments to hypertonic and hypersaline environments are multifold and include changes in the expression of photosynthesis-and respiration-related enzymes and the induction of several categories of stress proteins (Allakhverdiev et al 1999; De la Cerda et al 1997; Fulda and Hagemann 1995; Fulda et al 1990; Hagemann et al 1990; Jeanjean et al 1993; Jeanjean et al 1990). These changes can be followed using fluorescence and spectroscopic methods, which are capable of reporting on a variety of parameters ranging from the relative concentrations and activities of the photosystems to changes in the soluble pools of reduced substrate in the cytoplasm. This allows for real-time detection of several physiological changes in a non invasive manner.

Acquisition of tolerance is, in part, effected through the accumulation of the osmolyte glucosylglycerol, by de novo synthesis (Hagemann et al 1997a). Mutant analysis has provided insight regarding the role of two genes expressed in response to salt shock. *StpA* encodes glucosylglycerol-P phosphatase involved in the salt-induced accumulation of glucosylglycerol. *GgtA, B, C,* and *D* encode subunits of an ABC-type transporter, which permits the maintenance of high levels of solutes in the cytoplasm by recovering leaked glucosylglycerol (Hagemann et al 1996; Hagemann et al 1997b; Mikkat et al 1996). At this stage, clear functional homologs to the *Omp* system of genes have not been identified (nor have *EnvZ* homologs) in *Synechocystis* sp PCC6803, although multiple members of the conserved *OmpR* subfamily of regulatory molecules are present in the genomic sequence.

As mentioned above, many of these genes were identified through time-consuming random mutagenesis, screening, and cloning projects. Now with multiple genomic sequences available, it is possible to identify their orthologs in other cyanobacteria with sequence comparison algorithms such as Blast (Altschul et al 1997). It seems likely that this procedure will not identify novel proteins, yet it still remains one of the most useful tools when comparing two or more genomes. Moreover, this strategy is prone to difficulties in accurately predicting functions of the homologs. Predicted regulatory proteins such as the approximately 80 two-component histidine kinase proteins found in *Synechocystis* PCC6803 show very high deduced amino acid sequence similarity, but are activated by different environmental and physiological signals. Thus, the similarity analysis provides only a very superficial assessment of function (Throup et al 2000).

Another example in which sequence similarity may be confusing is in the case of $Na^+/H^+$ antiporters. Here, very high sequence similarity is observed, but each functions differently for ion homeostasis under different conditions. One transporter may be responsible for maintaining cytoplasmic ion concentrations under high-salt or low-pH conditions, whereas another is responsible for low-salt or high-pH conditions. Only through direct molecular and biochemical techniques is it possible to decipher how subtle differences in sequence determine functionality (Krulwich, 1983).

Nevertheless, genomic sequence data can be exploited to facilitate these direct approaches, as the genomic data can be used to create mutants in the interesting genes, with great efficiency in some cases as discussed below. One important point, in regard to the transporters, is the relationship between the number of paralogous genes in the genome and the biological niche of the organism. The cyanobacteria colonizing aquatic habitats that are more prone to fluctuations in ionic composition compared to those inhabiting marine environments may contain fewer ion homeostasis genes. For example, *Synechocystis* and the two *Nostocs* contain more $Na^+/H^+$ transporters than the *Synechococcus* or *Prochlorococcus* strains. It could be argued that because *Prochlorococcus* and *Synechococcus* are strictly marine and as marine conditions are largely invariant with respect to ionic composition, these species may not require as many ion transporters for maintaining ion homeostasis. This type of situation may, in turn, account for the apparently small genome size in the *Prochlorococcus* and *Synechococcus* strains.

The goal of one current project taken up by the authors of this work is to develop techniques to investigate the role of any or all genes of the genome of *Synechocystis* PCC6803 as they pertain to salt stress. We have begun by identifying targets of interest in two manners. The sequences of known salt-responsive genes in higher organisms such as *Saccharomyces* and *Arabidopsis* were used in a Blast search against the entire genome of *Synechocystis*. A list of homologs or potential targets was generated from the results of the Blast search. In addition, sub-and full-genome microarrays for differential gene expression were developed as described below. These arrays were used to screen total RNA from cultures before, and at various time points after, salt shock for differential expression. The resulting differentially expressed genes are also added to the list of potential target sequences, and those unaffected by salt shock were either removed from the list or put on a waiting list. The resulting list of genes was targeted for knockout mutagenesis using a PCR-mediated method developed to eliminate cloning steps. The resulting strains were then subjected to a battery of tests in a microtitre dish format to investigate their relative physiological fitness compared to wild-type strains. The information gathered in this manner is used to develop a database of the expression and physiological characteristics of each gene listed.

## II. DNA MICROARRAYS

With the growing availability of complete genomic sequences for scores of prokaryotic organisms, efforts have turned to the development of experimental approaches to measure transcription from a global perspective. DNA microarrays have emerged as a particularly effective tool for genome-wide transcript profiling, especially for studies examining eukaryotic organisms (DeRisi et al 1996; DeRisi et al 1997; Richmond et al 1999; Schena et al 1996). Information on the temporal patterns of accumulation and disappearance of transcripts for specific groups of genes is suggestive of cellular programs orchestrating these changes. A combination of knock-out mutagenesis of known or suspected regulatory loci coupled with microarray analysis should help elucidate the signal circuitry giving rise to programmatic changes in gene expression. Unfortunately, reports on the successful use of DNA microarrays for transcriptional profiles in prokaryotes have been sparse. As discussed below, this lag in the development of microarray technology for prokaryotes can be largely traced to technical issues associated with generating the high signal-to-noise ratio probes from non-poly (A) RNA (Hihara et al 2001). However, workable techniques are now available, as demonstrated by the recent report from the Blattner lab as well as experience in other labs (R. Burnap et al; A. Grossman et al, unpublished observations) (Richmond et al 1999).

DNA microarray technology uses microscopic, high-density arrays of DNA target elements immobilized to solid surfaces such as microscope slides. The DNA target elements typically represent specific gene sequences or subsequences that hybridize to cognate sequences in samples under investigation. There are currently two major types of DNA microarrays: oligonucleotide microarrays and DNA fragment microarrays. Oligonucleotide microarrays generally utilize in situ oligonucleotide synthesis techniques that directly build individual DNA targets on the surface of the array. Such an approach is epitomized by the proprietary manufacturing methods of the Affymetrix Corporation, a successful but extremely expensive implementation. Alternatively, oligos on the average of 60 to 70 bases in length can be printed onto the surface in the same manner as printing larger PCR products or clones. This second type of array, the DNA fragment microarray, was initially pioneered for *Saccharomyces* by Brown at Stanford University and for *Escherichia coli* by Blattner at the University of Wisconsin. These arrays consist of PCR-amplified fragments either from Expressed Sequence Tagged (EST) project libraries or from sequencing information. This latter approach is more broadly accessible to academic and commercial labs and has been used to produce full genomic and subgenomic microarrays for cyanobacteria.

Once constructed, microarrays are interrogated with fluorescently labeled DNA or RNA mixtures from biological sources, washed to remove background fluorescence, and scanned using laser confocal microscopy. A key advantage of the glass slide microarrays, compared with membrane-based, high-density arrays hybridized with radioactive probe, is the ability to hybridize more than one fluorescently labeled probe to the same microarray. Typically, this capability is used to compare control and treated (e.g., uninduced and induced) samples. This feature allows a direct qualitative comparison of hybridization intensities on the same DNA, thus controlling for potential disparities in a given immobilized DNA spot between two replica microarrays. The magnitude of the fluorescence emission intensities from the control and treated samples hybridized to a given DNA spot is then used to compute the corresponding hybridization signal ratio, theoretically reflecting the relative abundance of the mRNA species in each sample. Obviously, this raises questions about how well the cDNA product of the labeling procedure reflects the population of mRNA that it is meant to represent and to what extent the labeling reaction generates nonspecific background signal, which diminishes the sensitivity of the procedure. For reasons discussed below, these issues are especially acute for prokaryotic systems. What has probably been the greatest detractor in publication of microarray data is its reproducibility. Microarray hybridizations are very sensitive and are susceptible to even the slightest differences in protocol alterations. Most obviously, RNA purity can have a major impact on the final output. The production and hybridization of two target cDNA pools requires many steps at which error can be introduced. Automated machines and molecular biology kits developed for just such a purpose have recently been developed and have begun to hit the market, decreasing user interface and increasing reproducibility. The greatest impact on results might be the experiment itself. The extent to which apparently subtle differences in cell culture result in significant variations in global transcriptional patterns remains to be evaluated. Preliminary indications suggest these variations may be quite pronounced (Prof. H. Pakrasi, personal communication). Therefore, the preliminary data presented here should be considered tentative, and accordingly, treated with caution.

Improved labeling, hybridization, and normalization techniques could solve most of the current problems with analysis of prokaryotic gene expression with DNA microarrays. One of the most difficult challenges relates to the lack of polyadenylated mRNA. In the case of prokaryotic microorganisms, typical labeling conditions for production of cDNA involve the use of random oligos. The final result is a cDNA pool, which is rich in ribosomal RNA compliment. The abundance of ribosomal RNA nonspecifically binds across the surface of the array. This prevents the analysis of many genes present at low concentrations in the pool, including important regulatory proteins, which may or may not respond to a change in conditions. New techniques are being developed to circumvent such problems. One such attempt uses arbitrary primers, or a small nonrandom pool of specific primers capable of preferential binding to only mRNA sequences during cDNA production, thus eliminating the production of labeled ribosomal cDNA (Talaat et al 2000). Unfortunately, without a large pool of oligos, it is difficult at best to determine the amount necessary to gain an excess coverage of all transcripts and still not bind to rRNA. Alternatively, all oligos annealing to the 3′ end of the open reading frame used for the original amplification of fragments for the array could be pooled and used in the reverse transcription reaction. However, this is not possible if oligonucleotide arrays were produced.

## A. Cyanobacterial Microarrays

The first cyanobacterial microarrays, "Cyanochips," were produced for *Synechocystis* sp PCC6803 in Japan by a commercial arrangement.The microarrays were used by Ikeuchi et al (31) to follow global patterns of gene expression as a function of time in response to changing light regimens, and by the Murata group to investigate cold-responsive genes. The latter study included the parallel comparative analysis of wild-type and mutant cells that have a deletion in a histidine kinase gene previously shown to control the expression of cold-responsive lipid desaturase genes. The Ikeuchi work provides an evaluation of the sources and level of errors in their experiments, and they were able to generate data on a large fraction of genes. Their results agree with other published works for the most part, and have validated the use of microarrays for gene expression in cyanobacteria. Subgenomic DNA microarrays for *Synechocystis* sp PCC6803 have been used to explore salt stress and provide some interesting parallels to the studies of the Ikeuchi and Murata groups. During the early stages of the response, the photosynthetic complexes are repressed, followed shortly after acclimation with an accumulation to near prestress levels. During this acclimation phase, many events happen. The membrane becomes more rigid because of the expression and activity of lipid desaturase genes. Many chaperonins and dnaK proteins are also up-regulated. The production of compatible osmolytes is induced. This is not measurable using microarrays; however, as this is a posttranslational regulation event. Eventually the cellular ionic homeostatic environment is secured and under the protection of increased levels of reactive oxygen species (ROS) protecting enzymes such as superoxide dismutase and glutathione peroxidase, and with the help of many chaperonins, the photosynthetic apparatus is reintroduced to the thylakoid membrane and cellular proliferation resumes. Many of these events can be monitored through gene expression. Table 3 demonstrates the temporal expression of several representative genes after 1 and 5 hours exposure to salt stress relative to prestress expression levels. After exposure for 1 hour to salt shock, a large fraction of the induced genes belong to regulatory complexes. Four hours later, the effect of increased expression of regulatory proteins on the pool of RNA is seen as several transcripts encoding structural proteins are increased in number. Table 4 includes a larger list of genes whose expression is affected, either induced or repressed, after 5 hours acclimation to 684 mM salt. This list was developed from the analysis of several hybridizations between subgenomic arrays and (2) labeled total cDNA pools (control and experimental samples). In concert with the more well-annotated genes is the regulation of a number of uncharacterized open reading frames currently annotated only as "hypothetical". Table 5 lists the 10 most highly regulated hypothetical genes from the same subgenomic array experiments. The inclusion of hypothetical genes in our subgenome array arose only through their identification as homologs to salt-regulated genes from either *Saccharomyces* or *Arabidopsis*. This result confirms the method for selection criteria of subgenomic array constituents. Similar responses in the light adaptation experiment cannot be confirmed, however, as the results of all hypothetical genes were not reported.

Full genome arrays are being developed as part of a National Science Foundation project in the following manner. Each gene or sequence was used to determine primers that could PCR amplify products representing the entire open reading frame, but with a cut-off of 2000 bases in length, such that the longer genes have products that are truncated at 2000 bp. The amplification process has been diagrammed in Figure 3. Oligonucleotides were produced in individual tubes and transferred in cognate pairs of primer to 96-well format, allowing for high-throughput amplification and analysis. In addition to the gene-specific sequences, each oligo was designed to include universal ends, which allows second round amplifications using the first round products as template and universal oligos for priming. This second round amplification proved critical in maximizing yield

**Table 3** Representative Genes Induced After One and Five Hours Salt Shock from 2mM to 684mM NaCl in *Synechocystis* PCC6803 Determined Through Analysis Subgenomic DNA Microarrays

| Genes induced after 1-hour salt stress | | Genes induced after 5-hour salt stress | |
|---|---|---|---|
| Orf | Annotated function | Orf | Annotated function |
| slr0210 | Sensory transduction histidine kinase | slr2075 | 10kD chaperonin (groES) |
| sll1590 | Sensory transduction histidine kinase | ssl1633 | CAB/ELIP/HLIP superfamily |
| slr0395 | Transcriptional activator protein NtcB (ntcB) | sll1441 | Delta 15 desaturase (desB) |
| slr1324 | Hybrid sensory kinase | slr1171 | Glutathione peroxidase |
| slr0533 | Sensory transduction histidine kinase | sll0170 | DnaK protein (dnaK) |
| slr1871 | Transcriptional regulator | ssr1169 | Salt-stress induced hydrophobic peptide (ESI3) |
| sll0248 | Flavodoxin (isiB) | sll0430 | Heat shock protein (htpG) |
| slr1774 | Hypothetical protein | ssr2595 | High light inducible protein |
| ssl1633 | CAB/ELIP/HLIP superfamily | ssl0452 | Phycobilisome degradation protein Nb1A (nb1A) |

Four hundred eighty gene arrays were produced from PCR products representing 300 to 2000 base fragments from selected genes from the *Synechocystis* PCC6803 genome. Each product was verified on an agarose gel, purified, and printed on Clontech Type I slides in 3X SSC with a Cartesian Pixsys 6000 microarrayer. These slides were subject to UV crosslinking with 300 mJ of energy and subsequent hybridization to cy3 labeled cDNA from control or time 0 hr. and cy5 labeled cDNA from five-hour treatment of 684 mM NaCl shocked cells. cDNA pools synthesized from 20 Ug total RNA were labeled using 10 Ug random hexamers and Superscript II from BRL according to the manufacturer's specifications. Incorporation of the appropriate Cyanine dye (Amersham Pharmacia) was achieved through substitution with a 3:2 ratio of dTTP to cydUTP. The resulting products were purified on Microcon 30 concentrators (Amicon), then combined, mixed with 0.5 Ug denatured salmon sperm DNA and an equal volume of Sigma Perfect Hyb solution (Sigma Corp.). The hybridization was performed under a polypropylene coverslip at 65° C for 16–18 hrs in a 40 l volume. The array was washed once in 0.2 X SSC and 0.2% SDS, twice in 0.2 X SSC and once in 0.1X SSC. The slides were dried and scanned using a Scanarray 3000 from GSI Lumonics. The average signal ratio from each gene listed above is significantly different by two standard deviations from control spots whose in vitro transcribed RNA products were spiked in the experimental and control samples at the same concentration.

**Table 4** *Synechocystis* Genes Regulated by Salt Stress After 5 hours of Treatment Revealed Through Microarray Differential Expression Profiling

| Genes induced by 5 hours salt stress | | Genes repressed by 5 hours salt stress | |
|---|---|---|---|
| orf | gene designation | orf | gene designation |
| slr2075 | groES | slr0640 | sensory transduction histidine kinase |
| sll1514 | hspA | slr2076 | 60kD chaperonin 1 |
| sll0330 | 3-oxoacyl-[acyl-carrier protein] reductase | slr0074 | ABC transporter |
| ssl1633 | CAB/ELIP/HLIP superfamily | slr1198 | antioxidant protein |
| slr0797 | cadA | slr1856 | anti-sigma B factor antagonist |
| sll0912 | ABC transporter ATP binding protein | slr2067 | allophycocyanin a |
| sll1927 | ABC transporter ATP-binding protein | slr1986 | apcB |
| sll0698 | dfr | sll0928 | apcD |
| slr0449 | DNR protein | slr0335 | apcE |
| sll1383 | extragenic suppresso | slr1459 | apcF |
| sll0567 | fur | sll1317 | apocytochrome f |
| sll1085 | glpD or glyD | sll1874 | AT103 |
| slr1390 | ftsH | sll1326 | ATP synthase |
| slr1604 | ftsH | slr1329 | ATP synthase |
| sll1633 | ftsZ | sll1325 | ATP synthase |
| slr0008 | ctpA | slr1330 | ATP synthase |
| slr1751 | carboxyl-terminal protease | sll1327 | ATP synthase |
| slr0756 | kaiA | sll1323 | ATP synthase |
| slr0757 | kaiB1 | sll1324 | ATP synthase |
| sll0030 | cmpR | sll1028 | carbon dioxide concen. |
| slr1509 | Na+ ATPase subunit J | sll1029 | carbon dioxide concen. |
| slr0415 | Na+/H+ antiporter | sll1030 | carbon dioxide concen. |
| slr1665 | dapF | sll1031 | carbon dioxide concen. |
| sll1018 | pyrC | slr1860 | icfG |
| sll0766 | radC | sll1579 | cpcC |
| slr1041 | patA | sll1580 | cpcC |
| slr1594 | patA | slr2051 | cpcG |
| slr1392 | feoB | slr1379 | cydA or cyd |
| slr0298 | FraH protein homolog | sll0258 | cytochrome c550 |
| sll1423 | ntcA, ycf28 | sll1796 | cytochrome c553 |
| sll0779 | PleD gene product | slr2073 | hypothetical protein YCF50 |
| slr0302 | pleD | sll1479 | devB |
| slr1657 | pleD | slr0150 | ferredoxin (petF) |
| sll1592 | regulatory component | ssl0020 | ferredoxin (petF) |
| slr0531 | ggtD | sll1009 | frpC |
| slr0529 | ggtB | sll1342 | gap2 |
| sll1566 | ggpS | sll1626 | LexA repressor |

**Table 4** Continued

| Genes induced by 5 hours salt stress | | Genes repressed by 5 hours salt stress | |
|---|---|---|---|
| orf | gene designation | orf | gene designation |
| sll0306 | sigB | slr0750 | chlN |
| sll2012 | sigD | sll0247 | iron stress |
| slr1974 | GTP binding protein | sll0248 | isiB |
| slr1336 | H+/Ca2+ exchanger | sll0689 | Na+/H+ antiporter |
| slr1234 | protein kinase C inhibitor | slr1595 | Na+/H+ antiporter |
| sll1875 | ho2 | sll1818 | rpoA |
| ssl2542 | hliA | smr0003 | petM |
| slr1675 | hypA | sll1525 | phosphoribulokinase |
| sll1845 | hypothetical protein | sll1194 | photosystem II 12 kD |
| slr0208 | hypothetical protein | slr1181 | photosystem II D1 |
| sll0837 | hypothetical protein | sll0427 | photosystem II manganese |
| ssr3122 | hypothetical protein | slr0906 | photosystem II P680 |
| slr0210 | sensory transduction | sll1578 | phycocyanin a subunit |
| slr0311 | sensory transduction | sll1577 | phycocyanin b subunit |
| slr0533 | sensory transduction | sll1533 | pilT |
| slr0974 | initiation factor IF-3 | ssl0563 | psaC |
| sll0556 | nhaS6 | slr0737 | psaD |
| slr1414 | sensory transduction | sll0819 | psaF |
| sll1852 | nucleoside diphosphate kinase | smr0004 | psaI |
| slr1805 | sensory transduction | slr1808 | hemA |
| sll1221 | hoxF | slr1655 | psaL |
| ssr1386 | ndhL | ssr3451 | psbE |
| sll1226 | hoxH | slr1239 | pyridine nucleotide |
| sll2014 | sfsA | sll1292 | regulatory component |
| slr0095 | O-methyltransferase | slr0194 | ribose 5 phosphate I |
| slr1213 | umQ gene product | slr0009 | ribulose bisphosphat |
| slr1181 | psbA1 | slr0012 | ribulose bisphosphat |
| ssr1425 | ycf34 | ssr1768 | unknown protein |
| slr0353 | unknown protein | slr2057 | water channel protein |
| slr0377 | unknown protein | sll1057 | trx |
| slr0392 | unknown protein | slr1139 | trxA or trx |
| slr0581 | unknown protein | slr0623 | trxA |
| slr1899 | ureF | sll1784 | unknown protein |
| slr1963 | water-soluble carotenoid protein | sll1785 | unknown protein |
| sll1491 | WD-repeat protein | sll1281 | ycf9 |
| ssr2595 | high light inducible protein | sll0998 | ysR transcriptional |
| sll1624 | two-component response regulator | sll1746 | rpl12 |

The data presented here as a more complete list of salt regulated genes from our subgenomic array was derived from the same experiments described in Table 3.

**Table 5**   The Most Induced or Repressed Hypothetical Orfs from Transcriptional Profiles Using Subgenomic Arrays to Analyze Before and After 5-hour Exposures of *Synechocystis* PCC6803 to High Salt Shock.

| Hypothetical genes most induced by salt | Hypothetical genes most repressed by salt |
| --- | --- |
| slr1403 | sll1663 |
| sll0623 | sll1945 |
| slr1557 | slr0442 |
| sll1387 | sll0242 |
| sll1373 | sll1686 |

The data for this list was derived from the same experiments described in Table 3.

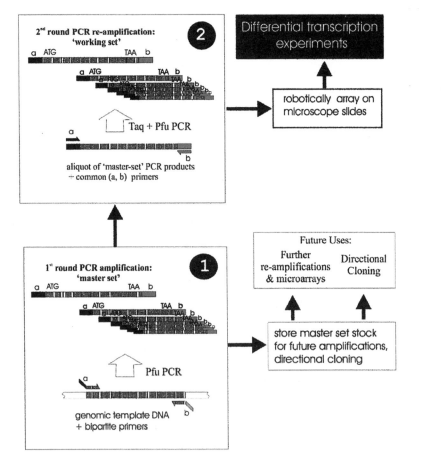

**Figure 3**   PCR amplification strategy used to develop full genome arrays.

without sacrificing accuracy. In addition, the universal portions of the oligos were designed to introduce restriction sites so that successfully amplified products could be digested with an endonuclease to directionally clone each of the products. Using universal oligos and template from first round amplifications produced elements of relatively equal and robust yield and had an unexpected leveling effect on the yield of the approximately 3200 genes being amplified. That is, the yields of products from the first round of amplification using the gene-specific sequences and chromosomal DNA template was comparatively much more heterogeneous than the yields on the second round of amplification using the universal primers. In the extreme case, some first round primers yielded no detectable product as judged by ethidium-stained genes, whereas as-good yield was obtained using the same poor-yielding first round product as template for the the second round reactions primed with the universal primers. Furthermore the yield of such reactions was comparable to the second round reactions based on first round template that had been produced in abundance. Amplified products were purified in 96-well format, then combined and transferred to 384-well plates under the appropriate conditions for robotic deposition on modified glass slides.

Subgenomic arrays are identical in concept to full-genomic arrays with one exception. They only contain a small number of genes. This makes it easier to analyze and coordinate the production of arrays on smaller scales, thus making them less expensive and more reliable. In addition, arrays can be printed with more replicates in the same space, allowing for better statistical averaging and more reliable data. Our subgenome arrays contain 480 *Synechocystis* genes from the networks of photosynthesis and respiration complexes, known stress-responsive genes, regulatory genes, and homologues to salt-induced genes from *Arabidopsis* and *Saccharomyces*.

As discussed previously, a determination of differential transcriptional profiles associated with abiotic stresses is expected to be a key component of an overall understanding of the adaptive mechanisms of cyanobacteria challenged with environmental stress. We are focusing on salt stress, paying close attention to the function, regulation, and protection of the photosynthetic apparatus. Already, cyanobacterial microarrays are being used by several laboratories to analyze global patterns of transcript abundance during time course experiments and are being used to monitor their adaptation to various abiotic stresses (Hihara et al 2001; Suzuki et al 2001). These studies are likely to be followed by comparative analyses of mutants with alterations in their stress response. Additionally, the microarrays have already begun to contribute to an understanding of the biogenesis and regulation of the photosynthetic mechanism. In accord with our project on the structure and assembly of the photosystem II (PSII) complex in cyanobacteria, we also envision using the microarrays to help dissect the processes governing the biogenesis and repair of the easily damaged and constantly turning-over PSII complex.

The results from our subgenomic arrays have already presented us with valuable information regarding the response of many cyanobacterial genes to salt stress. Table 3 lists genes whose expression was significantly regulated by the induction of a salt-shock response. In this example, total RNA from time zero and 3 hours after the addition of 684mM NaCl were labeled with fluorescent molecules cyanine 3(cy 3) and cyanine 5(cy 5), respectively, in a first-strand reverse transcriptase reaction. The resulting cDNA was hybridized to our subgenomic arrays overnight. The arrays were washed, dried, and scanned. Two false-colored images were overlaid and normalized using positive control spots whose transcripts were spiked into the labeling reaction at equal concentrations. Analysis of the resulting images produced results similar to previously described published works. Specifically, the repression of PSII genes early in the stress response is a protective measure against the production of reactive oxygen species. The induction of known salt-responsive genes such as *esi3* and *stpA* was observed. Interestingly, a few regulatory genes such as *sibG* and *patA* were also up-regulated. The results of these experiments has contributed to the targeted mutagenesis of several genes using a PCR-mediated procedure. In addition to those mentioned, mutants of several other up-regulated genes are now under close inspection (all Na+/H+ antiporters and several cation ATPases).

Other uses for DNA arrays besides measuring differential transcription include comparative genomics assays, including single nucleotide polymorphism analysis, and the development of arrays for clinical identification of strains (Cho and Tiedje 2001; Cummings and Relman 2000; Kane et al 2000; Malloff et al 2001; Salama et al 2000). As yet, however, no applications for cyanobacterial strains have been developed.

## B.  Mutational Analysis

Knock-out mutants have been used extensively by many researchers already and are a proven strategy. In the case of *Synechocystis* sp PCC 6803, gene knock-outs can be constructed by relying on this organism's natural competence and chromosomal transformation by integration of foreign DNA through homologous recombination. The vast majority of recombinants produced by this procedure occur by double crossover events in a manner at least superficially analogous to chromosomal transformation in *Saccharomyces*. Typical protocols require cloning of chromosomal DNA, followed by enzymatic insertion of an antibiotic-resistance cassette within the chromosomal fragment, transformation, and selection based on the antibiotic resistance provided by the cassette. The use of efficient procedures and cloning vectors specialized for the cloning of PCR products has greatly simplified the knock-out procedure, as does advanced knowledge of the restriction sites available for the disruption of the gene using an antibiotic-resistance cassette.

Alternatively, it is possible to use one of several PCR mediated knock-out strategies that require no cloning (Amberg et al 1995; Nikawa and Kawabata 1998). These strategies take advantage of double-crossover homologous recombination and the requirement for only short stretches ($> 200$ bp) of flanking homologous sequence on each end of an antibiotic cassette. Such strategies are much less time consuming and may be less expensive in the long run. In most cases, a final PCR product can be manufactured through fusion PCR reactions involving two flanking regions and a selectable marker gene. The resulting fragment is capable of transforming the wild-type strain and inserting the selectable marker directly into the targeted gene, either interrupting it or replacing it completely. Using the two most distal oligonucleotides designed for amplifying the two flanking regions, it is possible to screen mutants for complete loss of the wild-type genotype directly from cultured cells without purification of DNA (Howitt 1996). Coupling PCR-knockout strategies with whole-cell PCR screens to confirm gene deletion or disruption make these protocols very fast and efficient.

Having the capacity to clone each PCR product synthesized for array production, including virtually every identified gene in the genome, into one vector increases the capabilities of the study several-fold. Cloned products can be over-expressed in *E. coli*, and/or histidine-tagged to simplify purification. Site-directed mutagenesis is also a viable strategy taking advantage of the cloned products. Including this step in the aforementioned strategy increases the capabilities several-fold.

## III.  OUTLOOK FOR THE FUTURE

At the time of this writing, several additional cyanobacterial sequencing projects are underway, as well as the numerous other genome sequencing projects that will continue to be initiated and finished in the years to come. The subjects of the cyanobacterial sequencing projects include the genetically amenable marine strain *Synechococcus* sp PCC7002, the thermophilic species *S. vulcanus* (also referred to as *Thermosynechococcus elongatus*), the long-used *Synechoccus* sp PCC7942, and the evolutionarily interesting *Gloeobacter violaceous*. Gloeobacter remains largely unstudied, but from the little information available, it is clear that it exhibits unusual and ostensibly primitive cellular properties. This organism is interesting from the evolutionary perspective, as analysis of rRNA sequences place this organism near the root of the cyanobacterial phyletic radiation. Among other things, the features of the photosynthetic apparatus in this organism are most likely to hold clues as to the evolutionary transition from anoxygenic to oxygenic photosynthesis, and the availability of the full genome sequence will likely revolutionize this analysis. The analysis of the *Synechococcus vulcanus* genome also ought to provide interesting information because of its thermophilicity and because there are preliminary indications that it may hold some interesting gene clustering features. The genes for cytochrome *C*6 and cytochromes *C*550, heme proteins that serve distinct and nonoverlapping functions in photosynthesis, are not usually physically associated with each other in other cyanobacteria. However, not only are these genes found in tandem in the *S. vulcanus* genome, but an additional and apparently nonfunction cytochrome gene (pseudo-gene) is also found at this locus. This gene arrangement is highly suggestive of gene duplication and divergence events that started from unequal crossovers. The retention of the intact gene family cluster is intriguing because it raises the as yet unanswered question of whether additional ancient relics for other clustered paralogs are more common in this organism in comparison to those whose sequences have been finished. If this turns out to be the case, it will provide an unprecedented window on gene family evolution in the cyanobacteria. It is likely that as microarray analysis proceeds, the regulatory networks governing the coordinate expression of functional classes of proteins will be revealed. Given the large number of candidate regulatory proteins found in these organisms, this will remain an awesome task. In the case of *Synechocystis* sp PCC 6803, for example, there are more than 80 histidine kinase genes and some may participate in multiple regulatory pathways. This is evidenced by the finding that the histidine kinase designated HK33, functioning in the regulation of the lipid desaturation pathway, has also been found to govern the expression of high-light inducible proteins (HLIPs), which function as part of the high-light stress response mechanism. Consistent with this apparent multiple-function role, analysis of the HK33 protein has lead to the identification of multiple-sensing domains. Thus, the regulatory networks may prove to be highly ramified through the existence of multiple regulatory interactions.

This brings up the issue of how to go beyond the sequence data and evaluate the nature of protein-protein and protein-ligand interactions that underlie this network. This will require the analysis of the proteins through, perhaps, the proteomics approaches already used for other organisms. Since the advent of genomic sequence availability, several advancements have been made toward the development of technologies focusing on the expression and activity of proteins. Included in these are protein overexpression tools, improved purification methods using, for example, hexahistidine-tagging methods, 2-D gel electrophoresis, and MALDI-TOF characterization of protein complexes. The success of each of these technologies has required, or been enhanced by, the availability of sequence data. Through (matrix assisted laser desorption/ionization time of flight mass spectrometry), MALDI-TOF proteins, and fragments thereof, can be identified by comparing accurate experimental molecular masses with databases of deduced peptides, using computationally intensive algorithms. Although protein overexpression and 2-D gel electrophoresis have been available for some time, they can be technically difficult to master, so their use to this point has been limited, but rewarding. At least one group at the Kazusa DNA Research Institute has developed the techniques to use 2-D gels and some of their data are available on the Cyanobase website. Unfortunately, this single case is the only true example of proteomics research in a cyanobacterium that has been published to date. Expressing a protein modified to include a six amino acid linker (six histidines) allows for chromatographic purification along with any other protein(s) that might make up a complex. This practice is referred to as His-Tagging. Coupling two or more of these technologies, such as 2-D gel electrophoresis or six histidine tagging, with MALDI-TOF mass spectrometry sequencing has just begun to redefine the field. The ability to purify protein complexes and identify their constituents is a very powerful tool. Within this decade, it is likely that these technologies will move to the forefront of molecular biology as they become more widely understood, accepted, and used by cyanobacteriologists and plant scientists exploiting the cyanobacterial models and databases.

## REFERENCES

Allakhverdiev, S.I., Nishiyama, Y., Suzuki, I., Tasaka, Y., Murata, N. (1999). Genetic engineering of the unsaturation of fatty acids in membrane lipids alters the tolerance of Synechocystis to salt stress. Proc Natl Acad Sci USA 96, 5862–5867.

Altschul, S.F., Madden, T.L., Schaffer, A.A., Zhang, J., Zhang, Z., Miller, W., Lipman, D.J. (1997). Gapped BLAST and PSI-BLAST: A new generation of protein database search programs. Nucleic Acids Res 25, 3389-402.

Amberg, D.C., Botstein, D., Beasley, E.M. (1995). Precise gene disruption in Saccharomyces cerevisiae by double fusion polymerase chain reaction. Yeast 11, 1275–1280.

Bhaya, D., Schwarz, R., Grossman, A.R. (2000). Molecular Responses to Environmental Stress. In: The Ecology of Cyanobacteria. B.A. Whitton, M. Potts, eds. Netherlands: Kluwer Academic Publishers, pp 397–442.

Brahamsha, B. (1996). An abundant cell-surface polypeptide is required for swimming by the nonflagellated marine cyanobacterium Synechococcus. Proc Natl Acad Sci U S A 93, 6504–6509.

Brahamsha, B. (1999). Non-flagellar swimming in marine Synechococcus. J Mol Microbiol Biotechnol 1, 59–62.

Breton, J., Nabedryk, E., Leibl, W. (1999). FTIR study of the primary electron donor of photosystem I (P700) revealing delocalization of the charge in P700(+) and localization of the triplet character in (3)P700. Biochemistry 38, 11585–11592.

Chen, Y.B., Dominic, B., Zani, S., Mellon, M.T., Zehr, J.P. (1999). Expression of photosynthesis genes in relation to nitrogen fixation in the diazotrophic filamentous nonheterocystous cyanobacterium Trichodesmium sp. IMS 101. Plant Mol Biol 41, 89–104.

Chisholm, S.W., Olson, R.J., Zettler, E.R., Goericke, R., Waterbury, J.B., Welschmeyer, N.A. (1988). Chlorophyll b-containing marine picoplankton. Nature 334, 340–343.

Cho, J.C., Tiedje, J.M. (2001). Bacterial species determination from DNA-DNA hybridization by using genome fragments and DNA microarray. Appl Environ Microbiol 67, 3677–3682

Clark, M.S. (2000). Comparative genomics: An introduction: Sequencing projects and model organisms. In: Comparative Genomics. M.S. Clark, ed. Netherlands: Kluwer Academic Publishers, pp. 1–22.

Cummings, C.A., Relman, D.A. (2000). Using DNA microarrays to study host-microbe interactions. Emerg Infect Dis 6, 513–525.

De la Cerda, B., Navarro, J.A., Hervas, M., De la Rosa, M.A. (1997). Changes in the reaction mechanism of electron transfer from plastocyanin to photosystem I in the cyanobacterium Synechocystis sp. PCC 6803 as induced by site directed mutagenesis of the copper protein: Biochemistry 36, 10125–10130.

Debus, R.J., Campbell, K.A., Gregor, W., Li, Z.L., Burnap, R.L., Britt, R.D. (2001). Does histidine 332 of the D1 polypeptide ligate the manganese cluster in photosystem II? An electron spin echo envelope modulation study. Biochemistry 40, 3690–3699.

DeRisi, J., Penland, L., Brown, P.O., Bittner, M.L., Meltzer, P.S., Ray, M., Chen, Y., Su, Y.A., Trent, J.M. (1996). Use of a cDNA microarray to analyse gene expression patterns in human cancer. Nat Genet 14, 457–460.

DeRisi, J.L., Iyer, V.R., Brown, P.O. (1997). Exploring the metabolic and genetic control of gene expression on a genomic scale. Science 278, 680–686.

Diner, B.A., Force, D.A., Randall, D.W., Britt, R.D. (1998). Hydrogen bonding, solvent exchange, and coupled proton and electron transfer in the oxidation and reduction of redox-active tyrosine Y(Z) in Mn-depleted core complexes of photosystem II. Biochemistry 37, 17931–17943.

Fulda, S., Hagemann, M. (1995). Salt treatment induces accumulation of flavodoxin in the cyanobacterium Synechocystis sp PCC 6803. J Plant Physiol 146, 520–526.

Fulda, S., Hagemann, M., Libbert, E. (1990). Release of glucosylglycerol from the cyanobacterium Synechocystis spec. SAG 92.79 by hypoosmotic shock. Arch Microbiol 153, 405–408.

Gantt, E. (1994). Supramolecular membrane organization. In: The Molecular Biology of Cyanobateria. D. A. Bryant, ed. Netherlands: Kluwer Academic Publishers, pp. 119–138.

Golden, S.S. (1988). Mutagenesis of cyanobacteria by classical and gene transfer-based methods. Methods in Enzymology 167, 714–727.

Hagemann, M., Richter, S., Mikkat, S. (1997a). The ggtA gene encodes a subunit of the transport system for the osmoprotective compound glucosylglycerol in Synechocystis sp. strain PCC 6803. J Bacteriol 179, 714–720.

Hagemann, M., Richter, S., Zuther, E., Schoor, A. (1996). Characterization of a glucosylglycerol phosphate accumulating, salt sensitive mutant of the cyanobacterium Synechocystis sp. strain PCC 6803. Arch Microbiol 166, 83–91.

Hagemann, M., Schoor, A., Jeanjean, R., Zuther, E., and Joset, F. (1997b). The stpA gene form Synechocystis sp. strain PCC 6803 encodes the glucosylglycerol phosphate phosphatase involved in cyanobacterial osmotic response to salt shock. J Bacteriol 179, 1727–1733.

Hagemann, M., Woelfel, L., Krueger, B. (1990). Alterations of protein synthesis in the cyanobacterium Synechocystis sp. PCC 6803 after a salt shock. J Gen Microbiol 136, 1393–1399.

Hienerwadel, R., Boussac, A., Breton, J., Diner, B.A., Berthomieu, C. (1997). Fourier transform infrared difference spectroscopy of photosystem II tyrosine D using site-directed mutagenesis and specific isotope labeling. Biochemistry 36, 14712–14723.

Hihara, Y., Kamei, A., Kanehisa, M., Kaplan, A., Ikeuchi, M. (2001). DNA microarray analysis of cyanobacterial gene expression during acclimation to high light. Plant Cell 13, 793–806.

Howitt, C.A. (1996). Amplification of DNA from whole cells of cyanobacteria using PCR. Biotechniques 21, 32–34.

Jansson, C., Debus, R.J., Osiewacz, H.D., Gurevitz, M., Mcintosh, L. (1987). Construction of an obligate photoheterotrophic mutant of the cyanobacterium synechocystis 6803 inactivation of the psb-a gene family. Plant Physiol (Bethesda) 85, 1021–1025.

Jeanjean, R., Matthijs, H., Onana, B., Havaux, M., Joset, F. (1993). Exposure of the cyano-bacterium *Synechocystis* PCC6803 to salt stress induces concerted changes in respiration and photosynthesis. Plant Cell Physiol 34, 1073–1079.

Jeanjean, R., Onana, B., Peschek, G., Joset, F. (1990). Mutants of the cyanobacterium Synechocystis PCC6803 impaired in respiration and unable to tolerate high salt concentrations. FEMS Microbiol Lett 68, 125–130.

Kane, M. D., Jatkoe, T. A., Stumpf, C. R., Lu, J., Thomas, J. D., and Madore, S. J. (2000). Assessment of the sensitivity and specificity of oligonucleotide (50mer) microar-rays. Nucleic Acids Res 28, 4552–4557.

Kaneko, T., Sato, S., Kotani, H., Tanaka, A., Asamizu, E., Nakmura, Y., Miyajima, N., Hirosawa, M., Sugiura, M., Sasamoto, S., et al. (1996). Sequence analysis of the genome of the unicellular cyanobacterium *Synechocystis* sp. strain 6803. II. Se-quence determination of the entire genome and assignment of potential protein-coding regions. DNA Res 3, 109–136.

Kaneko, T., Tabata, S. (1997). Complete genome structure of the unicellular cyanobacter-ium *Synechocystis* sp. PCC 6803. Plant Cell Physiol 38, 1171–1176.

Kaplan, A., Schwarz, R., Lieman-Hurwitz, J., Ronen-Tarazi, M., Reinhold, L. (1994). Physiological and molecular studies on the response of cyanobacteria to changes in the ambient inorganic carbon concentration. In: The Molecular Biology of Cya-nobacteria. D. Bryant, ed. Netherland: Kluwer Academic Publishers, pp. 469–485.

Krulwich, T.A. (1983). Na$^+$/H$^+$ antiporters, Biochim Biophys Acta 726, 245–264.

Labarre, J., Chauvat, F., Thuriaux, P. (1989). Insertional mutagenesis by random cloning of antibiotic resistance genes into the genome of the cyanobacterium Synechocystis strain PCC 6803. J Bacteriol 171, 3449–3457.

Malloff, C.A., Fernandez, R.C., Lam, W.L. (2001). Bacterial comparative genomic hybrid-ization: A method for directly identifying lateral gene transfer. J Mol Biol 312, 1–5.

Marraccini, P., Bulteau, S., Cassier Chauvat, C., Mermet Bouvier, P., Chauvat, F. (1993). A conjugative plasmid vector for promoter analysis in several cyanobacteria of the genera Synechococcus and Synechocystis. Plant Mol Biol 23, 905–909.

Mikkat, S., Hagemann, M., Schoor, A. (1996). Active transport of glucosylglycerol is in-volved in salt adaptation of the cyanobacterium Synechocystis sp. strain PCC 6803. Microbiology 142, 1725–1732.

Mizuno, T., Kaneko, T., Tabata, S. (1996). Compilation of all genes encoding bacterial two component signal transducers in the genome of the cyanobacterium, Synechocystis sp. strain PCC 6803. DNA Res 3, 407–414.

Nikawa, J., Kawabata, M. (1998). PCR and ligation mediated synthesis of marker cassettes with long flanking homology regions for gene disruption in Saccharomyces cerevisiae. Nucleic Acids Res 26, 860–861.

Noguchi, T., Inoue, Y., Tang, X.S. (1999). Hydrogen bonding interaction between the primary quinone acceptor QA and a histidine side chain in photosystem II as revealed by Fourier transform infrared spectroscopy. Biochemistry 38, 399–403.

Partensky, F., Hess, W. R., Vaulot, D. (1999). Prochlorococcus, a marine photosynthetic prokaryote of global significance. Microbiol Mol Biol Rev 63, 106–127.

Poncelet, M., Cassier-Chauvat, C., Leschelle, X., Bottin, H., Chauvat, F. (1998). Targeted deletion and mutational analysis of the essential (2Fe-2S) plant-like ferredoxin in Synechocystis PCC6803 by plasmid shuffling [In Process Citation]. Mol Microbiol 28, 813–821.

Potts, M. (2000). Nostoc. In: The Ecology of Cyanobacteria. B. A. Whitton, M. Potts, eds. Netherlands, Kluwer Academic Publishers, pp. 465–504.

Richmond, C.S., Glasner, J.D., Mau, R., Jin, H., Blattner, F.R. (1999). Genome-wide expression profiling in Escherichia coli K-12. Nucleic Acids Res 27, 3821–3835.

Rocap, G. (2001). Personal Communication.

Salama, N., Guillemin, K., McDaniel, T.K., Sherlock, G., Tompkins, L., Falkow, S. (2000). A whole-genome microarray reveals genetic diversity among Helicobacter pylori strains. Proc Natl Acad Sci USA 97, 14668–14673.

Schena, M., Shalon, D., Heller, R., Chai, A., Brown, P.O., Davis, R.W. (1996). Parallel human genome analysis: Microarray-based expression monitoring of 1000 genes. Proc Natl Acad Sci USA 93, 10614–10619.

Schopf, J. (1993). Microfossils of the Early Archean Apex Chert: New evidence of the antiquity of life. Science, 640–646.

Shintani, D., DellaPenna, D. (1998). Elevating the vitamin E content of plants through metabolic engineering. Science 282, 2098–100.

Suzuki, I., Kanesaki, Y., Mikami, K., Kanehisa, M., Murata, N. (2001). Cold-regulated genes under control of the cold sensor Hik33 in Synechocystis. Mol Microbiol 40, 235–244.

Talaat, A.M., Hunter, P., and Johnston, S.A. (2000). Genome-directed primers for selective labeling of bacterial transcripts for DNA microarray analysis. Nat Biotechnol 18, 679–682.

Throup, J.P., Koretke, K.K., Bryant, A.P., Ingraham, K.A., Chalker, A.F., Ge, Y., Marra, A., Wallis, N.G., Brown, J.R., Holmes, D.J., et al. (2000). A genomic analysis of two-component signal transduction in Streptococcus pneumoniae, Mol Microbiol 35, 566–576.

Tomitani, A., Okada, K., Miyashita, H., Matthijs, H.C., Ohno, T., and Tanaka, A. (1999). Chlorophyll b and phycobilins in the common ancestor of cyanobacteria and chloroplasts. Nature 400, 159–162.

Urbach, E., Scanlan, D.J., Distel, D.L., Waterbury, J.B., Chisholm, S.W. (1998). Rapid diversification of marine picophytoplankton with dissimilar light-harvesting structures inferred from sequences of Prochlorococcus and Synechococcus (Cyanobacteria). Mol Evol 46, 188-201.

Vermaas, W.F.J., Williams, J.G.K., Rutherford, A.W., Mathis, P., Arntzen, C.J. (1986). Genetically engineered mutant of the cyanobacterium synechocystis 6803 lacks the photosystem ii chlorophyll-binding protein cp-47. Proc Natl Acad Sci USA 83, 9474–9477.

Waterbury, J.B., Watson, S.W., Valois, F.W., Franks D.G. (1986). Biological and ecological characterization of the marine unicellular cyanobacterium Synechococcus. Can Bull Fish Aquat Sci 214, 71–120.

Whitton, B., Potts, M. (2000). Introduction to the cyanobacteria. In: The Ecology of Cyanobacteria. P. Whitton, ed. Netherlands, Kluwer Academic Publishers, pp. 1–11.

Williams, J.G.K. (1988). Construction of specific mutations in Photosystem II photosynthetic reaction center by genetic engineering methods in Synechocystis 6803. Methods Enzymol 167, 766–778.

# 7

# Genomic Analysis of *Arabidopsis* Gene Expression in Response to a Systemic Fungicide

**Huey-wen Chuang, Tzung-Fu Hsieh, Manuel Duval, and Terry L. Thomas**
*Texas A&M University, College Station, Texas, U.S.A.*

## I.  INTRODUCTION

Fosetyl (aluminum ethyl phosphate) is a systemic fungicide used widely to control plant Oomycetes in the field. Typically, fungicides act by inhibiting fungus-specific biochemical pathways that are absent or less sensitive to the applied compound in treated plants (1). Although fosetyl is a common and effective fungicide in the field, the mechanism for its antifungal activity is still not fully elucidated. It is suggested that phosphonic acid, the in planta metabolite of fosetyl, is the active form of the compound and that it acts by interrupting phospholipid metabolism in the fungus (1). However, fosetyl has low antifungal activity in vitro; thus, an indirect mode of action has also been proposed. It has been further suggested that this compound acts indirectly against pathogens by stimulating host defense mechanisms, such as the hypersensitive response (HR) and the accumulation of phytoalexins (2, 3).

The possibility that plant natural defense response may play a role in disease control by chemicals has been proposed previously (4). By using mutants defective in different signal transduction pathways, Molina et al (5) demonstrated that fosetyl efficacy against pathogen was reduced in a systemic acquired resistance (SAR) compromised mutant blocked in salicylic acid (SA) accumulation. Systemic acquired resistance is an inducible resistance developed in plants after pathogen attack and is accompanied by the systemic expression of a group of genes called SAR (6). Salicylic acid is the apparent natural signaling molecule for the SAR signal transduction pathway (7, 8). Exogenous SA induces SAR and SAR gene expression (6). In addition to SA, the synthetic chemicals 2,6-dichloroisonicotinic acid (INA) and benzo(1,2,3)thiadiazole-7-carbothioic acid S-methyl ester (BTH) were shown to be potent activators of the same set of SAR genes induced by SA (6). Benzo(1,2,3)thiazide-7-carbothioic acid S-methyl ester and fosetyl act synergistically in increasing plant resistance to pathogen infection (5). This study suggests that fosetyl action may involve a complex signal transduction network in plants, and that host defense responses may play an important role in fosetyl's antifungal activity.

Fosetyl is a fungicide selected based on its fungicidal activity in vitro. However, Molina et al (5) demonstrated that fosetyl fungicidal activity was affected by plant defense signal transduction pathway(s). This suggests that, in addition to its direct fungicidal activity, fosetyl modulates expression of plant genes involved in pathogen defense, for example, fosetyl induces the SAR marker gene *PR-1* (5). To gain further insight into the relationship between fosetyl and plant defense response, we examined the effect of this fungicide on gene expression. We have combined SSH and cDNA microarrays to detect alterations in gene expression patterns of fosetyl-treated plants. Suppressive subtractive hybridization combines normalization and subtraction in a single step and selectively amplifies differentially expressed genes (9). DNA microarrays detect transcription changes in a large number of genes simultaneously (10–12). Recent studies have shown that combining these two technologies can be highly effective in isolating differentially expressed genes (13–15).

By using SSH and DNA microarrays in combination, we show that fosetyl is a potent inducer of plant defense responses, which leads to the activation of diverse defense-related genes. In many respects, the induction patterns of fosetyl resemble that of microbial elicitors and is shared by multiple defense signaling pathways.

## II.  RESULTS

### A.  Fosetyl Induction of PR-1 Expression

Molina et al (5) showed previously that fosetyl induced expression of *Arabidopsis PR-1*. In the Ws ecotype, maximum *PR-1* expression occurred at 5 g/L of fosetyl. To start this experiment, we determined an optimal fosetyl concentration for the CO-1 ecotype. *Arabidopsis* seedlings (4 weeks post-planting) were sprayed with 1 or 5 g/L of fosetyl; RNA was isolated 12, 24, 36, and 48 hours later. Plants treated with 5 g/L of fosetyl exhibited severe leaf lesions at every treatment time. In contrast, plants treated with 1 g/L of fosetyl remained healthy 48 hours after

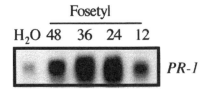

**Figure 1**   Kinetics of fosetyl *PR-1* induction; 20 μg of total RNA from fosetyl-treated *Arabidopsis* was loaded on each lane. Four-week-old *Arabidopsis* was treated with fosetyl for 12, 24, 36, and 48 hours by foliar spraying. Control plants were sprayed with water for 24 hours. The *PR-1* probe was prepared from a PCR product amplified from a seed cDNA library by primers flanking the coding sequence of *PR-1* gene (Uknes et al, 1992).

treatment. Therefore, the aerial parts of seedlings treated with 1 g/L of fosetyl were harvested for all subsequent time-course experiments. RNA gel blots containing total RNA from fosetyl treated seedlings (12–48 hours) were probed with *PR-1* cDNA to monitor gene induction by fosetyl in *Arabidopsis*. From this experiment, we determined that the induction of *PR-1* mRNA expression began after 12 hours and reached its highest level between 24 and 36 hours (Fig. 1). Based on this result, *Arabidopsis* seedlings treated with 1 g/L of fosetyl for 24 hours were used in subsequent SSH experiments.

## B.   Combination of Subtractive cDNA Library with DNA Microarray for Gene Discovery

To isolate genes responding to fosetyl treatment, SSH (Clontech) was used to enrich for differentially expressed sequences after fosetyl treatment. A subtractive cDNA library was constructed from the forward-subtracted cDNA. This library was screened on nylon filters with radioactively labeled forward and reverse subtractive cDNA probes. Of 1,536 clones screened, 192 showed differential hybridization signals and were selected for DNA microarray analysis.

The resulting array containing selected subtractive cDNAs was hybridized with fluorescently labeled cDNAs prepared from mRNA of fosetyl-treated and control plants with the use of reverse transcription. The fosetyl-treated cDNA was labeled with Cy3 and the control cDNA with Cy5. Hybridization was detected with a fluorescent scanner. In this array, nine clones have Cy3/Cy5 ratio ranging from 4.5 to 11 (Table 1).

We note in passing that there were 14 different clones with a "fosetyl-induced" reduction of two- to threefold. Two were analyzed in more detail. One encodes delta-9 desaturase-like protein (Accession # BAB02316.1), and the other is similar to an aconitase gene (Accession # T04820). In both cases, RNA gel

**Table 1**  Fosetyl-Induced Genes Identified by SSH

| Gene location | Gene identity | Accession number | Cy3/Cy5 ratio | Induction at 24 hr[a] | Maximum induction[a,b] |
|---|---|---|---|---|---|
| 1C6 | *PR-1* | P33154 | 11 | 6.7 | 9.7 |
| 1G6 | nsLTP | | 4.5 | 5.8 | 5.8 |
| 2C10 | *ATCYP1* | U32186 | 8.0 | 4.8 | 5.5 |
| C8 | *PR-2* | M90509 | 8.8 | 5.3 | 6.0 |
| 1E12 | *PR-2* | M90509 | 8 | ND[c] | ND |
| 1G11 | *PR-2* | M90509 | 8.3 | ND | ND |
| 2A8 | *PR-2* | M90509 | 6.3 | ND | ND |
| 2A12 | *PR-2* | M90509 | 6.5 | ND | ND |
| 2C8 | *PR-2* | M90509 | 9.9 | ND | ND |
| D8 | *HEL* | U01880 | ND | ND | 5.7 |
| C15 | unknown | | ND | ND | 2.6 |

[a]Induction at the 24-hour time point determined by RNA gel blot analysis.
[b]Maximum induction during time course from 12–48 hours.
[c]*ND*, Not determined.

blot analysis revealed reduced gene expression on fosetyl exposure, but the significance of repression was difficult to assess because of the low signals involved (data not shown).

Northern blot analysis was used to confirm that candidate fosetyl-induced genes are indeed differentially expressed in fosetyl treated plants and to provide a second measurement after fosetyl treatment (Fig. 2). Sequence analysis revealed that clone 1G6 is similar to a nonspecific lipid transfer protein (nsLTP) with 5.8-fold induction (Fig. 2 and Table 1). Nonspecific lipid transfer proteins are known to transfer a broad range of lipids between membranes in vitro (16). Several possible functions for nsLTPs have been proposed, including involvement in cuticle biosynthesis (17), pathogen defense mechanisms (18–21), and responses to abiotic stress (22, 23). Clone 2C10 is a cyclophilin (Cyp), *ATCYP1*; it was induced 4.2-fold. The main function of cyclophilin is in protein folding (24). Cyclophilin expression is induced by various environmental stresses (25). 1C6 is *PR-1*; it was induced by fosetyl nearly 10-fold. Six clones of *PR-2* were identified. One of them, C8, was selected for Northern blot analysis; it was induced 6-fold by fosetyl. *Arabidopsis* PR proteins, *PR-1* and *PR-2*, were previously known to be induced by pathogen attack (26), as well as by abiotic stresses such as mechanical wounding (27).

One unique feature of DNA microarray analysis is the use of two-color fluorescence labeling in one hybridization. This allows direct comparison of rela-

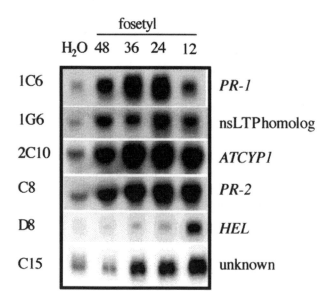

**Figure 2** RNA gel blot confirmation of fosetyl induction of genes identified by SSH. Experiment was conducted as described in Figure 1, except that the indicated gene probes were used.

tive mRNA abundance in two RNA populations. In this case, the Cy3/Cy5 ratio can be used to reflect gene induction levels by fosetyl. Comparing the Cy3/Cy5 ratio with fosetyl induction measured by Northern analysis, we found that Cy3/Cy5 ratios, with one exception, tended to be somewhat higher (up to 40%) than the gene induction level observed in the Northern blot analysis (Table 1). The observation of slight disparities in relative expression levels when Cy3/Cy5 and RNA gel blot analysis has been observed elsewhere (15, 28).

During the preliminary screening on nylon filters with the forward and reverse subtracted cDNAs, two cDNAs were apparently induced by fosetyl but were not included in the subtractive microarray. Sequence analysis revealed that cDNA D8 is a hevein-like protein (*HEL*). cDNA C15 has no significant sequence similarity to any known gene. Gene induction by fosetyl for these two cDNAs was confirmed by Northern blot hybridization (Fig. 2). D8 (*HEL*) and C15 were up-regulated by fosetyl 5.7-and 2.6-fold, respectively (Table 1). *HEL* was originally isolated by using tobacco acidic *PR-4* as probe and shown to be similar to the antifungal chitin-binding protein, hevein, from the rubber tree latex (29). *HEL* was shown to be induced by pathogen, insect feeding, and ethylene (27, 29). So far, no functional information on C15 is available.

## C. Combination of Subtractive cDNA Probes with an EST Microarray

Although it is usually fruitful to probe microarrays with fluorescent probes representing mRNA populations, there are compelling arguments (15, 30, 31) to suggest that this approach may not detect low abundance transcripts. To circumvent this possibility, subtractive cDNAs have been successfully used to detect low abundance transcripts owing to the enrichment step in SSH (32). In this study, we tested the feasibility of using subtractive probes to hybridize DNA microarrays. Specifically, subtractive cDNAs from the forward and reverse subtractions were used directly to hybridize EST microarrays.

Subtractive cDNAs from SSH were used to screen an EST array containing 2,500 *Arabidopsis* genes: 1,720 *Arabidopsis* cDNAs with known annotation (Table 2), 860 randomly picked cDNAs, two mouse cDNAs, and *PR-1* cDNA. Forward and reverse subtractive cDNAs were fluorescently labeled with Cy3 and Cy5, respectively, by random primer labeling and hybridized to this strategic EST array. To normalize the hybridization signal on the slide, mouse cDNAs

**Table 2**  Annotated ESTs in *Arabidopsis* Strategic Array

| Category | Number of ESTs in each category[a] | Category | Number of ESTs in each category[a] |
|---|---|---|---|
| GTP-binding | 45 | Transcription factor | 114 |
| Kinase | 302 | Storage protein | 11 |
| Helicase | 23 | Binding protein | 98 |
| Phosphatase | 66 | Cell cycle/division | 15 |
| Peroxidase | 29 | Auxin related | 40 |
| Transferase | 202 | ABA related | 21 |
| Genase | 195 | Stress related | 78 |
| Desaturase | 14 | Pathogen related | 8 |
| Thioesterase | 4 | Chaperone | 13 |
| Elongase | 1 | Extensin | 9 |
| Dismutase | 13 | Histone | 29 |
| Synthase | 168 | Resistance | 7 |
| Acyl-CoA | 17 | Calcium related | 22 |
| Acyl carrier protein (ACP) | 10 | Receptor | 68 |
| Steroid related | 9 | Transporter | 44 |
| Seed Specific | 6 | Unknown | 33 |
| Embryo specific | 6 | | |

[a]Total number of ESTs is 1,720.

were also included in labeling reaction, and scanning sensitivity for both channels was adjusted according to mouse cDNA signals so that the signal ratio of Cy3/Cy5 at two mouse clones was as close to 1.0 as possible. In this array, one mouse gene displayed a Cy3/Cy5 ratio of 0.93. Because of extremely low hybridization signal, the Cy3/Cy5 ratio of the second mouse cDNA was not measured. A positive control, *PR-1*, spotted in two locations in this array, had an average Cy3/Cy5 ratio of 3.9.

In our EST collection, 23 cDNAs exhibited Cy3/Cy5 ratios of 2.0 to 4.8 in three independent experiments. The Northern blot hybridization confirmed gene induction by fosetyl for 11 cDNAs with Cy3/Cy5 ratios greater than 2.5 (Table 3). Figure 3 shows the Northern blots confirming that these candidates were induced by fosetyl.

Sequence analysis revealed the identities of some fosetyl induced genes. A1-2, A1-3, A1-4 and B1-4 represented the same gene, *GST Atpm24.1*. This gene had an average fosetyl induction of 6.1-fold. B1-5 is another GST, ERD11; it is induced 3.1-fold by fosetyl. Glutathione-S-transferases are encoded by a multigene family; a primary GST function is to protect against oxidative stress (33). Different forms of GSTs have been shown to be induced by biotic and abiotic events (34, 35). Atpm24.1 was isolated in two separate experiments by its auxin-binding ability and inducibility by ethylene (36, 37). The biological

**Table 3**  Fosetyl-Induced Genes Identified by Screening Strategic Microarrays with SSH Probes

| Gene location | Gene identitity | Accession number | Cy3/Cy5 ratio | Induction at 24 hr[a] | Maximum induction[a,b] |
|---|---|---|---|---|---|
| A1-1 | Extensin (AtExt1) | U43627 | 3.2 | 2.0 | 3.4 |
| A1-2 | GST (Atmp24.1) | P46422 | 3.2 | 3.4 | 7.1 |
| A1-3 | GST (Atmp24.1) | P46422 | 3.2 | 2.5 | 5.0 |
| A1-4 | GST (Atmp24.1) | P46422 | 3.4 | ND[c] | ND |
| A2-1 | Unknown | | 3.6 | 1.6 | 1.9 |
| B1-2 | Calreticlulin (Crt3) | U66345 | 3.2 | 2.8 | 2.8 |
| B1-3 | Luminal Binding Protein (BIP) | D84414 | 3.5 | 3.6 | 3.6 |
| B1-4 | GST (Atmp24.1) | P46422 | 3.8 | 3.4 | 6.2 |
| B1-5 | GST (ERD11) | D17672 | 4.8 | 3.0 | 3.1 |
| Control | PR-1 | | 2.6;5.2 | 6.7 | 9.7 |

[a]Induction at the 24-hour time point determined by RNA gel blot analysis.
[b]Maximum induction during time course from 12–48 hours.
[c]*ND*, Not determined.

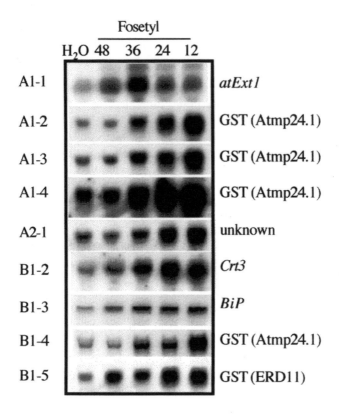

**Figure 3**  RNA gel blot confirmation of fosetyl induction of genes identified by screening DNA microarray with SSH probes. Experiment was conducted as described in Figure 1, except that the indicated gene probes were used.

function for this GST is still not defined. ERDl1 is a GST induced by desiccation (38).

A1-1 is similar to an *Arabidopsis* extensin gene, *atExt1*. It is induced 3.4-fold by fosetyl. Extensin, a hydroxyproline-rich glycoprotein, is an abundant protein in plant cell walls (39). Extensin expression is induced during host-pathogen interaction, ethylene treatment, and mechanical wounding (40–42). The *Arabidopsis* extensin, atExt, is normally expressed in root but is activated in leaves responding to abiotic inducers, such as wounding, application of jasmonic acid (JA), abscisic acid (ABA), and SA (43).

B1-3 is a luminal binding protein, *BiP*; it is induced 3.6-fold by fosetyl. Luminal binding protein (*BiP*), a member of the heat shock protein (hsp) 70 family, acts as a molecular chaperone in the lumen of the ER (44). The level of

BiP mRNA increases in response to protein misfolding in the lumen of the ER in tobacco (44). Expression of *BiP* was induced by wounding (45) and by pathogen attack (46). Expression of fosetyl-inducible BiP has been shown to be induced by chaotropic agents and heat shock (47).

B1-2 is a calcium-binding protein, calreticulin (*Crt3*), and is induced 2.5-fold by fosetyl. Crt3, another ER-binding protein, is believed to be the primary calcium carrier in the ER (48). Similar to *BiP*, *Crt* expression has been shown to be induced by various stimuli, including pathogen attack and hormone application (49). The fosetyl inducible Crt, *Crt3*, is abundantly expressed in reproductive organs. The biological function for this gene is not yet defined (50).

A2-1 is an unknown gene with 1.9-fold induction. Two genes, one encoding glucose-6-phosphoaldehydrogenase and another similar to the flowering-time gene *CONSTANS* (51), had Cy3/Cy5 ratios of 3.8 and 4.3, respectively, but did not show induction by Northern blot analysis. For these fosetyl-inducible ESTs, the Cy3/Cy5 ratios in the DNA microarray reflected the gene induction level in the Northern blot analysis at 24 hours (Table 3).

## III.  DISCUSSION

### A.  Discovery of Fosetyl Inducible Genes by SSH and DNA Microarray Technologies

In the first part of this study, subtractive cDNAs from SSH were verified by DNA microarray analysis. Six distinct genes with fosetyl-induction ratios ranging from 2.6 to 9.7 in Northern blot experiments were identified. These genes encode PR proteins *PR-1*, *PR-2*, and *PR-4*), a molecular chaperone (cyclophilin), and a non-specific lipid transfer protein (nsLTP) that might play an indirect role in plant defense.

In the second part of this work, we showed that the subtractive cDNAs from SSH can serve as reliable probes for detecting differentially expressed genes. Forward and reverse subtractive cDNAs were used as probes to screen *Arabidopsis* EST arrays. Six genes were identified that were up-regulated by fosetyl, including genes encoding antioxidants (GSTs), molecular chaperones (BiP and calreticulin), and a cell wall component (extensin). It has been suggested that first-strand cDNA probes representing complex mRNA are relatively insensitive to low abundance mRNAs (15, 31). Subtractive cDNA probes have been used to enhance detection of rare transcripts (31, 32). Here, we successfully used subtractive cDNA probes to identify differentially expressed genes in a cDNA microarray. The induction level of genes detected in the microarray by using subtractive probes tended to be lower than that of genes detected in the subtractive cDNA array by using mRNA probes (Tables 1 and 3). This demonstrates the high sensitivity of subtractive cDNA probes in detecting low abundance transcripts. Except for GST (Atmp24.1), which had an average induction level of 6.1, other fosetyl-inducible genes detected by the second approach displayed an induction level of 1.9 to 3.6, whereas induction of genes originally detected using complex probes ranged from 2.6 to 9.7 in Northern blot analysis.

Based on our results and those of others (13–15), it is clear that combining SSH and DNA microarrays results in a powerful gene discovery platform for isolating differentially expressed genes. The use of subtractive probes to screen the cDNA libraries provides a less time-consuming approach for isolating differentially expressed genes. Thus, when larger, genome-wide EST collections become available, the feasibility of using subtractive cDNAs as probes for DNA microarrays will provide a facile approach to monitor differential gene expression on a genome-wide scale.

## B.  Fosetyl May Mimic Microbial Elicitors by Inducing Plant Defense-Related Gene Expression

In host-pathogen interactions, elicitation by microbial pathogens is an essential step in inducing plant defense responses by activating various signal transduction cascades, such as protein phosphorylation, reactive oxygen species (ROS) and ion fluxes (52). For example, harpin, a bacterial elicitor, induces disease resistance and diverse plant defense genes, such as *PR-1*, *PR-2*, phenylalanine ammonia-lyase (*PAL*), *GST*, and anthranilate synthase (*ASA1*) (34, 53). Similarly, we showed that fosetyl induces a broad spectrum of plant response genes, including those coding for PR proteins (*PR-1*, *PR-2*, and *HEL*), antioxidants (*GSTs*), molecular chaperones (*BiP*, calreticulin, and cyclophilin), and other potential antimicrobial compounds such as *nsLTP* and extensin.

Although the expression of PR genes has been shown to be closely correlated to host-pathogen interaction, some of these genes are also induced by abiotic stress, such as mechanical wounding (27). Also, different forms of GSTs have been found to be induced by biotic and abiotic inducers (34, 35). Similar induction patterns were reported for other fosetyl-inducible genes, such as cyclophilin (25), and extensin (40–42). Thus, it is difficult to exclude the possibility that fosetyl may also exert an effect like abiotic stress to elicit diverse defense gene response. However, among these fosetyl-inducible genes, *HEL* was shown to be induced by pathogen attack, and recently it was demonstrated to be induced specifically by insect feeding, and not by abiotic stress, such as mechanical wounding or dehydration (27). Consequently, we propose that fosetyl mimics microbial elicitors in the activation of plant defense mechanisms. In this regard, it is noteworthy that Kiraly et al (54) observed that application of antibiotics with fungitoxic activity resulted in changes in pathogen-host interaction. Hypersensitive response was induced in compatible plants only when fungal pathogen was inhibited. It was proposed that molecules released from the dead fungus may trigger HR in host tissue. The pattern of fosetyl gene induction is similar to that of microbial elicitors; however, there is no fungus involved. This suggests that fungicide alone can be an inducer of plant defense response.

## C. Fosetyl Inducible Genes Are Shared by Multiple Signal Transduction Pathways

Fosetyl is an effective fungicide in controlling fungal pathogens in vivo. Although the mode of action of fosetyl is still unclear, our data provide strong evidence that fosetyl may interact with plant defense regulatory networks. In addition to SAR, induced systemic resistance (ISR) is another inducible resistance for protecting plants against pathogen attack (55, 56). There are distinct signaling molecules and sets of marker genes involved in both pathways. Salicylic accumulation is the signaling molecule in SAR and induces expression of the marker genes, *PR-1, PR-2,* and *PR-5.* Likewise, in the ISR signaling pathway, jasmonate (JA)/ ethylene are the signaling molecules that induce plant defense genes *PDF1,2, PR-3,* and *PR-4 (HEL)* in *Arabidopsis* (56). By using mutants defective in SA and ethylene/JA signaling transduction pathways, Molina et al (5) demonstrated that the effectiveness of fosetyl against fungal pathogens was affected by the SAR signal transduction pathway.

We showed that fosetyl induced PR proteins, *PR-1, PR-2,* and *HEL;* these are considered marker genes induced in both signal transduction pathways. There is additional evidence suggesting that fosetyl interacts with ethylene-dependent signaling pathways. For example, fosetyl-inducible GST, Atpm24.1, was isolated by its inducibility by ethylene (37). There is significant evidence indicating that there is cross-talk between the SA-, JA-, and ethylene-dependent signaling pathways. For example, SA can act as an antagonist to suppress JA-dependent defense gene expression (57, 58). Conversely, JA/ethylene acts synergistically with SA (7, 59). Futhermore, SA- and JA-dependent signaling pathways can be induced simultaneously and act additively to enhance plant resistance level for a broad spectrum of pathogens (60). In our case, the marker genes involved in both signaling pathways were induced by fosetyl. Thus, it is reasonable to suggest that fosetyl activates both signaling pathways, which leads to the induction of diverse defense-related genes.

The involvement of plant defense mechanisms in fungicide action has been long speculated (4, 61), but only limited data were available to support this hypothesis because of the restriction in monitoring gene expression by conventional methods. By combining SSH and DNA microarrays to detect differentially expressed genes on a genome-wide scale, we showed that the expression of a broad spectrum of defense-related genes was up-regulated in response to fosetyl treatment. Our data provide positive evidence for fosetyl activation of plant genes that are shared by multiple signal transduction pathways. They further suggest that fosetyl is a potent inducer of plant defense responses.

## IV.  EXPERIMENTAL PROCEDURES

### A.  Plant Treatment and Northern Blot Analysis

Four-week-old *Arabidopsis* seedlings (CO-1) were sprayed with 1 g/L of fosetyl (Aliette; Rhone-Poulenc) with 0.003% Tween 20. The control plants were sprayed with water containing 0.003% Tween 20. Plants treated for different times were harvested and total RNA was extracted (62). Briefly, frozen tissue was extracted with solution of 4% *p*-aminosalicylic acid (Sigma), 1% 1,5-naphthalenedisulfonic acid disodium salt (ACROS), and 50% water-saturated phenol. A chloroform extraction step was included. Ribonucleic acid was ethanol precipitated, dissolved in 4 mL of $H_2O$, and precipitated again with 2.5 mL of 8M LiCl.

Twenty µg of total RNA was separated on formaldehyde agarose gels and then transferred to nylon filters (Schleicher & Schuell) as described in Sambrook et al (62). Radioactive probes were synthesized by random primer labeling. Hybridization was performed in solution containing 50% formamide, 5X SSC, 50mM phosphate buffer (pH6.8), 5X Denhardts, 1% SDS, 5% dextransulfate, and 1 mg/mL sheared salmon sperm DNA, overnight at 45° C. Filters were washed with each of 5X SSC/0.1% SDS solution and 2X SSC/0.1% SDS solution for 15 minutes each at 65° C. The radioactive signal was quantitated by a phosphoimager (FUJI).

### B.  Suppressive Subtractive Hybridization

Four-week-old *Arabidopsis* seedlings were treated with fosetyl or water and harvested 24 hours later and used for total RNA preparation. mRNA from both tissues was purified with an Oligotex kit (Qiagen). Two µg of mRNA was used to perform SSH using the polymerase chain reaction (PCR) select cDNA subtraction kit (Clontech, catalog# K1804-1). Both forward (fosetyl-treated as tester and control as driver) and reverse (control as tester and fosetyl-treated as driver) subtractions were conducted. Subtractive hybridization and PCR amplification was performed after the manufacturer's protocol.

### C.  Preparation and Hybridization of Subtractive cDNA Microarray

To construct a subtractive library, the fosetyl-treated forward subtractive cDNA was ligated into pGEM-T vector (Promega). Individual clones were stored in

384-well microtiter plates. Preliminary screening was conducted on nylon filters. Duplicate macroarray filters were made from 384-well cultures and hybridized with radioactively labeled cDNAs of forward and reverse subtraction by random primer labeling. Clones with different hybridization intensities were selected for microarray printing.

The inserts of selected clones were amplified by PCR with primers flanking the cloning site of the vector. The PCR products were dissolved in 3X SSC and printed on poly-L-lysine–coated glass microscope slides (Sigma) using a DNA gridding robot (GeneMachines). Slides were processed following procedures described in DeRisi et al (10).

This subtractive cDNA microarray was hybridized with fluorescently labeled cDNA prepared from the mRNAs of fosetyl-treated and control plants by reverse transcription according to DeRisi et al (10) with Superscript reverse transcriptase (BRL). Fluorescent probes were purified on a P-30 column (BIORAD) and dried by SpeedVac. Labeled cDNA was dissolved in 15 μL of hybridization solution containing 2.5X SSC, 0.2% SDS, and 10 μg of sheared salmon sperm DNA. This probe mixture was hybridized with the DNA microarray at 62° C overnight. Arrays were washed with 2.5X SSC/0.2% SDS and 0.1X SSC/0.1% SDS solutions for 5 minutes each at room temperature. The hybridization signal was scanned with a ScanArray 3000 laser scanner (General Scanning Lumonics, Inc.) and quantitated with Genomic Solutions HDG software.

## D.  Preparation and Hybridization of an EST cDNA Microarray

The EST array contains 1,700 EST clones selected based on their sequence availability and partial annotation (Table 2), 860 randomly selected EST clones, two mouse cDNAs used for normalization, and *PR-1* cDNA used as a positive control. The inserts of EST clones and mouse cDNAs were amplified by PCR with vector-specific primers flanking cloning sites. The *PR-1* cDNA was isolated by PCR from a seed cDNA library with *PR-1* gene-specific primer sequences (26). Array printing and processing were as described above. This EST array was hybridized with forward and reverse subtractive cDNA probes fluorescently labeled, as described in Lashkari et al (11). Briefly, 1 μg of subtracted cDNA was labeled with Cy3- or Cy5-dUTP (Amersham), by random primer labeling. Twenty ng of mouse cDNA templates were added to this labeling reaction to provide an internal control for calibration. Array hybridization and analysis was performed as describe above.

## ACKNOWLEDGMENTS

We thank RhoBio for support for this research. We also acknowledge the Laboratory for Functional Genomics at Texas A&M University for access to the DNA microarray technology. We also appreciate critical comments and discussions from the members of the Thomas laboratory, especially Drs. Mona Damaj and John Ivy.

## REFERENCES

1.  LG Copping, HG Hewitt. Fungicides. In: Chemistry and mode of action of crop protection agents. Cambridge, UK: The Royal Society of Chemistry, 1998.
2.  DI Guest. Modification of defence responses in tobacco and capsicum following treatment with fosetyl-Al [Aluminium tris (o-ethyl phosphonate)]. Physiol Mol Plant Pathol 25:125–134, 1984.
3.  DI Guest. Evidence from light microscopy of living tissues that fosetyl-Al modifies the defence response in tobacco seedlings following inoculation by Phytophthora nicotianae var nicotianae. Physiol Mol Plant Pathol 29:251–261, 1986.
4.  D Cartwright, P Langcake, RJ Pryce, DP Leworthy. Chemical activation of host defence mechanisms as a basis for crop protection. Nature 267:511–513, 1977.
5.  A Molina, MD Hunt, JA Ryals. Impaired fungicide activity in plants blocked in disease resistance signal tranduction. Plant Cell 10:1903–1914, 1998.
6.  J Ryals, U Neuenschwander, M Willits, A Molina, H-Y Steiner, M Hunt. Systemic acquired resistance. Plant Cell 8:1809–1819, 1996.
7.  K Lawton, K Weymann, L Friedrich, B Vernooij, S Uknes, J Ryals. Systemic acquired resistance in Arabidopsis requires salicylic acid but not ethylene. Mol Plant Microbe Interact 8:863–870, 1994.
8.  K Lawton, L Friedrich, M Hunt, K Weymann, T Staub, H Kessmann, J Ryals. Benzothiadiazole induces disease resistance in Arabidopsis by activation of the systemic acquired resistance signal transduction pathway. Plant J 10:71–82, 1995.
9.  L Diatchenko, YC Lau, AP Campell, A Chenchik, F Moqadam, B Huang, S Lukyanov, K Lukyanov, N Gurskaya, ED Sverdlov. Suppression subtractive hybridization: A method for generating differentially regulated or tissue-specific cDNA probes and libraries. Proc Natl Acad Sci USA 93:6025–6030, 1996.
10. JL DeRisi, VR Jyer, PO Brown. Exploring the metabolic and genetic control of gene expression on a genomic scale. Science 278:680–686, 1997.
11. DA Lashkari, JL DeRisi, JH McCusker, AF Namath, C Gentile, SY Huang, PO Brown. Yeast microarray for genome wide parallel genetic and gene expression analysis. Proc Natl Acad Sci USA 94:13057–13062, 1997.
12. L Wodicka, H Dong, M Mittmann, MH Ho, DJ Lockhart. Genome-wide expression monitoring in Saccharomyces cerevisiae. Nat Biotechnol 15:1359–1367, 1997.
13. T Wang, D Hopkins, C Schmidt, S Silva, R Houghton, H Takita, E Repasky, SG Reed. Identification of genes differentially over-expressed in lung squamous cell carcinoma using combination of cDNA subtraction and microarray analysis. Oncogene 19:1519–28, 2000.
14. J Xu, JA Stolk, X Zhang, SJ Silva, RL Houghton, M Matsumura, TS Vedvick, KB Lesile, R Badaro, SG Reed. Identification of differentially expressed genes in human prostate cancer using subtraction and microarray. Cancer Res 60:1677–1682, 2000.

15. GP Yang, DT Ross, WW Kuang, PO Brown, RJ Weigel. Combining SSH and cDNA microarrays for rapid identification of differentially expressed genes. Nucleic Acids Res 27:1517–1523, 1999.
16. M Yamada. Lipid transfer proteins in plants and microorganisms. Plant Cell Physiol 33:1–6, 1992.
17. P Sterk, H Booij, G Schellekens, AV Kammen, SD Vries. Cell-specific expression of the carrot EP2 lipid transfer protein gene. Plant Cell 3:907–921, 1991.
18. JMM Caaveiro, A Molina, JM Gonzalez-Manas, P Rodriguez-Palenzuela, F Garcia-Olmedo, FM Goni. Differential effects of five types of antipathogenic plant peptides on model membranes. FEBS Lett 410:338–342, 1997.
19. F Garcia-Olmedo, A Molina, A Segura, M Moreno. The defensive role of nonspecific lipid-transfer proteins in plants. Trends Microbiol 3:72–174, 1995.
20. A Molina, A Segura, F Garcia-Olmedo. Lipid transfer proteins (nsLTPs) from barley and maize leaves are potent inhibitors of bacterial and fungal plant pathogens. FEBS Lett 316:119–122, 1993.
21. FRG Terras, IJ Goderis, FV Leuven, J Vanderleyden, BPA Cammue, WF Broekaert. In vitro antifungal activity of a radish (Raphanus sativus L.) seed protein homologous to nonspecific lipid transfer proteins. Plant Physiol 100:1055–1058, 1992.
22. M Trevino, M O'Connell. Three drought-responsive members of the nonspecific lipid-transfer protein gene family in Lycopersicon pennellii show different developmental patterns of expression. Plant Physiol 116:1461–1468, 1998.
23. MA Dunn, MA Hughes, L Zhang, RS Pearce, AS Quigley, PL Jack. Nucleotide sequence and molecular analysis of the low temperature induced cereal gene, BLT4. Mol Gen Genet 229:389–394, 1991.
24. CT Walsh, LD Zydowsky, FD Mckeon. Cyclosporin-A, the cyclophilin class of peptidylprolyl isomerases, and blockade of T-cell signal transduction. J Biol Chem 267:13115–13118, 1992.
25. J Marivet, P Frendo, G Burkard. DNA sequence analysis of a cyclophilin gene from maize: Developmental expression and regulation by salicylic acid. Mol Gen Genet 247:222–228, 1995.
26. S Uknes, B Mauch-Mani, M Moyer, S Potter, S Williams, S Dincher, D Chandler, A Slusarenko, ER Ward, JA Ryals. Acquired resistance in Arabidopsis. Plant Cell 4:645–656, 1992.
27. P Reymond, H Weber, M Damond, EE Farmer. Differential gene expression in response to mechanical wounding and insect feeding in Arabidopsis. Plant Cell 12:707–719, 2000.
28. SB Lee, K Huang, R Palmer, VB Truong, D Herzlinger, KA Kologuist, J Wong, C Paulding, SK Yoon, W Gerald, JD Oliner, DA Haber. The Wilms tumor suppressor WT1 encodes a transcriptional activator of amphiregulin. Cell 98:663–673, 1999.
29. S Potter, S Uknes, K Lawton, AM Winter, D Chandler, J DiMaio, R Novitzky, E Ward, J Ryals. Regulation of a hevein-like gene in Arabidopsis. Mol Plant Microbe Interact 6:680–685, 1993.
30. D Gerhold, T Rushmore, CT Caskey. DNA chips: Promising toys have become powerful tools. TIBS 24:168–173, 1999.

31. Z Wang, DD Brown. A gene expression screen. Proc Natl Acad Sci USA 88:11505–11509, 1991.

32. KA Lukyanov, MV Matz, EA Bogdanova, NG Gurskaya, SA Lukyanov. Molecule by molecule PCR amplification of complex DNA mixtures for direct sequencing: An approach to in vitro cloning. Nucleic Acids Res 24:2194–2195, 1996.

33. CB Pickett, AYH Liu. Glutathione S-transferases: Gene structure, regulation, and biological function. Annu Rev Biochem 58:743–764, 1989.

34. R Desikan, A Reynolds, JT Hancock, SJ Neill. Harpin and hydrogen peroxide both initiate programmed cell death but have differential effects on defence gene expression in Arabidopsis suspension cultures. Biochem J 330:115–120, 1998.

35. F Mauch, R Dudler. Differential induction of distinct glutathione-s-transferases of wheat by xenobiotics and by pathogen attack. Plant Physiol 102:1193–1201, 1993.

36. R Zettl, J Schell, K Palme. Photoaffinity labeling of Arabidopsis thaliana plasma membrane vesicles by 5-azido-[7-3H]indole-3-acetic acid: Identification of a glutathione S-transferase. Proc Natl Acad Sci USA 91:689–693, 1994.

37. J Zhou, PB Goldsbrough. An Arabidopsis gene with homology to glutathione S-transferase is regulated by ethylene. Plant Mol Biol 22:517–523, 1993.

38. T Kiyosue, K Yamaguchi-Shinozaki, K Shonozak. Characterization of two cDNAs (ERD11 and ERD13) for dehydration-inducible genes that encode putative glutathione S-transferases in Arabidopsis thaliana L. FEBS Lett 335:189–192, 1993.

39. GI Cassab, JE Varner. Tissue printing on nitrocellulose paper: A new method for immunolocalization of proteins, localization of enzyme activities and anatomical analysis. Cell Biol Int Rep 13:1147–1152, 1989.

40. DR Corbin, N Sauer, CJ Lamb. Differential regulation of a hydroxyproline-rich glycoprotein gene family in wound and infected plants. Mol Cell Biol 7:4337–4344, 1987.

41. JR Ecker, RW Davis. Plant defense genes are regulated by ethylene. Proc Natl Acad Sci USA 84:5202–5206, 1987.

42. AM Showalter, JN Bell, CL Cramer, JA Bailey, JE Varner. Accumulation of hydroxyproline-rich glycoprotein mRNA in response to fungal elicitor and infection. Proc Natl Acad Sci USA 82:6551–6555, 1985.

43. G Merkouropoulos, DC Barnett, AH Shirsat. The Arabidopsis extensin gene is developmentally regulated, is induced by wounding, methyl jasmonate, abscisic and salicylic acid, and codes for a protein with unusual motifs. Planta 208:212–219, 1999.

44. J Denecke, MHS Goldman, J Demolder, J Seurinck, J Botterman. The tobacco luminal binding protein is encoded by multigene family. Plant Cell 3:1025–1035, 1991.

45. A Kalinski, DL Rowley, DS Loer, C Foley, G Buta, EM Herman. Binding-protein expression is subject to temporal, developmental and stress-induced regulation in terminally differentiated soybean organs. Planta 195:611–621, 1995.

46. EPWM Jelitto-VanDooren, S Vidal, J Denecke. Anticipating endoplasmic reticulum stress: A novel early response before pathogenesis-related gene induction. Plant Cell 11:1935–1943, 1999.

47. N Koizumi. Isolation and responses to stress of a gene that encodes a luminal binding protein in Arabidopsis. Plant Cell Physiol 37:862–865, 1996.

48. TJ Ostwald, DH MacLennan. Isolation of a high affinity calcium-binding protein from sarcoplasmic reticulum. J Biol Chem 249:974–979, 1974.

49. J Denecke, LE Carlson, S Vidal, A-S Hoglund, B Ek, MJv Zeijl, KMC Sinjorgo, ET Palva. The tobacco homolog of mammalian calreticulin is present in protein complexes in vivo. Plant Cell 7:391–406, 1995.

50. DE Nelson, B Glaunsinger, HJ Bohnert. Abundant accumulation of the calcium-binding molecular chaperone calreticulin in specific floral tissues of Arabidopsis thaliana. Plant Physiol 114:29–37, 1997.

51. JJ Putterill, SE Ledger, K Lee, F Robson, G Murphy, G Coupland. The flowering-time gene CONSTANS and homologue CONSTANS LIKE 1 (accession nos. Y10555 and Y10556) exist as a tandem repeat on chromosome 5 of Arabidopsis (PGR97-077). Plant Physiol 114:396–396, 1997.

52. Y Yang, J Shah, D Klessig. Signal perception and transduction in plant defense responses. Gene Dev 11:1621–1639, 1997.

53. H Dong, TP Delaney, DW Bauer, SV Beer. Harpin induces disease resistance in Arabidopsis through the systemic acquired resistance pathway mediated by salicylic acid and the NIM1 gene. Plant J 20:207–215, 1999.

54. Z Kiraly, B Barna, T Ersek. Hypersensitivity as a consequence, not the cause, of plant resistance to infection. Nature 239:456–458, 1972.

55. CMJ Pieterse, JAvP SCM. van Wees, M Knoester, R Laan, H Gerrits, PJ Weisbeek, LC van Loon. A novel signaling pathway controlling induced systemic resistance in Arabidopsis. Plant Cell 10:1571–1580, 1998.

56. BPHJ Thomma, K Eggermont, IAMA Penninckx, B Mauch-Mani, R Vogelsang, BPA Cammue, WF Broekaert. Separate jasmonate-dependent and salicylate-dependent defense-response pathways in Arabidopsis are essential for resistance to distinct microbial pathogens. Proc Natl Acad Sci USA 95:15107–15111, 1998.

57. SA Bowling, JD Clarke, Y Liu, DF Klessig, X Dong. The cpr5 mutant of Arabidopsis expresses both NPR1-dependent and NPR1-independent resistance. Plant Cell 9:1573–1584, 1997.

58. IAMA Penninckx, BPHJ Thomma, A.Buchala, J-P Metraux, WF Broekaert. Concomitant activation of jasmonate and ethylene response pathways is required for induction of a plant defensin gene in Arabidopsis. Plant Cell 10:2103–2113, 1998.

59. Y Xu, P-F Chang, D Liu, ML Narasimhan, KG Raghothama. Plant defense genes are synergistically induced by ethylene and methyl jasmonate. Plant Cell 6:1077–1085, 1994.

60. S van Wees, ED Swart, JV Pelt, L van Loon, C Pieterse. Enhancement of induced disease resistance by simultaneous activation of salicylate- and jasmonate-dependent defense pathways in Arabidopsis thaliana. Proc Natl Acad Sci USA 97:8711–8716, 2000.

61. EWB Ward. Suppression of metalaxyl activity by glyphosate: Evidence that host defence mechanisms contribute to metalaxyl inhibition of Phytophthora megasperma f. sp. glycinea in soybeans. Physiol Plant Pathol 25:381–386, 1984.

62. J Sambrook, EF Fritsch, T Maniatis. Molecular Cloning—A Laboratory Manual. Cold Spring Harbor Laboratory Press, Cold Spring Harbor, NY, 1989.

# 8

# Identification of T-DNA Insertions in *Arabidopsis* Genes

**László Szabados**
*Institute of Plant Biology, Biological Research Center, Szeged, Hungary*

**Csaba Koncz**
*Max-Planck-Institute for Plant Breeding Research, Cologne, Germany*

## I. INTRODUCTION

During the last 2 decades, one of the most important developments in plant science was the exploitation of a model organism, *Arabidopsis thaliana*, which has attracted a critical mass of scientists and led to exceptional progress in plant genetics and molecular biology. This progress is illustrated by the isolation of hundreds of genes and mutants, creation of recombinant inbred lines, and sequencing of thousands of EST clones (1–7). Sequencing of the *Arabidopsis* genome recently has been completed, providing an essential source of information for gene isolation and characterization (8–13). The Internet-based comprehensive database Arabidopsis Information Resource (TAIR) offers a wealth of information for sequence analysis and retrieval (14). Extensive collections of *Arabidopsis* mutants and ecotypes are available for the research community in international seed stock centers (15).

Based on the availability of complete genome sequence, the next challenge for plant science is the determination of function of all *Arabidopsis* genes within the next 10 years (4,16–18). Fast advancing elucidation of gene functions and interactions in budding yeast serves as a model for similar efforts in plants (19, 20). Most current approaches in functional analysis of plant genes use knock-out and gain of function mutants (7, 21). In combination, genome-wide gene expression data are obtained by transcript profiling techniques using gene chips (22–25), whereas protein interaction and cellular localization studies are facilitated by yeast two-hybrid protein interaction studies and epitope-tagging approaches (26–28). An integral bioinformatic analysis of accumulating data generates new disciplines for computational genomics in support of experimental biology (29, 30).

A simple approach to identification of plant gene functions is the analysis of mutants that are characterized by well-defined phenotypes. Random mutagenesis, followed by either phenotypic screening or smart selection, is a common strategy to identify mutant alleles of a gene. Unfortunately, site-specific recombination-based targeted mutagenesis is not yet available to easily create gene disruptions in plants. Therefore, insertion mutagenesis techniques are used to saturate the genome with mutations. These techniques offer the advantage that besides creating mutations, the inserts in genes are marked by a known DNA tag, which facilitates their isolation and functional analysis. In *Arabidopsis*, large collections of mutants have been generated using the transferred DNA (T-DNA) of *Agrobacterium* Ti-plasmids and maize transposons as insertion elements (31–38). Currently, these T-DNA and transposon insertion mutant collections are exploited to develop forward and reverse genetic approaches by polymerase chain reaction (PCR)-based identification of *Arabidopsis* knock-out mutants (37, 39). Similar mutagenesis programs have also been initiated in maize (40) and rice (41–43). In this chapter, we discuss the technical aspects of insertion mutagenesis and identification of gene mutations by random sequencing of T-DNA insertion sites.

## II. pPCV-TYPE T-DNA VECTORS

Insertion mutagenesis is based on the integration of foreign DNA fragments into plant genes, which leads to elimination of gene functions. The T-DNA of *Agrobacterium* is a natural insertion element that was modified to construct versatile gene tagging vectors for insertion mutagenesis. One of these is represented by the pPCV-type T-DNA vectors (44). Outside of the T-DNA, these vectors carry replication and conjugation transfer origins of plasmid RK2 (ori$_v$ and ori$_T$, respectively), whereas within the T-DNA border they contain a pBR322-derived ColEl replication origin and a bacterial ampicillin/carbenicillin resistance gene. The latter two allow the isolation of T-DNA tags and flanking plant DNA sequences in *Escherichia coli* by plasmid rescue (32). To assist the identification of T-DNA tags in plant genes, gene-and promoter-trap vectors were constructed (Fig. 1) (45) that allow the detection of transcriptional activity at the T-DNA insertion sites by monitoring the activity of in situ gene fusions. The gene fusion vectors pPCV621 and pPCV6NFHyg carry a hygromycin resistance gene (46), as a selectable marker, and a promoterless neomycin phosphotransferase (*aph(3′)II*) reporter gene (either with or without ATG codon, respectively) at the right T-DNA border. Thus, if the T-DNA is integrated in proper orientation in a plant gene, the *aph(3′)II* reporter gene is activated by plant gene transcription. In case of pPCV6NFHyg T-DNA, the reporter gene carries no ATG, thus the activation of the *aph(3′)II* gene is detectable only if the T-DNA integration resulted in a plant gene-reporter gene fusion in a proper reading-frame. Other gene fusion and promoter trap vectors, including Tgus, TluxF and Tluc, carry respectively promoterless β-glucuronidase (*uidA*), fused bacterial luciferase (*luxF*), and firefly luciferase (*luc*) reporter genes at the right T-DNA border, respectively (47–50). Using the same principle, the activation of these reporter genes in fusion with tagged plant genes can be monitored by histochemical staining, quantitative enzyme measurement, and nondestructive in vivo detection of bioluminescence.

To induce dominant mutations by T-DNA–mediated activation of plant gene transcription, activation tagging vectors are used that carry either strong enhancers or promoters in the vicinity of T-DNA ends. A CaMV 35S (Cauliflower Mosaic Virus 35S RNA gene) promoter with four enhancer domains is thus fused to the right T-DNA border in the pTAc1 vector, whereas four CaMV 35S enhancers are present at the same position in the pT4E vector (45). pTac1 and pT4E are used to promote or enhance transcription of plant genes that are located either at the border or in close vicinity of T-DNA tags. Conditional, temporal, or spatial activation of plant genes is achieved by similar application of a dexamethasone-inducible CaMV 35S promoter (51) in the pTac16 vector.

## III. *ARABIDOPSIS* TRANSFORMATION SYSTEMS

The T-DNA of *Agrobacterium* Ti plasmid is transported into the nucleus of infected plant cells and integrated into the nuclear genome by illegitimate recombination (52–55). The T-DNA is an efficient mutagen as T-DNA integration occurs randomly without apparent preference for specific sequences and chromosomal loci (56, 57). To generate T-DNA insertion mutants in *Arabidopsis*, several *Agrobacterium*-mediated transformation methods were developed. The first mutant collections were created using tissue culture-based root transformation methods (32, 58, 59). This technique was perfected by transformation of cell suspensions that yield $10^4 - 10^5$ transformed cell colonies in a single experiment (60). This technology supports smart mutant selection approaches based on activation tagging (61); however, the regeneration of transgenic plants from transformed cell suspensions is laborious and handicapped by a quick loss of regeneration capacity of cultivated cells.

The development of in planta transformation methods made realistic to perform a saturation mutagenesis of *Arabidopsis*, with T-DNA (62). Variants of in planta transformation techniques include seed infection (31), vacuum infiltra-

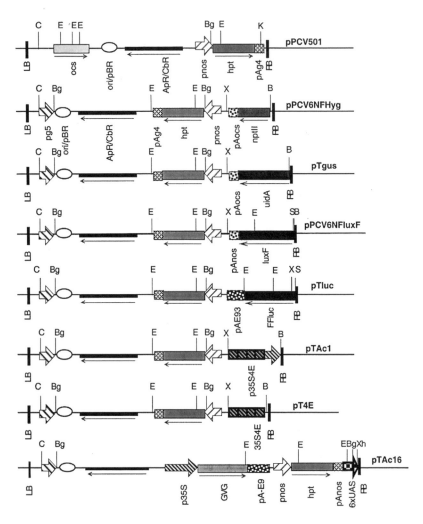

**Figure 1** Schematic map of pPCV T-DNA vectors developed for insertion mutagenesis. All vectors carry a pBR322-derived replication origin and bacterial ampicillin/carbenicillin resistance marker within the T-DNA borders (44). pPCV501: general tagging vector; pPCV6NFHyg: promoter trap gene fusion vector for generation of translational fusions with the *aph(3′)II* neomycin phosphotransferase reporter gene at the right T-DNA border; pPCV6NFluxF: promoter trap gene fusion vector carrying a fused bacterial luciferase reporter gene at the right T-DNA border; pTgus: promoter trap gene fusion vector with an *uidA* β-glucuronidase reporter gene at the right T-DNA border; pTluc: promoter trap gene fusion vector with a firefly luciferase reporter gene at the right T-DNA border; pTAc16: activation tagging vector with a dexamethasone-inducible CaMV35S promoter at the right border; pTAc1: activation tagging vector carrying a CaMV35S promoter with four enhancers at the right border; pT4E: activation tagging vector with four enhancers from the CaMV35S promoter.

tion of seedlings and adults plants (63) and flowers (64, 65). Using these methods, several collections of T-DNA insertion mutants were generated (36–38,59,66). To find an insertion all 26,000 *Arabidopsis* genes with a probability of 95% to 99%, it is estimated that about 120,000 to 600,000 T-DNA tagged *Arabidopsis* lines would be necessary (37, 67). This saturation level will probably soon be reached, as the currently available collections already include several hundred thousand lines (39). Therefore, using appropriate strategies, the identification of T-DNA and transposon insertions in most *Arabidopsis* genes is a realistic goal.

## IV. IDENTIFICATION OF INSERTION MUTATIONS

### A. Forward Genetics

A classic approach to mutant isolation is the phenotypic screening for loss-of-function insertion mutations followed by characterization of tagged genes. The success rate in the identification of specific mutants depends on the size of mutant collection and the efficiency of screening techniques used. The first characterized T-DNA tagged genes were identified by screening for mutations that caused easily observable changes in phenotype. These included, for example, the *GL1* (*glabrous 1*) gene that controls trichome differentiation (68), a mutant allele of *DWF1* gene causing dwarfism (69), the *CS/CH-42* gene controlling chlorophyll biosynthesis (70), the *AG1* (*AGAMOUS*) homeotic gene regulating flower organ identity (71) and several *EMB* gene mutations resulting in embryo lethality and abnormal embryo development (72). During the recent years, T-DNA insertion mutants not only helped to gain a better insight into signaling processes and regulation of development and metabolism, but also led to surprises, such as the *cpd* dwarf mutant, the analysis of which demonstrated that steroids play a role as essential hormones in plants as in animals (73). Nonetheless, the majority of T-DNA insertions do not appear to result in obvious mutant phenotypes because many metabolic and developmental functions are controlled by either small gene families

---

*Abbreviations*: *E*, *EcoR*I; *C*, *Cla*I; *Bg*, *Bgl*II; *X*, *Xba*I; *B*, *BamH*I; *Xh*, *Xho*I; *S*, *Sal*I; *LB*, left border; *RB*, right border; *pg5*, promoter of T-DNA gene 5; *ocs*, octopine synthase gene; *ori/pBR*, replication origin from plasmid pBR322; *Ap^R^/Cb^R^*, bacterial ampicillin/carbenicillin resistance marker; *pAg4*, 3' polyA signal sequence of T-DNA gene4; *hpt* hygromycin phosphotransferase gene; *pnos*, nopaline synthase promoter; *pAocs*, 3' polyA region of octopine synthase gene; *nptII*, neomycin phosphotransferase *(aph(3')II)* gene; *uidA*, β-glucuronidase gene; *FFluc*, firefly luciferace gene; *pAE93*, 3' polyA region of gene E93; *p35S4E*, CaMV35S promoter with 4xCaMV35S enhancer sequences; *35S4E*, 4xCaMV35S enhancer sequences; *p35S*, CaMV35S promoter; *GVG*, glucocorticoid responsive chimaeric transcription factor (51); *pAE9*, 3' polyA region of gene E9; *6xUAS*, CaMV35S promoter fused to 6x yeast UAS elements; *CaMV*, cauliflower mosaic virus.

or duplicated genes in *Arabidopsis* (13) Thus, a knock-out mutation in one gene may be compensated by another member of a gene family.

Activation tagging may help to overcome this problem by generation of dominant gene mutations that result from altered transcription of genes located in the vicinity of T-DNA tags (38). The application of activation T-DNA tagging is illustrated by the identification of *Arabidopsis* genes involved in cytokinin signal transduction (61), flower induction (74), and regulation of flowering time, disease resistance, and development (38).

### B.  Promoter and Enhancer Trap Mutagenesis

To specifically detect insertion of the T-DNA in functional plant genes, and thereby monitor transcriptional activity of plant genes in situ, promoter and enhancer trap mutagenesis techniques are used. As described above, this technique exploits promoterless reporter genes, such as *aph(3′)II* (75, 76), β-glucuronidase (*uidA*) (77) and luciferase (78), that are positioned close to the T-DNA border and used as gene fusion markers. Gene identification by promoter traps is based on the insertion of a promoterless reporter genes into transcribed regions of plant gene in proper orientation, which generate gene fusions and activate the transcription of reporter genes. As expected, activation of a promoter trap reporter gene indicates, in most cases, that the insertion is located in a functional plant gene (79, 80). However, cryptic promoter elements may also be detected by the reporter gene fusions (81, 82), which calls for a precise analysis of promoter trap lines. In the enhancer trap vectors, the reporter gene is fused to a minimal promoter (i.e., containing only TATA-box and transcription initiation sequences), which allows the detection of specific enhancer elements located in close vicinity of T-DNA or transposon tags (83). Enhancer traps could detect *cis*-acting regulatory elements independently of their orientation, increasing the probability of gene identification. The promoter and enhancer trap methods are exploited for detection of specific gene expression patterns in vivo using the β-glucuronidase, luciferase, and green fluorescent protein reporter genes (47, 78, 84, 85).

### C.  Reverse Genetics with PCR-Based Mutants Screens

Collections of T-DNA and transposon tagged mutants are exploited in sequence-based mutants screens. The individual tagged lines are arranged in larger pools from which DNA is prepared. As the sequence of insertion elements and all *Arabidopsis* genes is known, T-DNA insertions in any gene can be identified by PCR-amplification using pairwise combinations of gene-specific and T-DNA, or transposon specific primers (86). For T-DNA inserts, the left and right border-specific primers are combined either with 5′ or 3′ gene specific primers to search for PCR products that indicate the presence of a T-DNA tag in the gene of interest. Once a DNA pool resulted in a PCR product, DNA is prepared from each individual plant present in the pool and subjected to repeated PCR screen to identify a single individual carrying the tagged gene. Several research groups created PCR-screenable T-DNA and transposon tagged collections (37, 67, 87). The "Knockout Facility" at the University of Wisconsin (Madison) offers pooled DNA from more than 60,000 insertion lines, which is available as public PCR screening facility (88). The potential of PCR-based reverse genetics is well illustrated by recent identification of insertion mutants in phenylpropanoid biosynthesis genes (87), the actin act2-1 and act4-1 genes(89), or in members of the R2R3 Myb transcription factor gene family (90).

## D. Gene Identification by Random Sequencing of T-DNA Tags

Based on the availability of large insertion mutant collections that are estimated to carry a T-DNA tag at every 5 kb in the *Arabidopsis* genome, novel approaches aim at direct sequencing of PCR-amplified plant DNA sequences marked by T-DNA tags to create a collection of characterized gene mutations (60, 91, 92). Amplification of T-DNA insertion sites by inverse and tail PCR techniques allows facile isolation and automatic sequencing of tagged plant genomic sequences on a large scale. The T-DNA tagged genomic sequences can be identified and mapped by performing a simple homology search with known sequences of the *Arabidopsis* genome. Random sequencing of T-DNA tagged loci provides the advantage that the identification of insertions is not biased by limitation of any selection and screening method. When combined with gene fusion technology, this approach is also expected to help transcriptional profiling of each individual gene.

The ultimate goal of T-DNA end-sequencing projects is the generation of a comprehensive database for identification of mutations in all plant genes (88). The database of sequenced insertion sites will thus be an essential component of future developments that aim at functional characterization of all *Arabidopsis* genes. Random sequencing of transposon insertion sites to identify tagged genes in maize and rice also has been initiated recently (40–42).

## V. ISOLATION AND ANALYSIS OF T-DNA INSERTIONS

The procedures used for isolation of plant genes tagged by insertion elements were greatly improved during the past decade. Based on the presence of a plasmid

**Table 1**  PCR Program Used for Long-
Range Inverse PCR

| Steps | Temperature | Time |
|-------|-------------|------|
| 1 | 95° C | 2 min |
| | 1x | |
| 2 | 94° C | 15 sec |
| 3 | 65° C | 30 sec |
| 4 | 70° C | 8 min |
| | *35x* | *2–4 steps* |
| 5 | 40° C | hold |

replicon within the T-DNA, a plasmid rescue technique was developed for isolation of T-DNA inserts with flanking plant DNA sequences in *E. coli* (45). This
technique is now replaced by PCR-based methods, such as TAIL-PCR (93) and
inverse PCR (60, 78) that permit a high throughput amplification of T-DNA–
tagged genomic DNA fragments that can directly be used as templates for automatic DNA sequencing. Long-range inverse and tail PCR methods used for routine isolation of T-DNA–tagged plant genomic fragments are described below
(Table 1, Fig. 2).

## A.  Long-Range Inverse PCR

To prepare a substrate for inverse PCR (iPCR), the genomic DNA from a T-
DNA–tagged line is digested with a restriction endonuclease and circularized by
self-ligation (78). To increase the efficiency of amplification, we adapted a long-
range PCR amplification technique for iPCR (60). Two μg genomic DNA is
digested with a restriction enzyme (e.g., *EcoR*I, *Xba*I, or *Sal*I, depending on the
vector type used) that cleave once or twice within the T-DNA (Fig. 1). An aliquot
of digested DNA is tested on agarose gel, whereas the rest is self-ligated for 12
hours with 1 U of T4 DNA ligase in a reaction volume of 20 μl. Half the ligation
mix (0.5 μg DNA) is used as a template for PCR reaction. The T-DNA–specific
PCR primer pairs are selected according to the type of transformation vector and
restriction enzyme used for digestion of genomic DNA. The long-range PCR
reaction is performed in a volume of 50 μl with TaKaRa Ex-Taq polymerase as
recommended by the manufacturer. The PCR program used for Long-iPCR is
shown in Table 1. This long-range iPCR technique yields PCR fragments of 1
kb to 8 kb on an average 2/3 of reactions that can be gel purified and used directly
as a template for DNA sequencing.

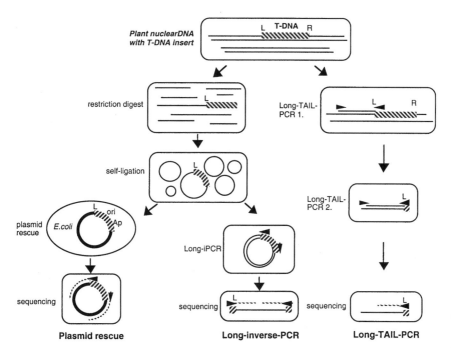

**Figure 2** Strategies used for isolation and sequencing of T-DNA insertion sites. Plasmid rescue: plant DNA is digested by a restriction enzyme and self-ligated. The ligation product, carrying a bacterial selection marker and replication origin within the T-DNA, is isolated by transformation of competent *E. coli*. Plasmid DNA is isolated and sequenced using T-DNA–specific sequencing primers. Long-range inverse-PCR (iPCR): plant DNA is digested and self-ligated as for plasmid rescue. Long-range iPCR is performed using T-DNA specific primers and PCR conditions favoring an extended elongation. Long-range TAIL-PCR: long-range TAIL-PCR is performed with undigested plant DNA template using a T-DNA specific primer and a short degenerated primer. The PCR product is diluted and a second PCR is performed with a nested T-DNA–specific primer and a degenerated primer. The PCR products are purified and sequenced with T-DNA specific sequencing primers.
Labels: L, left T-DNA border; R, right border; ⎯ plant DNA; ▧▧▧ T-DNA-derived DNA; ⎯⎯ PCR-derived DNA; ► T-DNA specific primer; ▶ degenerated primer; ·····-
sequencing.

## B.  Long Range TAIL-PCR

TAIL PCR permits the amplification of T-DNA inserts from undigested genomic DNA (93). The PCR amplification is performed by alternating high and low stringency annealing temperatures, using a combination of high $T_m$ T-DNA specific primers with low $T_m$ degenerate primers that can frequently anneal with plant genomic DNA (Table 2). The procedure includes two amplification steps with nested T-DNA–specific primers. The first step is performed in a volume of 10 µl of PCR mix that contains genomic DNA template, the first T-DNA specific primer, and a degenerate primer. After amplification, the volume of PCR reaction is increased to 100 µl, from which 1 µl is added to a second PCR reaction mix of 50 µl. The second PCR reaction is performed with the second nested T-DNA–specific primer and a degenerate primer. The PCR reactions are carried out in thermocycler 96-well microtiter plates with TaKaRa Ex-Taq™ polymerase. We routinely use a Perkin Elmer GeneAmp PCR System 9700 thermocycler, but the

**Table 2**  PCR Programs Used for Long-Range TAIL-PCR

| | PCR 1 | | | PCR 2 | |
|---|---|---|---|---|---|
| Steps | Temperature | Time | Steps | Temp/oC | Time |
| 1 | 95° C | 2 min | 1 | 94° C | 30 sec |
| | *1x* | | | 1x | |
| 2 | 94° C | 15 sec | 2 | 94° C | 15 sec |
| 3 | 65° C | 30 sec | 3 | 65° C | 30 sec |
| 4 | 70° C | 8min | 4 | 70° C | 8 min |
| | *5x* | *2–4 steps* | 5 | 94° C | 15 sec |
| 5 | 94° C | 15 sec | 6 | 65° C | 30 sec |
| 6 | 45° C | 2 min | 7 | 70° C | 8 min |
| 7 | 70° C | 8 min | 8 | 94° C | 15 sec |
| | *2x* | *5–7 steps* | 9 | 45° C | 30 sec |
| 8 | 94° C | 15 sec | 10 | 70° C | 8 min |
| 9 | 65° C | 30 sec | | *15x* | *2–10 steps* |
| 10 | 70° C | 8 min | 11 | 72° C | 7 min |
| 11 | 94° C | 15 sec | 12 | 4° C | Hold |
| 12 | 65° C | 30 sec | | | |
| 13 | 70° C | 8 min | | | |
| 14 | 94° C | 15 sec | | | |
| 15 | 45° C | 30 sec | | | |
| 16 | 70° C | 8 min | | | |
| | *12x* | *8–16 steps* | | | |
| 17 | 4° C | Hold | | | |

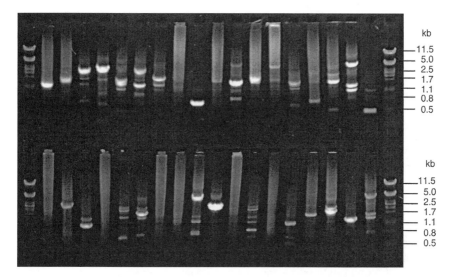

**Figure 3** Amplification of genomic DNA fragments flanking the T-DNA tags by long-range TAIL-PCR. In this typical experiment, one to three PCR fragments are obtained in 26 reactions from 36 samples. The fragment sizes range between 300 bp and 5500 bp.

protocol yields comparable results with other thermocyclers from MJResearch and Hybaid. Because of the use of an extended DNA synthesis time in the cycling program with the proofreading TaKaRa Ex-Taq™ polymerase, long PCR fragments can be reproducibly amplified. Sufficiently high yields can be obtained in the two-step procedure, which eliminates the need for a third PCR reaction (Fig. 3.).

## C. Primers Used for Long-Range iPCR and TAIL-PCR

Right border-specific primers for vector pPCV6NFHyg:
RB11: GTTTCGCTTGGTGGTCGAATGGGCAGGTAG
RB21: CCCGGCACTTCGCCCAATAGCAGCCAGTCC
Km21:CCCAGTCATAGCCGAATAGCCTCTCCACCC
Right border-specific primers for vector pTgus:
TG1:CGATACGCTGGCCTGCCCAACCTTTCGG
TG2:CATCGGCGAACTGATCGTTAAAACTGCCTGG
TG3:CGATCCAGACTGAATGCCCACAGG
Right border-specific primers for vector pPCV6NFLuxF:
LX1:CGTCTTCTGCTTGTCGAACCGGATGGGC

LX2:GGTGTGCGGCAGCAACATAAGGATTCCC
LX3:GGTGGCTGATAAGTGAGAAGGAAG
Left border-specific primers for all pPCV vectors:
LB11:CAGACCAATCTGAAGATGAAATGGGTATCTGGG
LB21:GTGAAGTTTCTCATCTAAGCCCCCATTTGG
LB41:TTCTCCATATTGACCATCATACTCATTGC
LB51: TCATACTCATTGCTGATCCATG
Primers used for T-DNA left border iPCR with *Eco*RI digestion of
pPCV6NFHyg, pTluc, pPCV6NFLuxF, pTgus, and pTAcl T-DNA inserts:
PR1:CACATTTCCCCGAAAAGTGCCACCTGACG
PR11: CCGCGCACATTTCCCCGAAAAGTGCCACC
PR2:CCTATAAAAATAGGCGTATCACGAGGCCC
Primer used for T-DNA left border iPCR with *Xba*I digested pPCV6NFHyg,
pPCV6NFLuxF, and pTgus T-DNA inserts:
PC3:CCTTGCGCCCTGAGTGCTTGCGGCAGC
Primers used for T-DNA right border iPCR with *Eco*RI digested pPCV6NFHyg,
pTluc, pPCV6NFLuxF, pTgus, and pTAcl T-DNA inserts:
EH11: GTCCTGCGGGTAAATAGCTGCGCCGATGG
EH21:CGTTATGTTTATCGGCACTTTGCATCGGC
Degenerate primers used for long-range TAIL-PCR (93):
AD1: NTCA(GC)T(AT)T(AT)T(GC)G(AT)GTT
AD2: NGTCGA(GC)(AT)GANA(AT)GAA
AD3: (AT)GTGNAG(AT)ANCANAGA

The size of PCR fragments generated by both long-range iPCR and TAIL-PCR ranges from a few hundred bp to 8 kb with an average fragment size of 2500 bp (Fig. 3). In most PCR reactions, amplification of a single fragment is observed, although agarose gel assays may reveal some additional faint nonspecific fragments. Therefore, the purification of PCR products from agarose gels is essential. The best sequencing results are obtained by electroelution of DNA fragments from excised agarose blocks.

## D.  DNA Sequence Analysis

Proper analysis of sequence data generated by large functional genomics projects requires extensive computation and application of new bioinformatics methods (29, 30, 94). Random sequencing of T-DNA insertion sites generates a vast amount of sequence information that has to be precisely processed to identify tagged plant genes. Sequence analysis is thus necessary to determine the exact position of T-DNA insertion sites and, if a gene was tagged, to obtain some useful predictions concerning the function of tagged genes. Further computing is required for establishment of a database that should display the most important

**Table 3** Internet Sites Used for Sequence Analysis of T-DNA Insertion Sites

| Sequence homology search | Address |
|---|---|
| NCBI, BLAST service | http://www.ncbi.nlm.nih.gov/blast/ |
| Arabidopsis Information Resource (TAIR), Blast service | http://www.arabidopsis.org/blast/ |
| EMBL Blast service | http://dove.embl-heidelberg.de/Blast2/ |
| *Gene identification programs* | |
| Genscan: prediction of complete gene structures | http://genome.dkfz-heidelberg.de/cgi-bin/ GENSCAN.genscan.cgi |
| NetPlantGene: predictions of Arabidopsis splice sites | http://www.cbs.dtu.dk/services/ NetPGene/ |
| GRAIL: Gene Recognition and Assembly Internet Link | http://compbio.ornl.gov/Grail-1.3/ |
| *Analysis of protein domain structure* | |
| SMART: Simple Modular Architecture Research Tool | http://smart.embl-heidelberg.de/ |
| BLOCKS: the Blocks WWW Server | http://blocks.fhcrc.org/blocks/ |
| CDD: Conserved Domain Database and Search Service | http://www.ncbi.nlm.nih.gov/Structure/ cdd/wrpsb.cgi |
| eMOTIF Search, Biochemistry, Stanford University | http://motif.stanford.edu/emotif/emotif- search.html |
| *Sequence analysis* | |
| Arabidopsis Information Resource (TAIR) | http://www.arabidopsis.org/ |
| Diverse protein and DNA sequence analysis | http://www.cbs.dtu.dk/services/ |
| Multiple sequence alignment | http://www.ebi.ac.uk/clustalw/ |
| PSORT WWW server: protein targeting signal prediction | http://psort.nibb.ac.jp/ |

data on the T-DNA insertions and corresponding tagged genes (Table 3, Fig. 4). Chromosomal positions of T-DNA insertions in the *Arabidopsis* genome are indicated by BlastN DNA homology searches (95) using the NCBI GenBank (96) and TAIR (Arabidopsis Information Resource) databases (14). The available gene annotation data often call for corrections. Therefore, a larger region of T-DNA–tagged genomic sequences are also processed in homology searches performed with the BlastX algorithm, which compares nucleotide sequences translated in all reading frames to amino acid sequences providing a prediction for coding regions (95). Gene prediction programs, such as GENSCAN, GRAIL, and Net-

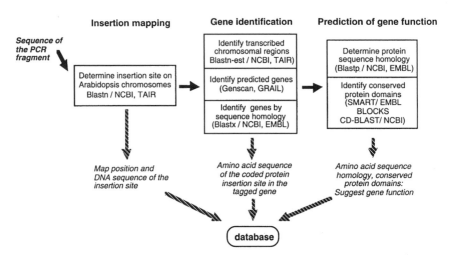

**Figure 4**   Main steps in sequence analysis. PCR-amplified genomic sequences are analyzed to identify the insertion sites in the *Arabidopsis* genome and to gain information about the presence of conserved sequences and motifs in the predicted protein products of identified genes. The localization of insertions is performed by BlastN nucleic acid sequence homology search; the identification of protein products, if any, is helped by BlastX homology search. GeneScan, Grail, and NetGene2 programs are used for gene predictions, and the obtained sequences are compared by BlastN with the EST database.

Gene2 further help to identify exon/intron boundaries, as well as ATG and STOP codons, for confirmation of available annotations (Table 3). In addition, the EST database provides information on sequenced cDNAs (expressed sequence tags, EST) that are useful to confirm the predicted positions of transcribed sequences. Thus, gene mutations can be safely identified if database searches reveal an overlap between EST clones and T-DNA insertion sites.

A combined use of gene prediction and sequence homology searches with the non redundant *Arabidopsis* genomic and EST database will often identify genes with yet unknown function. Therefore, a further mining of the database is useful to gain more information about the nature of predicted proteins encoded by the T-DNA–tagged genes. Protein homology searches performed with the Gapped-BLAST or PSI-Blast programs (95) in the nonredundant GenBank, EMBL, and SwissProt databases may reveal a conservation of amino acid sequences with other known proteins from different organisms. In fact, our experience shows that more than half the tagged *Arabidopsis* genes can be assigned with a putative function based on finding motif homologies. The modular protein motifs, *trans*-membrane segments, and cellular localization signals can then be further analyzed by Internet-based search programs (97). The analysis of domain structure of a predicted protein offers an alternative approach to gain information about putative gene functions using the SMART tool (Simple Modular Architecture Research Tool) of EMBL database (98), the BLOCKS server (99), the CD-Blast Search service (Conserved Domain Database and Search Service) of GenBank (96), and the eMotif Search facility at Stanford University (100). The obtained sequence analysis data are collected in a computer database. Usually, such a database combines data on the T-DNA insertion sites, predicted function and expression of tagged genes, and corresponding literature references to provide a useful inventory of *Arabidopsis* insertion mutations.

## VI. GENOMIC DISTRIBUTION OF T-DNA INSERTIONS IN *ARABIDOPSIS*

Direct sequencing of PCR-amplified fragments using T-DNA–specific primers yields, on average, a readable DNA sequence of 550 bp with iPCR and 450bp with TAIL-PCR for database searches. About 25% of all sequences obtained by PCR reveal only homology to the T-DNA owing to amplification of T-DNA tandem repeats. To remove these noninformative sequences, the PCR products are subjected to hybridization with T-DNA–specific probes before DNA sequencing.

To optimize and automate the technology, we have performed a T-DNA end-sequencing program producing more than 1,000 T-DNA insertion sites. T-DNA insertions were detected in random chromosomal positions, as observed earlier (56). Earlier reports suggested that T-DNA integration has preference into transcriptionally active or GC-rich chromosomal regions (52, 57). Mapping of T-DNA insertion sites (Table 4) showed that 24% of all insertions were localized

**Table 4**  Distribution of 1000 T-DNA Insertions in the *Arabidopsis* Genome

| Category | Insert no. |
| --- | --- |
| Promoter (within −600bp from ATG) | 160 |
| Coding region | 166 |
| Intron | 74 |
| Transcribed region | 38 |
| 3′ end (within 300bp from the STOP codon) | 48 |
| All predicted tagged genes | 486 |
| Unknown | 514 |
| *Total* | *1000* |

**Table 5**  Distribution of Known Motifs and Domains
Within the Predicted Proteins Sequences of Encoded by
T-DNA-tagged *Arabidopsis* Genes

| Functional Categories | Inserts | % |
| --- | --- | --- |
| Tagged genes | 521 | 100.0 |
| Putative protein | 356 | 68.3 |
| Ribosomal protein | 8 | 1.5 |
| Kinase | 42 | 8.1 |
| Receptor kinase | 10 | 1.9 |
| Phosphatase | 2 | 0.4 |
| Transcription factor | 12 | 2.3 |
| Transporter | 14 | 2.7 |
| HSP-like protein | 2 | 0.4 |
| Disease-related | 9 | 1.7 |
| Multidrug resistance | 3 | 0.6 |
| MADS box protein | 3 | 0.6 |
| ATPase | 4 | 0.8 |
| Helicase | 4 | 0.8 |
| RNA-binding protein | 4 | 0.8 |
| GTP-binding protein | 7 | 1.3 |
| Cytochrome P450 | 5 | 1.0 |
| Trans-membrane protein | 120 | 23.0 |
| Proteins with COIL domain | 18 | 3.5 |
| Proteins with RING domain | 8 | 1.5 |
| Proteins with LRR repeats | 8 | 1.5 |
| Proteins with ankyrin repeats | 4 | 0.8 |
| Proteins with WD40 domain | 1 | 0.2 |

in either exons or introns of predicted genes (excluding 3′ untranslated regions). These T-DNA insertions in transcribed sequences are predicted to result in knockout mutations. In addition, 16% of insertions were identified in promoter regions within 600 bp from the ATG translation initiation site. Six percent of T-DNA tags were found in 3′ untranslated sequences within 300 bp from the stop codon. These insertions in promoter and 3′-untranslated regions may alter gene activity by affecting transcription or RNA stability. Altogether, about half of all T-DNA insertions were localized within or near genes in *Arabidopsis*. Homology searches and domain-prediction analyses (Table 5) indicated that about 70% of these T-DNA tags were located in predicted genes with hypothetical protein products. However, a putative function could be assigned to about 50% of hypothetical proteins based on sequence similarity to either known genes and proteins from other organisms or well-defined protein domains.

## VII. CONCLUSIONS

The availability of the *Arabidopsis* genome sequence, a milestone in plant science, marks the next challenge, which is the determination of function of all plant genes. Characterization of mutants remains the most efficient strategy for functional genomics approaches that exploit *Agrobacterium* T-DNA and transposon insertions for saturation mutagenesis of the *Arabidopsis* genome. The currently available mutant collections are widely used for identification of T-DNA and transposon tags using PCR-based screening techniques. However, a new approach based on random sequencing of T-DNA tags is already emerging to identify insertion mutations in all *Arabidopsis* genes. New automated methods are in use to identify and map the position of tagged plant genes and predict the function of their protein products. In addition to a collection of tagged mutants, a comprehensive database on the location of insertion sites and tagged genes is being constructed to facilitate the work of plant scientists. In the near future, mutant alleles of plant genes may be identified in computer databases without a need for mutant identification programs, complementation analyses, genetic mapping, and PCR screens. As a side product, the database will also provide precise information on the integration preference of T-DNA and transposons unraveling the mechanism and regulation of T-DNA and transposon integration. Nevertheless, the exploitation of this gold mine remains a task of inventive researchers who wish to learn more about the function and regulation of genes, in higher plants.

## ACKNOWLEDGMENTS

This work was supported by Vitality Biotechnologies Ltd., Haifa, Israel, a Hungarian-German Cooperation Project, No. D-8/97 (DLR Ung-027-97), OTKA Grant No. T-029430, OMTB-00496/2002..

## REFERENCES

1. GP Rédei. *Arbidopsis* as a genetic tool. Annu Rev Genet 9:111–127, 1975.
2. EM Meyerowitz. *Arabidopsis thaliana*. Annu Rev Genet 21:93–111, 1987.
3. T Newman, FJ de Bruijn, P Green, K Keegstra, H Kende, L McIntosh, J Ohlrogge, N Raikhel, S Somerville, M Thomashow, E Retzel, C Somerville. Genes galore: A summary of methods for accessing results from large-scale partial sequencing of anonymous *Arabidopsis* cDNA clones. Plant Physiol 106:1241–1255, 1994.
4. DW Meinke, JM Cherry, C Dean, SD Rounsley, M Korneef *Arabidopsis thaliana*: A model plant for genome analysis. Science 282:662–682, 1998.

5. FM Ausubel. *Arabidopsis* genome: A milestone in plant biology. Plant Physiol 124:1451–1454, 2000.

6. V Walbot. A green chapter in the book of life. Nature 408:794–795, 2000.

7. R Wambutt, G Murphy, G Volckaert, et al. Progress in *Arabidopsis* genome sequencing and functional genomics. J Biotechnol 78:281–92, 2000.

8. A Theologis, JR Ecker, CJ Palm, NA Federspiel, S Kaul, O White, et al. Sequence and analysis of chromosome 1 of the plant *Arabidopsis thaliana*. Nature 408:816–820 (2000).

9. X Lin, S Kaul, S Rounsley, et al. Sequence and analysis of chromosome 2 of the plant *Arabidopsis thaliana*. Nature 402:761–768, 1999.

10. M Salanoubat, K Lemcke, M Rieger, et al. Sequence and analysis of chromosome 3 of the plant *Arabidopsis thaliana*. Nature 408:820–822, 2000.

11. K Mayer, C Schüller, R Wambutt, et al. Sequence and analysis of chromosome 4 of the plant, *Arabidopsis thaliana*. Nature 402:769–777, 1999.

12. S Tabata, T Kaneko, Y Nakamuro, et al. Sequence and analysis of chromosome 5 of the plant *Arabidopsis thaliana*. Nature 408:823–826, 2000.

13. The Arabidopsis Genome Initiative. Analysis of the genome sequence of the flowering plant *Arabidopsis thaliana*. Nature 408:796–815, 2000.

14. E Huala, AW Dickerman, M Garcia-Hernandez, D Weems, L Reiser, F LaFond, D Hanley, D Kiphart, M Zhuang, W Huang, LA Mueller, D Bhattacharyya, D Bhaya, BW Sobral, W Beavis, DW Meinke, CD Town, C Somerville, SY Rhee. The Arabidopsis Information Resource (TAIR): A comprehensive database and web-based information retrieval, analysis, and visualization system for a model plant. Nucleic Acids Res 29:102–105, 2001.

15. RL Scholl, ST May, DH Ware. Seed and molecular resources for *Arbidopsis*. Plant Physiol 124:1477–1480, 2000.

16. N Terryn, P Rouzé, M Van Montagu. Plant genomics. FEBS Lett 452:3–6, 1999.

17. C Sommerville. The twentieth century trajectory of plant biology. Cell 100:13–25, 2000.

18. J Chory, JR Ecker, S Briggs, M Caboche, GM Coruzzi, D Cook, J Dangl, S Grant, ML Guerinot, S Henikoff, R Martienssen, K Okada, NV Raikhel, CR Somerville, D Weigel. National Science Foundation-Sponsored Workshop Report: The 2010 Project functional genomics and the virtual plant. A blueprint for understanding how plants are built and how to improve them. Plant Physiol 123:423–426, 2000.

19. SG Oliver, MK Winson, DB Kell, F Baganz. Systematic functional analysis of the yeast genome. TIBTECH 16:373–377, 1998.

20. P Ross-Macdonald, PS Coelho, T Roemer, S Agarwal, A Kumar, R Jansen, KH Cheung, A Sheehan, D Syrnoniatis, L Umansky, M Heidtman, FK Nelson, H Iwasaki, K Hager, M Gerstein, P Miller, GS Roeder, M Snyder. Large-scale analysis of the yeast genome by transposon tagging and gene disruption. Nature 402:413–418, 1999.

21. PSR Coelho, A Kumar, M Snyder. Genome-wide mutant collections: Toolboxes for functional genomics. Curr Opin Microbiol 3:309–315, 2000.

22. R Schaffer, J Landgraf, M Perez-Amador, E Wisman. Monitoring genome-wide expression in plants. Curr Opin Biotechnol 11:162–167, 2000.

23. CA Harrington, C Rosenow, J Retief. Monitoring gene expression using DNA microarrays. Curr Opin Microbiol 3:285–291, 2000.
24. T Zhu, X Wang. Large-scale profiling of the *Arabidopsis* transcriptome. Plant Physiol 124:1472–1476, 2000.
25. G Sherlock, T Hernandez-Boussard, A Kasarskis, G Binkley, JC Matese, SS. Dwight, M Kaloper, S Weng, H Jin, CA Ball, MB Eisen, PT Spellman, PO Brown, D Botstein, JM Cherry. The Stanford Microarray Database. Nucleic Acids Res 29:152–115, 2001.
26. A Ferrando, R Farras, J Jasik, J Schell, C Koncz. Intron-tagged epitope: A tool for facile detection and purification of proteins expressed in *Agrobacterium-* transformed plant cells. Plant J 22:553–560, 2000.
27. G Walter, K Büssow, D Cahill, A Lueking, H Lehrach. Protein arrays for gene expression and molecular interaction screening. Curr Opin Microbiol 3:298–302, 2000.
28. I Xenarios, E Fernandez, L Salwinski, XJ Duan, MJ Thompson, EM Marcotte, D Eisenberg. DIP: The Database of Interacting Proteins: 2001 update. Nucleic Acids Res 29:239–241, 2001.
29. S Tsoka, CA Ouzounis. Recent dvelopments and future directions in computational genomics. FEBS Lett 480:42–48, 2000.
30. D Lonsdale, M Crowe, B Arnold, BC Arnold. Mendel-GFDb and Mendel-ESTS: Databases of plant gene families and ESTs annotated with gene family numbers and gene family names. Nucleic Acids Res 29:120–122, 2001.
31. KA Feldmann, MD Marks. *Agrobacterium*-mediated transformation of germinating seeds of *Arabidopsis thaliana*: A non-tissue culture approach. Mol Gen Genet 208:1–9, 1987.
32. C Koncz, N Martini, R Mayerhofer, Z Koncz-Kálmán, H Körber, GP Rédei, J Schell. High-frequency T-DNA-mediated gene tagging in plants. Proc Natl Acad Sci USA 86:8467–8471, 1989.
33. AM Bhatt, T Page, EJ Lawson, C Lister, C Dean. Use of Ac as an insertional mutagen *Arabidopsis*. Plant J 9:935–945, 1996.
34. RA Martienssen. Functional genomics: Probing plant gene function and expression with transposons. Proc Natl Acad Sci USA 95:2021–2026, 1998.
35. AF Tissier, S Marillonnet, V Klimyuk, K Patel, MA Torres, G Murphy, JD Jones. Multiple independent defective suppressor-mutator transposon insertions in *Arabidopsis*: A tool for functional genomics. Plant Cell 10:1841–1852, 1999.
36. D Bouchez, F Granier, N Bouché, M Caboche, N Bechtold, G Pelletier. T-DNA insertional mutagenesis for reverse genetics in *Arabidopsis*. In: 10th International Conference on Arabidopsis Research, Univ. Melbourne, Australia, 1999, p 4–11.
37. PJ Krysan, JC Young, MR Sussman. T-DNA as an insertional mutagen in *Arabidopsis*. Plant Cell 11:2283–2290, 1999.
38. D Weigel, JH Ahn, MA Blázquez, JO Borevitz, SK Christensen, C Fankhauser, C Ferrándiz, I Kardailsky, EJ Malancharuvil, MM Neff, JT Nguyen, S Sato, Z-J Wang, Y Xia, RA Dixon, MJ Harrison, CJ Lamb, MF Yanofsky, J Chory. Activation tagging in *Arabidopsis*. Plant Physiol 122:1003–1013, 2000.
39. S Parinov, V Sundaresan. Functional genomics in, *Arabidpopsis*: Large-scale in-

sertional mutagenesis complements the genome sequencing project. Curr Opin Biotechnol 11:157–161, 2000.

40. S Hanley, D Edwards, D Stevenson, S Haines, M Hegarty, W Schuch, KJ Edwards. Identification of transposon-tagged genes by the random sequencing of Mutator-tagged DNA fragments from Zea mays. Plant J 22:557–566, 2000.

41. DH Jeong, JS Jeong, S Lee, KH Jung, SH Jun, JW Lee, CH Kim, S Jang. T-DNA insertional mutagenesis in rice. In: 6th international Congress of Plant Molecular Biology, Quebec, Canada, 2000, p S01–35.

42. JS Jeong, S Lee, KH Jung, SH Jun, DH Jeong, J Lee, C Kim, S Jang, K Yang, J Nam, K An, MJ Han, RJ Sung, HS Choi, JH Yu, JH Choi, SY Cho, SS Cha, SI Kim, G An. Technical advance: T-DNA insertional mutagenesis for functional genomics in rice. Plant J 22:561–570, 2000.

43. H Hirochika, A Miyao, M Yamazaki, R Hirochika, GK Agrawal, S Takeda, K Abe, T Watanabe. Knockout rice for the functional analysis of genes. In: 6th International Congress of Plant Molecular Biology, Quebec, Canada, 2000, p S01–33.

44. C Koncz, J Schell. The promoter of the TL-DNA gene 5 controls the tissue-specific expression of chimaeric genes carried by a novel type of *Agrobacterium* vector. Mol Gen Genet 204:383–396, 1986.

45. C Koncz, N Martini, L Szabados, M Hrouda, A Bachmair, J Schell. Specialized vectors for gene tagging and expression studies. In: Plant Molecular Biology Manual. S. Gelvin, (Ed.) 1994, pp B2/1–22.

46. C Waldron, EB Murphy, JL Roberts, GD Gustafson, SL Armour, SK Malcolm. Resistance to hygromycin B. Plant Mol Biol 5:103–108, 1985.

47. RA Jefferson, TA Kavanagh, MW Bevan. GUS fusions: b-Glucuronidase as a sensitive and versatile gene fusion marker in higher plants. EMBO J 6:3901–3907, 1987.

48. C Koncz, WHR Langridge, O Olsson, J Schell, AA Szalay. Bacterial and firefly luciferase genes in transgenic plants: Advantages and disadvantages of a reporter gene. Dev Genet 11:224–232, 1990.

49. CD Riggs, MJ Chrispeels. Luciferase reporter gene cassettes for plant gene expression studies. Nucl Acid Res 15:8115, 1987.

50. CD Riggs, DC Hunt, J Lin, MJ Chrispeels. Utilization of luciferase fusion genes to monitor differential regulation of phytohemagglutinin and phaseolin promoters in transgenic tobacco. Plant Sci 63:47–57, 1989.

51. T Aoyama, N-H Chua. A glucocorticoid-mediated transcriptional induction system in transgenic plants. Plant J 11:605–612, 1997.

52. R Mayerhofer, Z Koncz-Kálmán, C Nawrath, G Bakkeren, A Crameri, K Angelis, GP Rédei, J Schell, B Hohn, C Koncz. T-DNA integration: A mode of illegitimate recombination in plants. EMBO J 10:697–704, 1991.

53. S De Buck, A Jacobs, M Van Montagu, A Depicker. The T-DNA sequences of T-DNA junctions suggest that complex T-DNA loci are formed by a recombination process resembling T-DNA integration. Plant J 20:295–304, 1999.

54. SB Gelvin. *Agrobacterium* and plant genes involved in T-DNA transfer and integration. Annu Rev Plant Physiol Plant Mol Biol 51:223–256, 2000.

55. J Zupan, TR Muth, O Draper, P. Zambryski. The transfer of DNA from *Agrobacterium tumefaciens* into plants: A feast of fundamental insights. Plant J 23:11–28, 2000.

56. R Azpiroz-Leehan, KA Feldmann. T-DNA insertion mutagenesis in *Arabidopsis*: Going back and forth. Trends Genet 13:152–156, 1997.

57. A Barakat, P Gallois, M Raynal, D Mestre-Ortega, C Sallaud, E Guiderdoni, M Delseny, G Bernardi. The distribution of T-DNA in the genomes of transgenic *Arabidopsis* and rice. FEBS Lett 471:161–164, 2000.

58. D Valvekens, M van Montagu, M van Lijsebettens. *Agrobacterium tumefaciens*-mediated transformation of *Arabidopsis thaliana* root explants by using kanamycin selection. Proc Natl Acad Sci USA 85:5536–5540, 1988.

59. C Koncz, K Németh, GP Rédei, J Schell. T-DNA insertional mutagenesis in *Arabidopsis*. Plant Mol Biol 20:963–976 1992.

60. J Mathur, L Szabados, S Schaefer, B Grunenberg, A Lossow, E Jonas-Straube, J Schell, C Koncz, Z Koncz-Kalman. Gene identification with sequenced T-DNA tags generated by transformation of Arabidopsis cell suspension. Plant J 13:707–16, 1998.

61. T Kakimoto. CKI1, a histidine kinase homolog implicated in cytokinin signal transduction. Science 274:982–985, 1996.

62. AF Bent. *Arabidopsis* in-planta transformation. Uses, mechanisms, and prospects for transformation of other species. Plant Physiol 124:1540–1547, 2000.

63. N Bechtold, J Ellis, G Pelletier. In planta *Agrobacterium*-mediated gene transfer by infiltration of adult *Arabidopsis thaliana* plants. C R Acad Sci Paris 316:1194–1199, 1993.

64. N Bechtold, G Pelletier. In planta *Agrobacterium*-mediated transformation of adult *Arabidopsis thaliana* plants by vacuum infiltration. Methods Mol Biol 82:259–266, 1998.

65. SJ Clogh, AF Bent. Floral dip: A simplified method for *Agrobacterium*-mediated transformation of *Arabidopsis thaliana*. Plant J 16:735–743, 1998.

66. D Weigel, JH Ahn, MA Blázquez, JO Borevitz, SK Christensen, C Fankhauser, C Ferrándiz, I Kardailsky, EJ Malancharuvil, MM Neff, JT Nguyen, S Sato, Z-J Wang, Y Xia, RA Dixon, MJ Harrison, CJ Lamb, MF Yanofsky, J Chory. Activation tagging in *Arabidopsis*. Plant Physiol 122:1003–1013, 2000.

67. D Bouchez, H Höfte. Functional genomics in plants. Plant Physiol 118:725–732, 1998.

68. MD Marks, KA Feldmann. Trichome development in *Arabidopsis thaliana*: I. T-DNA tagging of the Glabrous I gene. Plant Cell 1:1043–1050, 1989.

69. KA Feldmann, MD Marks, ML Christianson, RS Quatrano. A dwarf mutant of *Arabidopsis* generated by T-DNA insertion mutagenesis. Science 243:1351–1354, 1989.

70. Koncz, R Mayerhofer, Z Koncz-Kálmán, C Nawrath, B Reiss, GP Rédei, J Schell. Isolation of a gene encoding a novel chloroplast protein by T-DNA tagging in *Arabidopsis thaliana*. EMBO J 9:1337–1346, 1990.

71. MF Yanofsky, H Ma, JL Bowman, GN Drews, KA Feldmann, EM Meyerowitz. The protein encoded by the *Arabidopsis* homeotic gene agamous resembles transcription factors. Nature 346:35–39, 1990.

72. D Errampalli, D Patton, L Castle, L Mickelson, K Hansen, J Schnall, K Feldmann, D Meinke. Embryonic lethals and T-DNA insertion mutagenesis in *Arabidopsis*. Plant Cell 3:149–157, 1991.

73. M Szekeres, K Nemeth, Zs Koncz-Kalman, D Mathur, A Kauschmann, T Altmann, GP Rédei, F Nagy, J Schell, C Koncz. Brassinosteroids rescue the deficiency of CYP90, a cytochrome P450, controlling cell elongation and de-etiolation in *Arabidopsis*. Cell 85:171–182, 1996.

74. I Kardailsky, VK Shukla, JH Ahn, N Dagenais, SK Christensen, JT Nguyen, J Chory, MJ Harrison, D Weigel. Activation tagging of the floral inducer FT. Science 286:1962–1965, 1999.

75. C André, D Colau, J Schell, M Van Montagu, JP Hernalsteens. Gene tagging in plants by y T-DNA insertion mutagen that generates *APH(3′)II*-plant gene fusions. Mol, Gen Genet 204:512–518, 1986.

76. TH Teeri, L Herrera-Estrella, A Depicker, M Van Montagu, ET Palva. Identification of plant promoters in situ by T-DNA-mediated transcriptional fusions to the npt-II gene. EMBO J 5:1755–1760, 1986.

77. S Kertbundit, H De Greeve, F Deboeck, M Van Montagu, JP Hernalsteens. *In vivo* random β-glucuronidase gene fusions in *Arabidopsis thaliana*. Proc Natl Acad Sci USA 88:5212–5216, 1991.

78. C Jiang, WHR Langridge, AA Szalay. Identification of genes in vivo by tagging with T-DNA border linked luciferase genes followed by inverse polymerase chain reaction amplification. Plant Mol Biol Rep 10:345–361, 1992.

79. E Babiychuk, M Fuangthong, M Van Montagu, D Inze, S Kushnir. Efficient gene tagging in *Arabidopsis thaliana* using a gene trap approach. Proc Natl Acad Sci USA. 94:12722–12727, 1997.

80. PS Springer. Gene traps: Tools for plant development and genomics. Plant Cell 12:1007–1020, 2000.

81. PR Fobert, H Labbe, J Cosmopoulos, S Gottlob-McHugh, T Ouellet, J Hattori, G Sunohara, VN Lyer, BL Miki. T-DNA tagging of a seed-coat-specific cryptic promoter in tobacco. Plant J 6:567–577, 1994.

82. L Ökrész, C Máthé, É Horváth, C Koncz, L Szabados. T-DNA trapping of a cryptic promoter identifies an ortholog of highly conserved SNZ growth arrest response genes in *Arabidopsis*. Plant Sci 138:217–228, 1998.

83. L Campisi, Y Yang, Y Yi, E Heilig, B Herman, AJ Cassista, DW Allen, H Xiang, T Jack. Generation of enhancer trap lines in *Arabidopsis* and characterization of expression patterns in the inflorescence. Plant J 17:699–707, 1999.

84. J Haseloff, KR Siemering, DC Prasher, S Hodge. Removal of a cryptic intron and subcellular localization of green fluorescent protein are required to mark transgenic *Arabidopsis* plants brightly. Proc Natl Acad Sci U S A 94:2122–2127, 1997.

85. J Haseloff, EL Dormand, AH Brand. Live imaging with green fluorescent protein. Methods Mol Biol 122:241–259, 1999.

86. RG Winkler, KA Feldmann. PCR-based identification of T-DNA insertion mutants. Methods Mol Biol 82:129–236, 1998.

87. E Wisman, U Hartmann, M Sagasser, E Baumann, K Palme, K Hahlbrock, H Saedler, B Weisshaar. Knock-out mutants from an En-1 mutagenized *Arabidopsis thaliana* population generate phenylpropanoid biosynthesis phenotypes. Proc Natl Acad Sci USA 95:12432–12437, 1998.

88. MR Sussman, RM Amasino, JC Young, PJ Krysan, S Austin-Phillips. The Arabi-

dopsis Knockout Facility at the University of Wisconsin-Madison. Plant Physiol 124:1465–1467, 2000.

89. EC McKinney, N Ali, A Traut, KA Feldmann, DA Belostotsky, JM McDowell, RB Meagher. Sequence-based identification of T-DNA insertion mutations in *Arabidopsis*: Actin mutants act2-1 and act4-1. Plant J 8:613–622, 1995.

90. RC Meissner, H Jin, E Cominelli, M Denekamp, A Fuertes, R Greco, HD Kranz, S Penfield, K Petroni, A Urzainqui, C Martin, J Paz-Ares, S Smeekens, C Tonelli, B Weisshaar, E Baumann, V Klimyuk, S Marillonnet, K Patel, E Speulman, AF Tissier, D Bouchez, JJ Jones, A Pereira, E Wisman, M Bevan. Function search in a large transcription factor gene family in *Arabidopsis*: Assessing the potential of reverse genetics to identify insertional mutations in R2R3 MYB genes. Plant Cell 11:1827–1840, 1999.

91. S Parinov, M Sevugan, D Ye, WC Yang, M Kumaran, V Sundaresan. Analysis of flanking sequences from *Dissociated* insertion lines: A database for reverse genetics in *Arabidopsis*. Plant Cell 11:2263-2270, 1999.

92. L Szabados, I Kovács, E Ábrahám, A Oberschall, R Nagy, L Zsigmond, I Krasovskaya, I Kerekes, A Ben-Haim, C Koncz. Identification of T-DNA insertions in *Arabidopsis* genes: The first *Arabidopsis* genome project in Hungary. 6th Congress ISPMB, Quebec, Canada, 2000, p S01–83.

93. YG Liu, N Misukawa, T Oosumi, RF Whittier. Efficient isolation and mapping of *Arabidopsis thaliana* T-DNA insert junctions by thermal asymmetric interlaced PCR. Plant J 8:457–463, 1995.

94. AD Baxevanis. The molecular biology database colection: An updated compilation of biological database resources. Nucleic Acids Res 29:1–10, 2001.

95. SF Altschul, TL Madden, AA Schaffer, J Zhang, Z Zhang, EW Miller, DL Lipmann. Gapped blast and PSI-blast: A new generation of protein database search prorams. Nucl Acids Res 25:3389–3402, 1997.

96. DL Wheeler, DM Church, AE Lash, DD Leipe, TL Madden, JU Pontius, GD Schuler, LM Schriml, TA Tatusova, L Wagner, BA Rapp. Database resources of the National Center for Biotechnology Information. Nucleic Acids Res 29:11–16, 2001.

97. K Nakai, P Norton. PSORT: A program for detecting sorting signals in proteins and predicting their subcellular localization. Trends Biochem Sci 24:34–36, 1999.

98. G Stoesser, W Baker, A van den Broek, E Camon, M GarciaPastor, C Kanz, T Kulikova, V Lombard, R Lopez, H Parkinson, N Redaschi, P Sterk, P Stoehr, MA Tuli. The EMBL nucleotide sequence database. Nucleic Acids Res 29:17–21, 2001.

99. JG Henikoff, EA Greene, S Pietrokovski, S Henikoff. Increased coverage of protein families with the blocks database servers. Nucleic Acids Res 28:228–230, 2000.

100. JY Huang, DL Brutlag. The EMOTIF database. Nucleic Acids Res 29:202–204, 2001.

# 9
# Genomic Approaches for Studying Gene Families in Plants

**Cathal Wilson, Balázs Melikant, and Erwin Heberle-Bors**
*Vienna Biocenter, University of Vienna, Vienna, Austria*

## I. INTRODUCTION

One of the most striking facts to emerge from the fully sequenced *Arabidopsis* genome is the large amount of gene duplication that has occurred during evolution (1). This has involved both tandem gene duplications (17% of the genes are arranged in tandem arrays) and segmental duplications (approximately 60% of the genome). The degree of gene duplication and the number of gene families with more than five members is higher than in other multicellular eukaryotes whose genomes have been fully sequenced (*Drosophila melanogaster*, *Caenorhabditis elegans*). The availability of the sequence of the entire *Arabidopsis* genome provides a wealth of information for studies on genome evolution and structure and has practical implications for genomic analyses that are currently used in studying plant biology. In particular, sequence conservation within gene families should be a consideration when using approaches such as insertional knock-outs and microarray analysis. We address some of these issues using the mitogen-activated protein (MAP) kinase gene family as an example.

## II. MAP KINASE SIGNALING

Mitogen-activated protein kinases are important signaling molecules that respond to a vast array of extra- and intracellular signals and activate a variety of cellular

279

responses. A signature of a MAP kinase signaling pathway is the presence of a three-component module composed of a MAP kinase that is activated by an upstream kinase (MEK), which in turn is activated by a MAPKKK (2). Mitogen-activated protein kinases respond to extracellular triggers, such as environmental stress and growth factors (3), and are involved in a number of developmental programs (4, 5). Activated MAP kinases may translocate to the nucleus and phosphorylate transcription factors, or target cytosolic proteins such as other kinases or cytoskeletal-associated proteins. A characteristic feature of MAP kinases is that activation occurs by phosphorylation of threonine and tyrosine residues (6). The phosphorylated amino acid residues are closely spaced as a TXY motif in the activation loop. The tyrosine residue of the TEY motif is located 10 amino acids before a highly conserved glutamate residue of kinase subdomain VIII. The TXY motif is a feature that may be regarded as distinguishing members of the MAP kinase superfamily (7). A large number of MAP kinases have been identified in mammals (8). The three major types of MAP kinases in mammals are ERK, JNK/SAPK, and p38; they have TEY, TPY and TGY motifs, respectively. More divergent members of the MAP kinase family, some containing a TDY motif, have also been reported (7).

## III.  THE *ARABIDOPSIS* MAP KINASE FAMILY

To date, seven Arabidopsis MAP kinase cDNA sequences have been published (9). All the encoded proteins contain a TEY motif before subdomain VIII. In addition, two kinases with a TDY motif (ATMPK8, accession AC069551, and ATMPK9, accession AB020749) have been described. The availability of the entire *Arabidopsis thaliana* genome now makes it possible to identify all members of the MAP kinase family. Protein sequences, domains, and motifs can be used to perform an in silico screen of sequence databases to identify the full complement of a gene family within a genome (10, 11). Such an analysis identified at least 1,533 transcriptional regulators in *Arabidopsis* (10). The Arabidopsis Genome Initiative (1) reported the presence of about 20 MAP kinase genes in *Arabidopsis*. We analyzed the entire *Arabidopsis* genome available at TIGR (http://www.tigr.org/tdb/ath1/htmls), TAIR (http://www.arabidopsis.org/home.html), and MIPS (http://mips.gsf.de/proj/thal) using BLAST searches with the region around the TEY motif, the already reported MAP kinase sequences, or with a PROSITE (http://www.expasy.ch/prosite) motif (F-x(10)-R-E-x(72,86)-R-D-x-K-x(9)-C) reported as being specific for MAP kinases using the GCG program. This analysis identified 25 protein sequences with a TEY, TDY, or TPY motif at the position found in MAP kinases. Further BLAST searches using these sequences identified two other sequences with a TPY motif. No TGY motif was found in the position characteristic of MAP kinases. Three of the TEY-

(A)
```
ATMPK1    GLARASNT---KGQFMTEYVVTRWYRAPE
ATMPK2    GLARTSNT---KGQFMTEYVVTRWYRAPE
ATMPK7    GLARTSQG---NEQFMTEYVVTRWYRAPE
ATMPK11   GLART------YEQFMTEYVVTRWYRAPE
ATMPK5    GLARTTS----ETEYMTEYVVTRWYRAPE
ATMPK16   GLARTSN----ETEIMTEYVVTRWYRAPE
ATMPK4    GLARTKS----ETDFMTEYVVTRWYRAPE
ATMPK20   GLARTKS----ETDFMTEYVVTRWYRAPE
ATMPK10   GLARTTS----DTDFMTEYVVTRWYRAPE
ATMPK3    GLARPTS----ENDFMTEYVVTRWYRAPE
ATMPK6    GLARVTS----ESDFMTEYVVTRWYRAPE
ATMPK18   GLARATP----ESNLMTEYVVTRWYRAPE
ATMPK9    GLARVSFNDAPTAIFWTDYVATRWYRAPE
ATMPK19   GLARVSFNDAPTAIFWTDYVATRWYRAPE
ATMPK8    GLARVSFNDAPTAIFWTDYVATRWYRAPE
ATMPK21   GLARVAFNDTPTAIFWTDYVATRWYRAPE
ATMPK14   GLARVSFNDTPTTVFWTDYVATRWYRAPE
ATMPK22   GLARVAFNDTPTTVFWTDYVATRWYRAPE
ATMPK17   GLARVAFNDTPTTIFWTDYVATRWYRAPE
ATMPK15   GLARVSFTDSPSAVFWTDYVATRWYRAPE
          ****              *:**.********
```

(B)
```
ATMPK1    YENRIDALRTLRE -(x)- RDLKPGNLLVNANC
ATMPK2    FENRIDALRTLRE -(x)- RDLKPGNLLVNANC
ATMPK7    FENRVDALRTLRE -(x)- RDLKPGNLLVNANC
ATMPK11   FENRIDALRTLRE -(x)- RDLKPGNLLVNANC
ATMPK5    FDNKVDAKRTLRE -(x)- RDLKPSNLLLNSNC
ATMPK16   FDNRVDAKRTLRE -(x)- RDLKPSNLVLNTNC
ATMPK4    FDNIIDAKRTLRE -(x)- RDLKPSNLLLNANC
ATMPK20   FGNIIDAKRTLRE -(x)- RDLKPSNLLLNANC
ATMPK10   FDNIIDAKRTLRE -(x)- RDLRPSNVLLNSKN
ATMPK3    FDNHMDAKRTLRE -(x)- RDLKPSNLLLNANC
ATMPK6    FDNKIDAKRTLRE -(x)- RDLKPSNLLLNANC
ATMPK18   FDNTIEAKRTLRE -(x)- RDLKPSNLLLSTQC
ATMPK9    FDHISDATRILRE -(x)- RDLKPKNILANADC
ATMPK19   FDHISDATRILRE -(x)- RDLKPKNILANADC
ATMPK8    FEHVSDATRILRE -(x)- RDLKPKNILANADC
ATMPK21   FEHVSDATRILRE -(x)- RDLKPKNILANADC
ATMPK14   FEHVSDALRILRE -(x)- RDLKPKNILANANC
ATMPK22   FEHISDALRILRE -(x)- RDLKPKNILANANC
ATMPK17   FEHISDAARILRE -(x)- RDLKPKNILANANC
ATMPK15   FEHVSDAIRILRE -(x)- RDLKPKNILANADC
           *  *  ***           **  *  *
PROSITE   F  -x(10)- RE-x(77)-RDxK -x(9)-  C
```

**Figure 1** Alignment of the 20 *Arabidopsis* MAP kinases. (**A**), the region surrounding the TXY motif. (**B**), matches to the MAP kinase PROSITE motif.

motif kinases (one of which [MHK] was previously reported [12] but not identified as a putative MAP kinase) and the four TPY-motif kinases are very divergent at the sequence level from the other identified kinases, many of which have been identified as bona fide MAP kinases. Pairwise identities between these seven kinases and the MAP kinases are similar to or less than the cdc2 protein kinase. The three TEY-motif kinases and the four TPY kinases were, therefore, not considered for further analysis.

The remaining 12 TEY-motif kinases are ATMPK1-7 (9) and ATMPK10, 11, 16, 18, and 20 (Fig. 1). Curiously, the ATMPK5 genomic sequence shows a deletion in the ATP binding site and differs in this respect from the ATMPK5 cDNA (9). We could not identify any sequence in the *Arabidopsis* genome database that had 100% identity with the ATMPK5 cDNA.

Eight proteins were found with a TDY motif (Fig. 1). These include ATMPK8 and 9, and the others have been called ATMPK14, 15, 17, 19, 21, and 22. ATMPK9 was previously reported as a cDNA sequence (accession AB020749). From the genomic sequence, it appears that this was only a partial sequence, and the ATMPK9 protein is much longer. No stop codon is present upstream of the reported ATMPK9 cDNA sequence.

## IV. PHYLOGENETIC ANALYSIS

The domain structure of the remaining 20 MAP kinases was compared by aligning the kinase domains (13). The MAP kinases as a whole show differences in their amino and carboxy terminal extensions (Table 1). The TDY motif MAP kinases have very long carboxy terminal extensions, whereas ATMPK8, ATMPK9, and ATMPK19 also have very long amino terminal extensions. To compare the phylogenetic relatedness, we chose only the kinase domains from the 20 MAP kinase protein sequences and aligned them using CLUSTALW (http://www.ebi.ac.uk/clustalw). A phylogenetic tree constructed using CLUSTALX (14) (viewed with TreeView [15], version 1.6.1) identifies three major groups within the MAP kinase family (Fig. 2). Group 1 contains ATMPK1, 2, 7, 11; Group 2 can be divided into Group 2a (ATMPK3, 6, 18) and Group 2b (4, 5, 10, 16, 20); Group 3 contains the TDY-motif kinases. The TDY motif kinases show approximately 50% identity with the MAP kinases from the other groups. All the MAP kinases except ATMPK1 and ATMPK10 contain the PROSITE motif (F-x(10)-R-E-x(72,86)-R-D-x-K-x(9)-C) (Table 1 and Fig. 1). A further indication of the relatedness of the members of each group is that the introns are located at exactly the same position in each member of a group (data not shown), except for ATMPK5, where the first intron is displaced because of the above-mentioned deletion.

**Table 1** MAP Kinases in *Arabidopsis thaliana*

| ATMPK | Chr. | MIPS Accession | TXY motif | Prosite | N Term. | C Term. |
|-------|------|----------------|-----------|---------|---------|---------|
| 1  | I   | AT1g10210 | E | F—Y     | 38  | 52  |
| 2  | I   | AT1g59580 | E | +       | 38  | 58  |
| 3  | III | AT3g45640 | E | +       | 44  | 47  |
| 4  | IV  | AT4g01370 | E | +       | 49  | 48  |
| 5  | IV  | AT4g11330 | E | +       | 49  | 48  |
| 6  | II  | AT2g43790 | E | +       | 69  | 48  |
| 7  | II  | AT2g18170 | E | +       | 38  | 50  |
| 8  | I   | AT1g18150 | D | +       | 110 | 149 |
| 9  | III | AT3g18040 | D | +       | 138 | 197 |
| 10 | II  | AT2g46070 | E | K—R,C-N | 81  | 46  |
| 11 | IV  | AT4g36450 | E | +       | 38  | 46  |
| 14 | III | AT3g14720 | D | +       | 26  | 283 |
| 15 | II  | AT2g01450 | D | +       | 22  | 210 |
| 16 | I   | AT1g07880 | E | +       | 39  | 45  |
| 17 | II  | AT2g42880 | D | +       | 19  | 291 |
| 18 | III | AT3g59790 | E | +       | 66  | 49  |
| 19 | I   | AT1g73670 | D | +       | 96  | 196 |
| 20 | I   | AT1g01560 | E | +       | 46  | 44  |
| 21 | V   | AT5g19010 | D | +       | 31  | 252 |
| 22 | I   | AT1g53510 | D | +       | 19  | 300 |

The chromosome (Chr.) on which each MAP kinase (ATMPK) is located and the MIPS accession number is given. In the column marked TXY motif, the amino acid in the X position is shown. Perfect matches to the PROSITE motif (F-x(10)-R-E-x(72,86)-R-D-x-K-x(9)-C) are indicated by +, while the amino acid deviations from the motif in ATMPK1 and ATMPK10 are given. The lengths of the amino (N) and carboxy (C) terminal extensions (outside the kinase domains) are also shown.

## V. FUNCTIONAL REDUNDANCY

A potentially powerful approach to identify the function of any protein in *Arabidopsis* is to knock out the respective gene by insertional mutagenesis. It is apparent from the foregoing analysis of the MAP kinase gene family, which is applicable to any gene family, that the presence of highly related genes might limit such an approach because of functional redundancy. The knock-out of one of a set of related genes that are expressed within the same cells or tissues might not be expected to result in a phenotype. In this event, it may be necessary to produce multiple knock-outs of the members of a multigene family.

   A recently described knock-out of the MAP kinase gene ATMPK4 led to a strong phenotype. The plants had a dwarf phenotype and exhibited constitutive

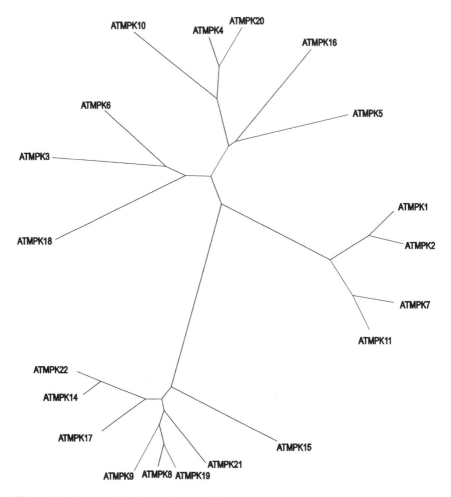

**Figure 2**   Unrooted phylogenetic tree of the 20 MAP kinase proteins constructed using the CLUSTALX program (14). Only the kinase domain from each protein was compared.

systemic acquired resistance (16). ATMPK4, on chromosome IV, has a highly related gene, ATMPK20 on chromosome I (Fig. 2 and Table 1). These two genes encode proteins that are 88.3% identical at the amino acid level. It appears, there-fore, that functional redundancy is not a problem in this case. A phenotype may be observed in a knock-out if similar proteins have different intracellular locations. Analyzing the protein sequences of ATMPK4 and ATMPK20 by PSORT (a com-puter program for the prediction of protein localization sites in cells; http://psort.-

nibb.ac.jp) showed a high score (0.909) for ATMPK4 localization to the stroma of the chloroplast, whereas the highest score (0.650) for ATMPK20 suggested a cytoplasmic location. None of the other group 2 proteins was predicted to locate to the stroma of the chloroplast. Another possibility is that ATMPK4 and ATMPK20 are differentially expressed. The high degree of sequence conservation within the ORF is not mirrored in the 5′ untranslated region (UTR). ATMPK4 and ATMPK20 are 83.7% identical at the nucleotide level over the entire open reading frame, but only 48.9% identical over the first 1000 bp of the 5′ UTRs. More distantly related MAP kinase genes, such as ATMPK1 and ATMPK6, are only 58.7% identical at the nucleotide level over the entire reading frame but 46.2% identical over the first 1000 bp of the 5′ UTRs. The absence of a high degree of sequence conservation in the 5′ UTRs was also found for other closely related MAP kinase genes. Therefore, it is possible that such closely related genes are differentially expressed in the plant. Although promoter elements may be retained within the 5′ UTR of related genes, the large number of changes that have occurred provide the possibility for changes in gene expression patterns. An interesting question for the future will be whether highly similar MAP kinases that show different expression profiles have taken on different functional roles because of co-expression of different interacting partners and substrates in the respective cells/tissues.

## VI. MICROARRAY ANALYSIS OF GENE FAMILIES

An important consideration is to establish the tissue-specific and stimulus-induced expression profiles of a given gene family. Plant MAP kinase genes may show tissue-specific expression and can be transcriptionally induced after stimulation (17–19). This has also been observed using microarray analysis (20, 21). The highly related nature of some gene family members may present a problem in this analysis. Most plant microarrays used to date are collections of expressed sequence tags (ESTs) and, as has been noted (22), might not distinguish between highly conserved sequences.

We took the DNA sequence (minus introns) of the kinase domains (13) from groups 1 and 2 and compared them using CLUSTALW and the ALIGN Query at the GENESTREAM network server (http://www2.igh.cnrs.fr/bin/align-guess.cgi). The degree of identity between the sequences is shown in Table 2. As can be readily seen, a number of these sequences show a very high level of sequence identity. Sequences with more than 70% identity over a stretch of 200 bp might be expected to cross hybridize (23). Clearly, there are a number of the MAP kinase sequences reported in Table 2 that meet this criterion.

One possibility would be to clone the 3′-untranslated region of all the genes to spot on a microarray, as these regions are less conserved. However, this re-

**Table 2**  Percent Nucleotide Sequence Identities of MAP Kinase Genes from Group 1 and Group 2

| ATMPK | 1 | 2 | 3 | 4 | 5 | 6 | 7 | 10 | 11 | 16 | 18 | 20 |
|---|---|---|---|---|---|---|---|---|---|---|---|---|
| 1 | 100 | | | | | | | | | | | |
| 2 | 85.2 | 100 | | | | | | | | | | |
| 3 | 64.7 | 63.8 | 100 | | | | | | | | | |
| 4 | 64.8 | 63.8 | 66.8 | 100 | | | | | | | | |
| 5 | 63.2 | 65.8 | 69.1 | 71.7 | 100 | | | | | | | |
| 6 | 63.7 | 65 | 71.2 | 69.1 | 65.1 | 100 | | | | | | |
| 7 | 74.6 | 75.4 | 64.6 | 63.5 | 65.3 | 62.4 | 100 | | | | | |
| 10 | 62 | 61.4 | 68 | 76.5 | 69.2 | 65.9 | 59.6 | 100 | | | | |
| 11 | 72.8 | 74.4 | 64.8 | 63.2 | 64.1 | 62.2 | 78.8 | 63.7 | 100 | | | |
| 16 | 63.3 | 65.3 | 66.4 | 70.6 | 70 | 69.1 | 61.6 | 67.1 | 63.6 | 100 | | |
| 18 | 64.1 | 63.1 | 68.3 | 66.1 | 63.4 | 78.1 | 61 | 66.2 | 61.8 | 66.3 | 100 | |
| 20 | 63.2 | 64.1 | 66.7 | 86.3 | 72.6 | 68.1 | 60.2 | 77.4 | 62.9 | 70.8 | 66.2 | 100 |

Only the kinase domain (less introns) from each gene was compared.

quires a good deal of time and effort. A simpler approach would be to use oligonucleotide arrays, which provides the possibility of being both selective and flexible in the analysis of gene family expression profiles.

We designed oligonucleotide microarrays to represent a large number of the *Arabidopsis* MAP kinase or MAP kinase-like gene family. Oligonucleotide design was performed using the phylogenetic tree of the different family members and aligning the DNA sequences of the different subgroups. This enabled the synthesis of specific oligonucleotides for each gene. The oligonucleotides were from 19 to 26 bases long and designed such that they had a $T_m$ of 60°C or greater. These oligonucleotide sequences were compared against each other and against the entire *Arabidopsis* genome sequence to check for specificity. None of the oligonucleotides had more than 75% identity to any other sequence in the *Arabidopsis* genome. Cross-hybridization may occur between oligonucleotides with between 75% and 80% sequence similarity (24). These characteristics of the oligonucleotides allow for discrimination and specificity for each MAP kinase member.

In initial trials, oligonucleotides were designed for the 5′ and 3′ ends of each gene. However, it was found that hybridization to the 5′ oligonucleotide was variable and weak compared to the 3′ oligonucleotide, presumably because of the absence of full-length cDNAs in the hybridization probe. The use of 5′ oligonucleotides on the array in a screen of many different tissues and under many different conditions to analyze the expression profile of the MAP kinase genes would require that there always be good synthesis of full-length labeled cDNAs, something that is not always easy to control. Hybridization of a new set of microarrays containing only the oligonucleotide from the 3′ UTR or from the end of the open reading frame (ORF), spotted in replicate, indicated that such arrays are able to identify tissue-specific expression of MAP kinase genes (unpublished results).

This illustrates the flexibility of using these oligonucleotide arrays. Additional genes, or new probes for the same gene, can be easily represented on a chip by the simple synthesis of additional oligonucleotides. To date, we have used a limited number of gene oligonucleotides on the array to study gene family expression, but this can be easily scaled up. Modular chips can be synthesized and used in different combinations, depending on requirements and previous results. For example, the entire panoply of molecules from MAP kinase signaling pathways (receptors, G-proteins, MAPKKKs, MAPKKs, and MAP kinases) can be unified on a single chip, or with different elements on separate chips that may be combined, depending on results that show their involvement in a particular experimental system. Such an approach for studying gene family expression can complement genome-wide cDNA microarray analysis in which specificity may be a problem.

Full genome sequences provide valuable information regarding genome organization, evolution, and gene and protein structure and relatedness. As described here, interrogation of such sequence databases should be useful for the rational design of experiments to address the function of genes belonging to multigene families.

## ACKNOWLEDGMENTS

The design, synthesis, and analysis of the oligonucleotide arrays were performed in conjunction with VBC Genomics, Bioscience Research GmbH, Vienna, Austria. This work was supported by the Austrian Fonds zur Förderung der wissenschaftlichen Forschung, project number P14751-GEN. We thank Dr. Martin Grabner (EMBnet node, Austria) for help with data mining.

## REFERENCES

1. The Arabidopsis Genome Initiative. Analysis of the genome sequence of the flowering plant Arabidopsis thaliana. Nature 408:796–815, 2000.
2. C Widman, S Gibson, MB Jarpe, GL Johnson. Mitogen-activated protein kinase: Conservation of a three-kinase module from yeast to human. Physiol Rev 79:143–180, 1999.
3. MH Cobb. MAP kinase pathways. Prog Biophys Mol Biol 71:479–500, 1999.
4. AR Nebreda, A Porras. p38 MAP kinases: Beyond the stress response. Trends Biochem Sci 25:257–260, 2000.
5. S Noselli, F Agnès. Roles of the JNK signaling pathway in Drosophila morphogenesis. Curr Opin Genet Devel 9:466–472, 1999.
6. MJ Robinson, MH Cobb. Mitogen-activated protein kinase pathways. Curr Opin Cell Biol 9:180–186, 1997.
7. Y Miyata, E Nishida. Distantly related cousins of MAP kinase: Biochemical properties and possible physiological functions. Biochem Biophys Res Comm 266:291–295, 1999.
8. D Kültz. Phylogenetic and functional classification of mitogen- and stress-activated protein kinases. J Mol Evol 46:571–588, 1998.
9. T Mizoguchi, N Hayashida, KT Yamaguchi-Shinozaki, H Kamada, K Shinozaki. ATMPKs: A family of plant MAP kinases in Arabidopsis thaliana. FEBS Lett 336:440–444, 1993.
10. JL Riechmann, J Heard, G Martin, L Reuber, C Jiang, J Keddie, L Adam, O Pineda, OJ Ratcliffe, RR Samaha, R Creelman, M Pilgrim, P Broun, JZ Zhang, D Ghandehari, BK Sherman, G Yu. Arabidopsis transcription factors: Genome-wide comparative analysis among eukaryotes. Science 290:2105–2110, 2000.
11. G Blanc, A Barakat, R Guyot, R Cooke, M Delseny. Extensive duplication and reshuffling in the Arabidopsis genome. Plant Cell 12:1093–1102, 2000.
12. TV Moran, JC Walker. Molecular cloning of two novel protein kinase genes from Arabidopsis thaliana. Biochem Biophys Acta 1216:9–14, 1993.
13. SK Hanks, T Hunter. The eukaryotic protein kinase superfamily: Kinase (catalytic) domain structure and classification. FASEB J 9:576–596, 1995.
14. JD Thompson, TJ Gibson, F Plewniak, F Jeanmougin, DG Higgins. The ClustalX windows interface: Flexible strategies for multiple sequence alignment aided by quality analysis tools. Nucl Acids Res 24:4876–4882, 1997.
15. RDM Page. TREEVIEW: An application to display phylogenetic trees on personal computers. Comput Appl Biosci 12:357–358, 1996.
16. M Petersen, P Brodersen, H Naested, E Andreasson, U Lindhart, B Johansen, HB Nielsen, M Lacy, MJ Austin, JE Parker, SB Sharma, DF Klessig, R Martienssen, O Mattsson, AB Jensen, J Mundy. Arabidopsis MAP kinase 4 negatively regulates systemic acquired resistance. Cell 103:1111–1120, 2000.
17. V Voronin, A Touraev, H Kieft, AAM van Lammeren, E Heberle-Bors, C Wilson. Temporal and tissue-specific expression of the tobacco ntf4 MAP kinase. Plant Mol Biol 45:679–689, 2001.

18. T Mizoguchi, K Irie, T Hirayama, N Hayashida, K Yamaguchi Shinozaki, K Matsumoto, K Shinozaki. A gene encoding a mitogen-activated protein kinase kinase kinase is induced simultaneously with genes for a mitogen-activated protein kinase and an S6 ribosomal protein kinase by touch, cold and water stress in Arabidopsis thaliana. Proc Natl Acad Sci USA 93:765–769, 1996.

19. S Seo, H Sano, Y Ohashi. Jasmonate-based wound signal transduction requires activation of WIPK, a tobacco mitogen-activated protein kinase. Plant Cell 11:289–298, 1999.

20. PM Schenk, K Kazan, I Wilson, JP Anderson, T Richmond, SC Somerville, JM Manners. Coordinated plant defense responses in Arabidopsis revealed by microarray analysis. Proc Natl Acad Sci USA 97:11655–11660, 2000.

21. P Reymond, H Weber, M Damond, EE Farmer. Differential gene expression in response to mechanical wounding and insect feeding in Arabidopsis. Plant Cell 12: 707–719, 2000.

22. Y Ruan, J Gilmore, T Conner. Towards Arabidopsis genome analysis: Monitoring expression profiles of 1400 genes using cDNA microarrays. Plant J 15:821–833, 1998.

23. CS Richmond, JD Glasner, R Mau, H Jin, FR Blattner. Genome-wide expression profiling in Escherichia coli K-12. Nucl Acids Res 27:3821–3835, 1999.

24. MD Kane, TA Jatkoe, CR Stumpf, J Lu, JD Thomas, SJ Madore. Assessment of the sensitivity and specificity of oligonucleotide (50mer) microarrays. Nucl Acids Res 28:4552–4557, 2000.

# 10

# Industrialization of Plant Gene Function Discovery

**Douglas C. Boyes, Mulpuri V. Rao, Susanne Kjemtrup, and Keith R. Davis**
*Paradigm Genetics, Inc., Research Triangle Park, North Carolina, U.S.A.*

**Robert Ascenzi**
*BASF Plant Science L.L.C., Research Triangle Park, North Carolina, U.S.A.*

**Andreas Klöti**
*ETH Zurich, Zurich, Switzerland*

## I. INTRODUCTION

The past decade has witnessed a revolution in agriculture arising from our ability to apply the tools of molecular biology to plant breeding and cultivation. The realization that these molecular tools can be used to enhance the nutritive value of food and produce stress-and pathogen-resistant crops in an environmentally friendly manner has led to the whole-genome sequencing of various plant species, including *Arabidopsis thaliana* and rice. Understanding the function of all the genes required for normal plant growth and development will be a major challenge for plant biologists during the next decade. The efficient completion of this task hinges on the development of rapid and reliable methods for altering the expression of each gene, followed by a consistent and systematic approach for identifying corresponding phenotypes. The process of assigning functions to novel genes has been performed on a small scale for decades, but the means and initiative to perform this work on a large scale in the context of the entire genome is a recent development. In building on the nomenclature of the structural genomics revolution, large-scale systematic efforts to determine gene function are typically classified under the heading of "functional genomics."

In this chapter, we summarize the tools and technologies available for developing an industrialized plant functional genomics platform and discuss some of the potential benefits to be realized by using this approach. The first section of this chapter provides a survey of plant model systems and underscores the wealth of genetic information available as a starting point for this endeavor. The second section reviews the breadth of technologies available to generate plants with altered gene expression. This is followed by a framework description of a high-throughput platform for performing an extensive phenotypic analysis of mutant arabidopsis plants. Finally, we discuss the power and significance of the functional genomics approach with regard to the development of improved agrochemical compounds and crop germplasm.

## II.  PLANT MODEL SYSTEMS FOR FUNCTIONAL GENOMICS

Two species have emerged as the primary plant model systems for functional genomics. First, the small mustard plant *Arabidopsis thaliana*, closely related to the crop plants canola, cabbage, and cauliflower, serves as a model for dicot plants (http://www.arabidopsis.org). The monocot, rice (*Oryza sativa*), is a major food crop and serves as a second model for the world's most important food crops like corn, wheat, barley, and sorghum (http://genome.cornell.edu/rice).

Genome sequencing projects have been completed recently, for arabidopsis in the public domain (1), and for rice in the private sector (2). Both arabidopsis and rice were chosen for sequencing because they have comparably compact genomes of approximately 125 Mb and 430 Mb, respectively, with little interspersed repetitive DNA. Tens of thousands of expressed sequence tags (ESTs) have been sequenced and stored in public databases. With this wealth of molecular information, both arabidopsis and rice are ideal models for functional genomics in plants. In addition, both species are diploid with straightforward genetics. This is in marked contrast to many crop species, which exhibit polyploid genetics. When compared with their polyploid relatives, diploid model organisms are expected to contain fewer functionally redundant genes, making it easier to uncover phenotypes through single-locus mutagenesis strategies. Both arabidopsis and rice have a reasonably short generation time of 8 or 12 to 14 weeks, respectively, are self-fertile, and produce hundreds to thousands of seeds per individual.

An important criterion for many plant functional genomics approaches is that the target plant species be transformed easily. Both arabidopsis and rice can be genetically transformed very efficiently. Arabidopsis germline transformation can be achieved by dipping flowering plants in *Agrobacterium tumefaciens* suspensions (3). This process requires very little labor and is amenable to high-throughput implementation. Genetic transformation of rice, although more laborious and requiring tissue culture, can be accomplished efficiently by particle bombardment, as well as by *Agrobacterium*-mediated methods. Because rice is not only a model plant, but also a very important food crop for a large part of the human population, information gained from this model system can be applied directly to potential product development. Additionally, tens of thousands of rice varieties exist that can be tapped as sources of naturally occurring genetic variation. Similarly, hundreds of wild-type accessions of arabidopsis have been collected and provide a rich source of genetic variants that can be exploited for functional analysis (4).

Arabidopsis and rice are ideal model plants for the study of basic angiosperm biology. However, additional model systems are required for the identification and characterization of genes controlling certain specialized functions. For example, arabidopsis neither establishes symbiotic relationships with nitrogen fixing *Rhizobia*, nor is it typically colonized by beneficial arbuscular mycorrhizal fungi. Therefore, arabidopsis is not a suitable model plant for studying the genes involved in these plant/microbe interactions. In addition, arabidopsis is not an ideal system for studying specific natural biosynthetic pathways, such as isoflavonoids and sesquiterpenoids. Two legumes have emerged as model plants for some of these aspects, *Medicago truncatula* (5; http://www.noble.org) and *Lotus japonicus* (6). Both have relatively small, diploid genomes (400–500 Mb), short generation times (around 14 weeks), are self-fertile, and can be efficiently transformed using *Agrobacterium tumefaciens*. *Medicago truncatula*, like arabidopsis, can be genetically transformed by infiltration of flowers with *Agrobacterium* suspensions (7).

For more basic cellular processes and metabolic pathways common to most plants, including cell differentiation, cell division, and photosynthesis, simple plant organisms suffice as model systems. One such organism is the unicellular green alga, *Chlamydomonas reinhardtii*, which has been the subject of genetic, biochemical, and cytological analyses for more than 50 years. It has a genome size of around 100 Mb, tens of thousands of EST sequences are publicly available, and efficient genetic transformation of the nucleus and the chloroplast can be accomplished using a variety of methods (8–10). The moss *Physcomitrella patens* has a relatively simple lifecycle containing a major haploid gametophyte stage. A very appealing feature of *P. patens* is that transgenes can be integrated at reasonable frequency in its nuclear genome by homologous recombination (11). This precise knock-out of specific genes is unique to higher plants, and is therefore a very valuable tool for functional genomics.

## III. MUTANT GENERATION TECHNOLOGIES

The most straightforward means of defining the function of a novel gene begins with the identification of a plant with a mutation in the gene of interest. A role for the mutated gene is then elucidated through a comparison with wild-type plants. This genetic variation can occur naturally, but more commonly is the result of a specific mutagenesis strategy. Single gene mutations can be generated using targeted or random approaches. Random mutagenesis is suitable for gene function discovery based on a forward genetics strategy, in which a biological screen is used to identify mutant plants exhibiting a desired phenotype. This approach requires follow-up analysis to identify the mutated gene responsible for the phenotype. In contrast, a systematic reverse genetics approach, in which expression of a specific gene is altered resulting in an unknown phenotype, is conducted most readily using a targeted mutagenesis approach. The following section provides an overview of mutagenesis strategies that are available currently and discusses the suitability of each for random or targeted approaches in a functional genomics enterprise.

## A.  Homologous Recombination

The exchange of covalent linkages between DNA molecules in regions of highly similar sequences is referred to as homologous recombination (12). Homologous recombination has been exploited as a routine method of targeting insertions into genes in yeast, bacteria, and mammals; however, its utility in plants has yet to be realized. Recent studies with arabidopsis suggest that targeted homologous recombination is possible (13), albeit at a low frequency (1 in 1,000,000). Approaches incorporating double-strand DNA breaks show a modest increase in the frequency of homologous recombination events (14,15). If perfected, this technology may offer widespread advantages in generating both "knock-out" and subtle "knock-in" mutant alleles (13). Until then, alternative mutagenesis strategies will likely remain at the forefront.

## B.  Radiation and Chemical Mutagenesis

Several radiation sources and chemical treatments have been used to induce heritable mutations in plants. Numerous forms of radiation, including X-rays, γ-rays, fast neutrons, and ultraviolet (UV) rays have been demonstrated to induce mutations in several plant species (16). X-ray treatments often induce large deletions in the genome, whereas the effects of ionizing radiation are less severe, limited primarily to point mutations or small deletions. As a result, ionizing sources are generally preferred, with γ-rays and fast neutrons used most commonly. Several chemical mutagens are also effective in creating mutants in arabidopsis and other plant species. The most preferred chemical is ethylmethane sulfonate (EMS; 16).

A major advantage of these mutagenesis mechanisms is that genome saturation can be achieved relatively easily, allowing the recovery of multiple mutant alleles for each locus under study. Once the genetic map position of a selected mutation is determined, the gene can be cloned using conventional strategies, such as chromosome walking. This undirected mutagenesis strategy is best suited for model plants such as arabidopsis, to which dense genetic maps with many visible and molecular genetic markers are available.

Mutagenesis by radiation or chemical means is by nature nontargeted and traditionally, its use has been restricted to forward genetic screens. Recently, however, the use of chemical mutagenesis in a reverse genetic approach has become a possibility. Henikoff and colleagues have developed a robust and automated reverse genetic genome-scale functional analysis by combining chemical-induced point mutations with rapid mutational screening to discover induced lesions (17). In brief, this technique (TILLING) uses the DNA from an EMS mutagenized plant population and a screen for mismatches in heteroduplexes created by melting and annealing of heteroallelic DNA with a denaturing HPLC. The presence of a heteroduplex PCR product is detected in the chromatogram as an extra peak, which can be recovered and sequenced to characterize the mutant gene. TILLING is well suited for large-scale screening and can be adapted easily to any organism that can be heavily mutagenized, even those that lack extensive genetic tools. Furthermore, TILLING also has the potential to directly generate phenotypic variants in agricultural crops without introducing a foreign gene into a plant genome.

## C. Insertional Mutagenesis

Insertional mutagenesis refers to the disruption of genes with known mobile DNA elements. In most cases the insertion results in a loss-of-function phenotype; however, rare gain-of-function insertional mutants have been documented (18). The insertion disrupting the gene serves as a "tag" that can be used as a molecular probe to clone the gene.

Although they are ubiquitous in plants, transposable elements from only a few species such as maize and snapdragon have been characterized in detail. Autonomous genetic elements such as *Activator* (Ac) and *Suppressor-mutator* (*Spm*), are capable of activating a family of nonautonomous *Dissociation* (*Ds*) and *defective Spm* (*dSpm*) elements, respectively. These transposon systems have been engineered and used widely to generate tagged mutants in several heterologous plant species, including arabidopsis and tobacco (19–21). The transposon mutagenesis strategy has been enhanced through the addition of a reporter gene cassette into the mobile element, facilitating the isolation of genes with interesting expression patterns. Such lines can also be used to mark certain cell types for developmental studies (22).

Similar in concept to transposon mutagenesis, T-DNA mutagenesis is based on the disruption of genes by the transfer of DNA by *Agrobacterium tumefaciens*. This technique has been used in several plant species, and the ease with which arabidopsis can be transformed has made it the greatest beneficiary of this mutagenesis approach (23–25).

## D.  Activation Tagging

The major drawback in loss-of-function mutant screening is that it is usually
unable to produce a detectable phenotype in cases where a gene is a member of
a functionally redundant family. Genome sequencing of arabidopsis, rice, and
other plant species revealed the existence of many duplicated genes. The identification of gain-of-function phenotypes offers an alternative for defining a mutant
for these genes (26). The traditional approaches to induce gain-of-function mutations include chromosome rearrangements, transposons, or ectopic overexpression of the gene. One novel approach to induce gain-of-function mutations includes construction of a T-DNA vector with four copies of an enhancer element
from the constitutively active promoter of the cauliflower mosaic virus (CaMV)
*35S* gene. These enhancer elements are believed to cause transcriptional activation of nearby endogenous genes (27). This activation method may be advanta-
geous over standard ectopic overexpression, as it may more faithfully reflect the
pattern and timing of endogenous gene expression (26). Once a mutant of interest
is identified, the activated gene can be cloned easily by plasmid rescue. The 1 in
1,000 frequency of dominant morphological mutants obtained using this approach
was found to be similar to that observed in a transposon mutagenized population
(26).

The insertional mutagenesis strategies discussed above result in a high percentage of stable, single-locus insertions and, through standard hybridization or
PCR-based techniques, allow relatively straightforward identification of the disrupted gene. These approaches are effective in generating random mutations in
plant genomes and are well suited for both forward and reverse genetic functional
genomic approaches. Based on the genome size of arabidopsis, it is estimated
that a collection of more than 100,000 individual insertional lines are required
for the saturation of the genome (28, 29). Thus, successful reverse genetics approaches typically involve a sample pooling strategy before the initial round of
amplification to increase the efficiency and throughput of the process of identifying the desired insertion line. Deconvolution of the pool to a single transgenic
line containing the desired insertion requires several subsequent rounds of PCR.
Once the mutant line is isolated, it must be outcrossed to wild-type remove any
T-DNA or transposon insertions that are unlinked to the target insertion locus.
The resulting line is then available for phenotypic characterization.

## E.  Antisense

Antisense technology has been used to assess gene function in plants since shortly
after the initial report of Izant and Weintraub (30). This report demonstrated
the effect of antisense RNA expression on the inhibition of endogenous gene
expression. Typically, antisense transgenes are integrated in the plant genome

and the in vivo transcribed antisense RNAs interact with their endogenous target mRNAs, resulting in the suppression of gene expression. Several factors such as T-DNA copy number, the secondary structure of the antisense RNA and other external factors modulate the effectiveness of antisense inhibition (31, 32). As the antisense effect is gene specific and independent of ploidy level, it has become a valuable method for creating targeted mutants.

## F.  Site Specific Gene Silencing

Antisense RNA suppression can be inconsistent, often requiring the analysis of many independent transformation events to identify ones with sufficient suppression of target gene expression. To overcome these limitations, several novel strategies of site-specific gene silencing have been developed and are being evaluated for use in high-throughput functional genomics platforms.

### 1.  Virus-Induced Gene Silencing

After the initial report on the ability of a virus vector carrying a fragment of phytoene desaturase (*PDS*) to suppress the expression of endogenous *PDS* mRNA, Baulcombe and colleagues developed a highly efficient technology that exploits virus vectors to induce gene silencing in plants on a large scale (33). Recombinant viral vector constructs carrying a fragment of a plant host gene are introduced into plant cells either by directly inoculating the infectious DNA on the plant or by *A. tumefaciens*-mediated transfer of DNA. The mechanism by which viral elements induce gene silencing is not well understood. Current research indicates that plants specifically recognize viral nucleic acids and induce gene silencing by an RNA-mediated defense mechanism (33, 34). A major advantage of virus-induced gene silencing (VIGS) is that it targets the homologs of a multigene family exhibiting 20% to 25% sequence divergence. Thus, this technique may be useful in revealing phenotypes in cases where the target gene is a member of large family of functionally redundant genes.Virus-induced gene silencing appears to be independent of position effect, and suppresses gene expression consistently. This is significant in that it minimizes the number of lines that need to be analyzed to find one with the desired level of endogenous gene suppression. Although most genes appear susceptible to VIGS, most viruses cannot penetrate meristematic zones of plants. Thus, the utility of VIGS in the functional analysis of genes expressed exclusively in meristems may be problematic.

### 2.  Homology-Dependent Gene Silencing

The presence of multiple copies of a particular sequence in a genome often results in a type of epigenetic inactivation termed homology-dependent gene silencing (HDGS; 35). Paradoxically, in HDGS, an increase in gene copy number leads to decreased gene expression. Two distinct types of HDGS have been distinguished: transcriptional gene silencing (TGS) and posttranscriptional gene silencing (PTGS). Transcriptional gene silencing mechanisms are most likely mediated by sequence-specific de novo methylation of nuclear DNA, whereas PTGS postulates sequence-specific degradation of aberrant RNAs (36). The sequence specificity and dominant character of HDGS can be exploited to silence all members of gene families that share high sequence similarity. Furthermore, PTGS based on 5′ untranslated regions or TGS based on promoter homology can be used to inactivate individual alleles or specific gene family members with unique sequences in these regions.

### 3.   Chimeric Oligonucleotide-Induced Gene Silencing

A strategy currently being used to induce gene silencing in animal systems uses self-complementary chimeric oligonucleotides (COs) composed of DNA and 2′-O-methyl RNA to target and mutate genes in vivo (37, 38). Extending the applicability of COs to mutate genes in plants, May and colleagues have reported that the creation of self-complimentary COs can create stable, site-specific targeted mutations in a plant nuclear gene (39). Although the precise mechanism by which COs induce gene-specific mutations in vivo is not known, it has been hypothesized that CO-induced mutations are dependent on DNA repair enzymes that recognize mismatches between the targeted gene and the CO that was designed to create a mismatch (40). A major advantage of this technique, if perfected, is that it will provide an opportunity to create mutations in multiple genes simultaneously. For instance, a mutation useful for selection could be generated simultaneously with the inactivation of an unlinked target gene. Subsequently, the selectable marker could be segregated away, leaving the desired target gene mutation in an otherwise wild-type genetic background. As with VIGS, direct gene knock-outs by COs would be of value to inactivate specific members of plant multigene families and could also be used to alter untranslated or nontranscribed regions of a gene.

### G.   Natural Variation

In arabidopsis, the majority of work characterizing phenotypes resulting from induced mutations has been conducted using a limited number of inbred strains (ecotypes), such as Landsberg *erecta* (Ler), Columbia (Col), and Wassilewskija (Ws). Focusing the analysis on a small number of well-characterized ecotypes has clear advantages for the coordination of map-based cloning efforts, but these ecotypes represent just a small fraction of arabidopsis genetic diversity found throughout the world. Indeed, surveys of inbred ecotypes have revealed considerable variation in traits, such as resistance to biotic and abiotic stress factors, flowering time, plant size, seed size, seed dormancy, phosphate uptake, water-use efficiency, wax compositions and several enzymatic activities (4, 41). Exploitation of this genetic variation can be used as a complementary tool for the functional analysis of the arabidopsis genome otherwise being conducted by generating induced mutants. The identification of natural variation is strongly dependent on high quality genetic maps and statistical methods to locate quantitative trait loci (QTLs). The availability of highly efficient molecular markers such as RFLPs, CAPSs, and AFLPs, together with the complete genome sequence and the high frequency of polymorphisms observed between ecotypes, will allow a rapid and accurate linkage mapping and eventual cloning of these loci. Exploitation of natural variation will facilitate the discovery of functions for genes with phenotypes that cannot be induced by conventional mutagenesis strategies. In addition, an analysis of alternative allelic variants of well-characterized genes may lead to the discovery of new or modified functions for those genes.

## IV. PHENOTYPIC CHARACTERIZATION OF MUTANTS

Many of the mutagenesis strategies detailed in the preceding section are amenable to high-throughput implementation. The success of these strategies is clearly defined and can be assessed readily through sequence analysis of the resulting mutation. It is a greater challenge, however, to identify a phenotype resulting from the specific gene mutation. Indeed, this process is typically the rate-limiting step in a systematic functional genomics platform.

Detailed analysis of the recently completed genome sequence of arabidopsis identified roughly 25,500 genes. Of these, nearly 30% have no assigned cellular function, and only 9% have been characterized experimentally (1). Those with no defined cellular function pose the greatest challenge, because this group contains no molecular or experimental basis for the design of a phenotypic screen. After development of a plant mutant for the assigned gene, the investigator is faced with determining which traits of the plant to measure. Because a single gene can affect any aspect of plant development or physiology, the investigative task becomes intrinsically daunting. Another difficult task involves the analysis of genes having a functional classification based primarily on sequence similarity with genes of known cellular or enzymatic activity. Although this preliminary classification may offer some guidance in experimental design, the large number of different genes encompassed by each class does little to simplify the task of a more comprehensive functional assignment (e.g., at the whole plant level). Indeed, even those genes that fall into the well-characterized category pose a challenge. Even if these genes were identified in a forward genetic approach on the basis of a positive outcome in a phenotypic screen, it remains desirable to determine if the mutation impacts other, unexpected aspects of the plant as well.

Many factors may preclude the identification of a phenotype resulting from a specific gene mutation. Biological effects, such as the presence of a functionally redundant gene, can obscure a phenotype. To some extent this problem can be overcome through the selection of an appropriate mutagenesis strategy. In many cases, if a functionally redundant gene exists, it will be a structural homolog of the gene under investigation. Thus, in model systems where the complete genome sequence is available, it is possible to generate strains affected in the expression of multiple gene family members that may be functionally redundant. Another alternative is to rely on a gain-of-function mutant generation strategy, such as activation tagging (see section III. D). A second factor precluding the identification of a phenotype is the fact that many genes have evolved highly specialized functions. Mutants carrying defects in these genes often show no, or very subtle phenotypes when grown under optimal conditions. Discovery of a phenotype for these mutants is only possible when assayed under the environmental condition(s) in which function of the gene is normally required.

The development of a systematic, high-throughput phenotypic screening platform is dependent on two criteria for success. First, data collected during the analysis need to reflect a broad range of traits representing the entire developmental timeline of the organism. This accomplishes two things. It allows the same process to serve as foundation for the analysis of all mutants and eliminates any bias that might otherwise be imparted as a result of predictive or previous functional assignment of the genes under study. It also results in a complete dataset for each gene under study. Second, phenotypic data collected from each mutant must be directly comparable. In much of the plant scientific literature, phenotypic data have been collected with reference to the chronological age of the subject plant. However, the rate of development can be affected greatly by environmental conditions as well as the very mutations under study. Thus, a data collection process based strictly on chronological age of the plant can result in a situation in which plants of vastly different developmental states are the subjects of comparison. In these cases, it becomes difficult to discriminate phenotypic abnormalities caused by genetic differences from those that simply reflect a difference in the developmental state of the plants being compared. The incorporation of a reference to growth or developmental stage in all data collection steps can be used to generate a robust dataset that will allow a more direct comparison of data from disparate sources. Indeed, similar methods for data collection are commonplace for other model systems including *D. melanogaster* (42) and *C. elegans* (43).

## A.  Collection of Phenotypic Data Based on Growth Stage

We have developed a platform for conducting high-throughput phenotypic analysis of arabidopsis mutants. The cornerstone of this platform is a data collection

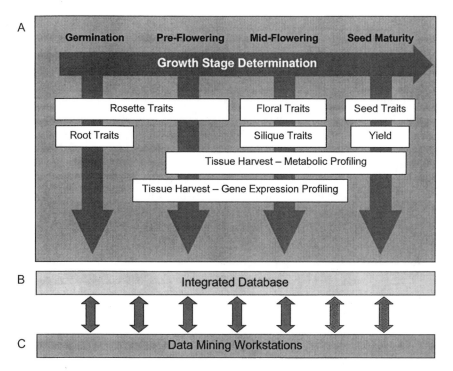

**Figure 1** Generalized data collection, integration and analysis model. **(A)** The main *horizontal arrow* represents a continuous measurement platform from which growth stages are determined. At key landmark growth stages (*vertical arrows*), specified data collection modules are invoked. These modules may include detailed morphological analyses relevant for that growth stage or the harvesting of tissue for various molecular or biochemical analyses. All sample tracking and data collection activities in **(A)** occur within the context of a computerized Laboratory Information Management System (LIMS). Resulting data are integrated into a common database structure **(B)**, which is then available for mining through a series of analysis tools and workstations **(C)**.

strategy based on series of distinct developmental growth stages that encompass the entire lifecycle of the plant.

A useful way to visualize data collection based on growth stages is as a continuum of measurements onto which modules of additional detailed data collection steps can be integrated at the appropriate growth stages (Fig. 1). Depending on the research goal, new data collection modules can be added or deleted. These modules may consist of detailed morphological analyses (e.g., leaf area, floral structure) at a specific growth stage, or the harvest of specific tissues for a variety of molecular or biochemical analysis. Examples of the latter include the analysis of global gene expression profiles using microarrays, and the analysis of metabolite profiles using analytical biochemistry techniques, such as mass spectrometry. The industrialization of this process relies heavily on customized computer systems, which track the samples throughout the process, drive the data collection activities, and provide an interface and tools for the analysis of the data. Figure 1 presents a framework model for the collection, integration, and mining of morphological, developmental, molecular, and biochemical data.

Growth stage definitions for agronomically important crops and weeds have been established previously and are used frequently in conventional breeding programs. The BBCH scale is one such example and was developed by a consortium of agricultural companies (BASF, Bayer, Ciba-Geigy and Hoechst) for the express purpose of uniform communication within the agricultural community (44). We have adapted the BBCH scale for arabidopsis, as outlined in Table 1. Congruent with the BBCH scale, there are 10 principal growth stages that extend across the entire developmental timeframe of arabidopsis. For descriptive purposes, each principle growth stage is subdivided based on secondary developmental events germane to the principal stage. The secondary growth criteria are also assigned numbers 0 through 9. For example, seed germination is principal growth stage 0, and the first developmental step in seed germination—seed imbibition—has been assigned the number 1. Hence, the seed imbibition event is growth stage 0.1.

The adaptability of the BBCH scale becomes more apparent during principal growth stage 1—leaf development. Leaf development is a longer and more complex event. For this reason, the major growth characteristic—number of rosette leaves—has been further subdivided to detail the number of emerging leaves. As an example, growth stage 1.10 is defined as the period when ten rosette leaves have emerged. This open-ended structure ensures that a late-flowering mutant that continues to produce rosette leaves before setting an inflorescence can still be classified easily (e.g., a 1.19 plant would have 19 rosette leaves).

## B.  High-Throughput Phenotypic Analysis of Arabidopsis

The foundation of the data collection model adapted for arabidopsis consists of a series of core measurements that determine growth stage. The growth stages 1.00, 1.04, 1.10, 6.00, 6.50, 6.90, and 9.70 have been chosen as developmental landmarks that trigger additional data collection events. These events include the collection of detailed quantitative and qualitative morphological data, and in some cases, tissue harvests for coordinated gene expression and metabolic analysis. The following is a description of our data collection model for arabidopsis (Fig. 2).

The stages of germination up to the four rosette-leaf stage (growth stage 1.04) of development are best observed by growing plants in agar medium on

**Table 1** *Arabidopsis* Growth Stages for Phenotypic Analysis

| Stage | Description |
|---|---|
| Principal growth | Seed germination |
| 0.10 | Seed imbibition |
| 0.50 | Radicle emergence |
| 0.70 | Hypocotyl and cotyledon emergence |
| Principal growth | Leaf development |
| 1.0 | Cotyledons fully opened |
| 1.02 | 2 rosette leaves > 1 mm in length |
| 1.03 | 3 rosette leaves > 1 mm in length |
| 1.04 | 4 rosette leaves > 1 mm in length |
| 1.05 | 5 rosette leaves > 1 mm in length |
| 1.06 | 6 rosette leaves > 1 mm in length |
| 1.07 | 7 rosette leaves > 1 mm in length |
| 1.08 | 8 rosette leaves > 1 mm in length |
| 1.09 | 9 rosette leaves > 1 mm in length |
| 1.10 | 10 rosette leaves > 1 mm in length |
| 1.11 | 11 rosette leaves > 1 mm in length |
| 1.12 | 12 rosette leaves > 1 mm in length |
| 1.13 | 13 rosette leaves > 1 mm in length |
| 1.14 | 14 rosette leaves > 1 mm in length |
| Principal growth | rosette growth |
| 3.20 | Rosette is 20% of final size |
| 3.50 | Rosette is 50% of final size |
| 3.70 | Rosette is 70% of final size |
| 3.90 | Rosette growth complete |
| Principal growth | Inflorescence emergence |
| 5.10 | First flower buds visible |
| Principal growth | Flower production |
| 6.00 | First flower open |
| 6.10 | 10% of flowers to be produced have |
| 6.30 | 30% of flowers to be produced have |
| 6.50 | 50% of flowers to be produced have |
| 6.90 | Flowering complete |
| Principal growth | Silique filling |
| Principal growth | Silique ripening |
| 8.00 | First silique shattered |
| Principal growth | Senescence |
| 9.70 | Senescence complete; ready for seed |

*Source*: Adapted from Ref. 44.

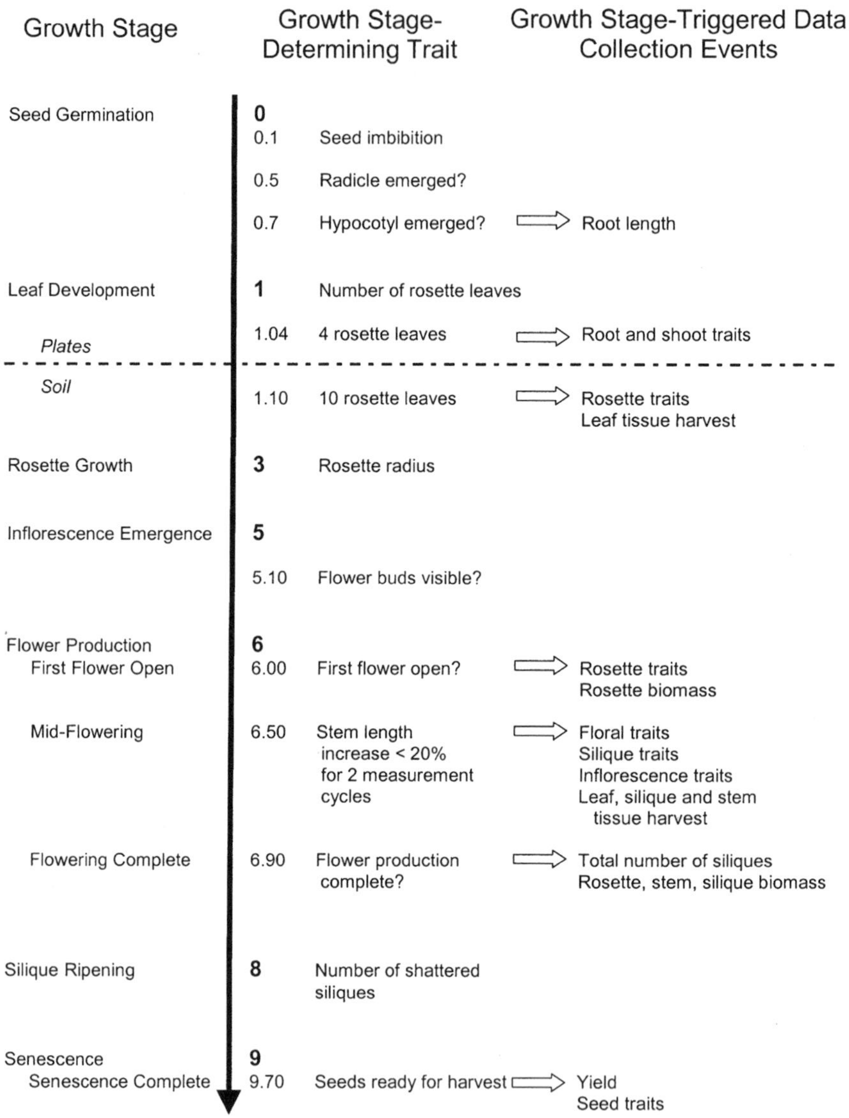

| Growth Stage | Growth Stage-Determining Trait | | Growth Stage-Triggered Data Collection Events |
|---|---|---|---|

| Seed Germination | **0** | | |
| | 0.1 | Seed imbibition | |
| | 0.5 | Radicle emerged? | |
| | 0.7 | Hypocotyl emerged? | ⇨ Root length |
| Leaf Development | **1** | Number of rosette leaves | |
| *Plates* | 1.04 | 4 rosette leaves | ⇨ Root and shoot traits |
| *Soil* | 1.10 | 10 rosette leaves | ⇨ Rosette traits / Leaf tissue harvest |
| Rosette Growth | **3** | Rosette radius | |
| Inflorescence Emergence | **5** | | |
| | 5.10 | Flower buds visible? | |
| Flower Production / First Flower Open | **6** / 6.00 | First flower open? | ⇨ Rosette traits / Rosette biomass |
| Mid-Flowering | 6.50 | Stem length increase < 20% for 2 measurement cycles | ⇨ Floral traits / Silique traits / Inflorescence traits / Leaf, silique and stem tissue harvest |
| Flowering Complete | 6.90 | Flower production complete? | ⇨ Total number of siliques / Rosette, stem, silique biomass |
| Silique Ripening | **8** | Number of shattered siliques | |
| Senescence / Senescence Complete | **9** / 9.70 | Seeds ready for harvest | ⇨ Yield / Seed traits |

**Figure 2**  A description of the arabidopsis model for data collection based on the modular model in Figure 1. The vertical *arrow* represents the developmental continuum of an arabidopsis plant. Growth stages assayed in the plate-based early analysis platform are outlined above the dotted line and those assayed in the soil-based analysis platform are described below that line. Primary growth stages are given in bold.

plates oriented vertically such that root structure as well as early leaf development can be evaluated. In this early development analysis, primary observations are taken at 1- to 2-day intervals to determine if a growth stage has been achieved. The achievement of a specified growth stage is recorded by the computer system as an affirmative answer to a specific query. For instance, an answer of "yes" to the question "Have the cotyledons and hypocotyl emerged from the seed coat?" indicates that the plant has reached stage 0.7. Once growth stage 0.7 is reached, the ensuing round of data collection is initiated by determining if growth stage 1.0 has been reached, and so on. As a landmark growth stage, stage 1.0 triggers a series of periodic data collection steps, including a qualitative assessment of seedling size, shape, and color, as well as root length and number of rosette leaves. The early phenotypic analysis is complete when root length reaches 6 cm or on day 14 after sowing, whichever occurs first.

Data describing the subsequent stages of development, through senescence and seed maturity (growth stage 9.70), are collected from plants grown on soil. In this assay, the development of the arabidopsis plant is analyzed from growth stage 1.04 (four rosette leaves) until growth stage 9.7 (senescence complete), a period that spans approximately 70 days. Not all of the growth stages in the BBCH scale are amenable for use as real-time data collection triggers. In some cases, analysis of collected data may show a correlative relationship that will serve as a working definition for that stage. For example, the BBCH scale defines stage 6.50 as the point when 50% of flowers have been produced, a measurement that is difficult to assess, as the final number produced by the plant is not known a priori. In the course of conducting a baseline analysis of the Col-0 ecotype, we found a correlation of stage 6.50 with a decrease in the rate of stem elongation (Fig 3A). Consequently, our working definition of stage 6.50 is a stem height increase of less than 20% for two consecutive measurement cycles. Stage 9.7 offers a similar growth stage-defining problem. Although the BBCH definition of stage 9.7 is senescence complete and seed ready for harvest, the timing with which this occurs differs along the inflorescence of each plant. This frequently results in seed loss caused by dehiscence (shatter) of the oldest siliques before the younger ones are ready for harvest. Therefore, to reduce seed loss and variation in yield from silique shatter, the intact inflorescence is collected at stage 6.9 (flower production complete) and dried. Stage 9.7 is then defined as the period when those siliques are dry enough for harvest.

Through the incorporation of the modifications described above, we have developed growth stages 1.10, 6.00, 6.50, 6.90, and 9.70 as triggers for secondary data collection in the soil-based assay. Stage 1.10 triggers the qualitative assessment of rosette traits. Stage 1.10 is typically one of the last stages before the transition to flowering and, therefore, is a source of tissue for the analysis of preflowering gene expression and metabolite composition. At stage 6.00 (first flower open), the rosette has typically reached its maximum size. Therefore, this

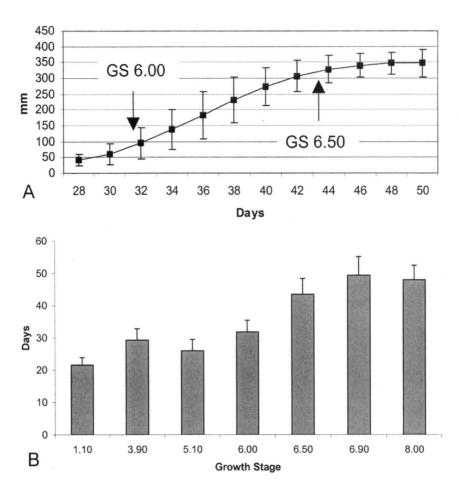

**Figure 3** Baseline analysis of Col-0 (**A**) Temporal data derived from the stem length measurement. *Arrows* indicate the timing of stages 6.00 and 6.50. (**B**) Vertical bars represent the average number of days required to reach each growth stage indicated. The error bars are standard deviation. Numbers are an average of approximately 300 wild-type Col-0 plants.

stage is well suited for the analysis of quantitative rosette traits such as leaf area and biomass. At stage 6.50 (midflowering), the inflorescence exhibits a continuum of floral and fruit development. Thus, an analysis of flower and fruit structure, as well as an assessment of reproductive success on a per-fruit basis can be performed readily at this stage. Stage 6.50 is also an appropriate stage for tissue collection to assess postflowering gene expression and metabolic profiles of specific organs. Stage 6.90 marks the completion of flowering and provides an opportunity to determine the final plant height, number of fruits produced, and the biomass of leaves, stems, and fruits. At stage 9.70 seeds are harvested and yield is assessed. Once seeds are available, image analysis can be used to quantify seed size and shape. The overall modular nature of this platform allows the addition or subtraction of harvests, or other data collection activities, depending on the ultimate goal of the process.

An analysis of data collected from the arabidopsis wild-type Col-0 ecotype shows that achievement of specific growth stages is reproducible (Fig 3B). This statistically robust baseline provides a solid foundation for the development of alternative conditions in which the growth stage measurements can be performed. For example, the effects of a variety of stress conditions on plant development can be measured and compared to the baseline. In one model stress assay, plants are continuously subject to stress conditions (e.g., cold or heat treatment) throughout the entire measurement period. Alternatively, the growth stage method can be used as a way to synchronize plants. A stress applied at a specific growth stage establishes a uniform developmental period from which to compare results.

The previous discussion centered on the suitability of growth stage analysis in arabidopsis for wild-type, mutant, and stress analysis. By extension, growth stage analysis can also be established for other model plants and crops. With the recent completion of the rice genome sequence, development of a similar platform for the analysis of rice mutant phenotypes is a likely extension of this phenotypic analysis strategy. Indeed, rice was used as one of the models for development of the BBCH scale.

## V. CURRENT APPLICATIONS OF INDUSTRIALIZED FUNCTIONAL GENOMICS

Plant functional genomics research should lead to rapid improvements in global crop production, nutrition, and human health. Indeed, the main goal of plant genomics is to understand how plants themselves function, so that they may be modified in ways that benefit our expanding population. Throughout history, each discovery of a new technology to modify plants has led to improvements in human welfare. For example, the domestication of plants by prehistoric humans through plant selection resulted in the transition from a hunter-gatherer lifestyle to an agrarian lifestyle. The resulting sedentary way of life eventually gave rise to civilizations (45). The discovery of genetics in the 20th century likewise led to dramatic improvements in plant breeding and, ultimately, to the "green revolution." In the last 20 years, it has become possible to "engineer" plants by adding (by transformation) one or more useful genes from another species which may not otherwise be transferred naturally. Currently, the major limiting step in harnessing the power of plant biotechnology is the dearth of gene function information. Therefore, gene discovery through functional genomics strategies will accelerate the pace of plant improvement in years to come.

The two main areas for plant improvement are in agronomic traits (input) and quality traits (output). Agronomic traits include those that increase yield, for example through a reduction in the use of herbicides and pesticides or increased tolerance to environmental stresses. Quality traits are those related to the nutritional value, tensile strength, or the other inherent qualities of the plant product. The industrialization of gene function discovery will allow scientists to ultimately improve both types of traits.

The modern generation of pest-resistant plants and herbicide-tolerant plants has increased yield while reducing chemical inputs. One example of such a modified plant is glyphosate-resistant soybeans. A report sent to the U. S. House Committee on Science (46) concluded that the use of Roundup Ready soybeans decreased the use of herbicides by 40% and saved farmers $30/Ha. Herbicide-tolerant crops will also allow farmers to adopt no-till agriculture, a technique that limits erosion and soil loss.

In its current early state, plant functional genomics can quickly lead to the discovery of novel herbicides and the simultaneous development of crops resistant to these new herbicides. Herbicides function through an interaction with proteins essential for life. These proteins are called herbicide targets. A useful herbicide target is one that is required for viability but is distinct from proteins in non-pest organisms. One way to discover new targets is through a combination of comparative and functional genomics. Gene sequences can be compared within a genome and across genomes to identify plant genes that are likely to be essential yet distinct from animal genes. This comparative work can be complemented by a more direct experimental approach in which essential genes would be identified by virtue of a lethal phenotype when "knocked out." Many of the mutant generation strategies described above (section III) could be used to obtain lines that can be screened for lethality. Screening can most simply be accomplished by a mutagenesis strategy such as T-DNA insertion followed by observing lethal phenotypes in homozygous plants and examining developing T2 seeds. Alternatively, a system in which known genes are knocked out by antisense or gene silencing can be used in conjunction with a transactivation system (47). In such a system, an antisense construct corresponding to a putative target gene is present in a latent state until combined with a transgene designed to drive expression of the target transgene. The transgenes can then be combined by sexual crossing and the resulting F1 population screened, as described above. This strategy provides an added advantage in that one is able to immediately know the identity of the target gene, unlike with chemical or insertional mutagenesis. A validated target gene could then be expressed and the protein screened in vitro against diverse chemical libraries of up to a million compounds in a high-throughput platform (48). Any hits resulting from such an assay would then lead to in vivo microscreening. Synthesis of the lead compound followed by greenhouse and field-testing would be the final steps in product development.

Another area for crop improvement is increasing yield through enhanced tolerance to environmental stress. Boyer (49), on the basis of comparing optimal yields with typical yields, has estimated that major crops in the United States achieve only 20% of their full genetic potential. The remaining 80% is lost to environmentally imposed stresses. In addition, marginal lands in other parts of the globe restrict the extent of agriculture. Traditional breeding efforts have had limited success with enhancing stress tolerance. This is most likely the result of the genetic and physiological complexities of plant stress responses. Understanding the processes underlying the stress response will undoubtedly open new avenues for obtaining stress-tolerant crops. Even with our limited knowledge, genetic engineering of plants with stress-related genes has already shown great potential. One of the more promising results involves expression of the gene encoding DREB1A/CBF3 under control of the stress-inducible promoter of *RD29A* (50). Transgenic arabidopsis lines in this study were notably more tolerant to imposed salinity, freezing, and drought stresses, whereas there was a minimal effect on growth during nonstressful conditions. These and other results suggest that DREB1A is a master regulator of dehydration-related stress responses. Another example that illustrates the potential of genomics to improve stress tolerance is the demonstration that moderate overexpression of a homologous cDNA encoding a $Na+/H+$ antiporter in arabidopsis confers tolerance to salt (51). This result suggests that glycophytes (species not naturally salt tolerant) possess the genes to become salt tolerant and that changes in the regulation of those genes is sufficient to produce a salt tolerant phenotype. In fact, morphological change in plants and animals is thought to be driven transcriptional regulation by changes in promoter sequences (52, 53). It is likely that non-morphological evolution (e.g. physiological changes) may also be caused by alterations in gene regulation.

One may envision a straightforward means of comprehensively identifying and then using genes involved in tolerance to stress or uptake and utilization of nutrients based on the premise that environmental adaptation occurs at the level of gene regulation. An experimental approach, whereby genes may be misregulated (either under- or overexpressed) can be used to identify those involved in the environmental condition of interest. In this scenario, a prioritized list of genes or all genes in model plants such as arabidopsis can be cloned in both sense and antisense orientations. Transformed plants could then be analyzed under defined conditions, enabling one to identify both tolerant and hypersensitive mutants. By this approach, one should be able to identify genes involved in plant environmental adaptations, and in some cases, even genes that can be used to directly engineer desirable agronomic traits.

## VI. CONCLUSIONS AND FUTURE PERSPECTIVES

The pace of gene function discovery will rapidly accelerate as a result of (1) increasing functional knowledge of a greater percentage of model genomes; (2) more comprehensive knowledge of each individual gene; (3) improving genomics tools (such as bioinformatics). The industrialization of gene function assignment described earlier in this chapter should advance genomics in the three areas stated above through its broad yet consistent and integrated approach to gene function discovery. That DREB1A controls the dehydration "regulon" was discovered by molecular genetics techniques from the last decade. We believe a new genomics approach can reveal similar regulons specific to salinity stress, water-deficit stress, and high-temperature stress. Coordinately regulated genes can be unveiled by use of gene expression profiling. In addition, mutations in genes yielding similar phenotypes may be involved in the same pathway. Ultimately, plant scientists will have a "wiring diagram" of model plants through their entire lifecycles. As part of this conceptual diagram, similar but smaller networks of gene interactions will be associated with specific nonstandard environmental conditions. This virtual plant will certainly aid applied scientists interested in plant improvement, as well as enrich our understanding of the diversity of life and, ultimately, interactions between plants and other organisms.

As the genomes of model plant systems, such as arabidopsis, rice, and *Medicago truncatula* become better defined, the focus for subsequent detailed research will shift. New genomics tools will be used to study the biology of more recalcitrant plants with intrinsic value or having unique adaptations. Important examples include wheat, cotton, commercially valuable trees, and medicinal plants. Corn, wheat, and other cereal crops, despite having large genomes, are important because grain is the major component of the human diet in the developing world. Understanding the specific properties of these cereal genomes should help with eventual engineering of these crops for food and renewable resource production. Likewise, the genomes of woody and medicinal plants are likely to provide insight on the production of secondary metabolites.

The future applications of plant functional genomics-derived knowledge are diverse and will impact both input and output traits. It is through improvement of the input traits that maximal production could be achieved in a sustainable way to meet the demands of a growing population. Specific input traits not detailed in the previous section include improved photosynthesis, disease resistance, and nutrient acquisition. The latter two applications will increase yield and decrease chemical inputs in agriculture. In the realm of quality-related traits, the focus should be on improving nutrition, engineering foods with therapeutic or medicinal qualities ("pharmafoods"), and providing renewable resources.

Biotechnology has the potential to decrease the use of environmentally harmful agrochemicals (such as pesticides and fertilizers) by introducing sources of pathogen resistance and improving nutrient acquisition. Functional genomics has already contributed to dissecting mechanisms of plant pathogen resistance. Using DNA microarrays, Maleck and colleagues (54) have identified a common promoter element in genes of the systemic acquired resistance regulon. The next step is to examine the function of transcription factors associated with this element. One can envision that moderate and inducible expression members of the plant-specific transcription factor family found by Maleck et al. (54), followed by a pathogen screen, would yield a discovery similar to the master regulator of dehydration, DREB1A.

Another clear area for genomics is in nutrient acquisition, including both macro- and micronutrients. Narang and colleagues (55) recently provided data that illustrate how a functional genomics strategy can be used to increase nutrient uptake efficiency. In this study, various arabidopsis ecotypes were found to vary in their efficiency of phosphate uptake. While their roots differed both physiologically and morphologically, the $Km$ values of the phosphate transporters were found to be the same in the most divergent ecotypes. However, the $I_{max}$ was greater in those ecotypes exhibiting greater phosphate uptake efficiency. This suggests that the observed variation is the result of differential regulation of the genes in the ecotypes, not the coding sequence of the genes. If this is the case, then a strategy similar to the one described for the functional genomics of stress tolerance can be used to identify genes involved in nutrient uptake and, possibly, utilization.

One issue not yet addressed is the functional genomics of nontranscribed DNA. Although a smaller problem than transcriptome genomics, little is known about how chromosomal structure influences high-fidelity gene expression and inheritance. One great advance would be the development of plant artificial chromosomes, which could be engineered to contain entire pathways (involved in such adaptations as synthesis of important secondary metabolites or resistance to pathogens) and transferable into any plant of interest. Therefore, the study of nontranscribed regions, including centromeres, telomeres, and putative nuclear subdomain targeting regions, is an area of plant genomics that should not be neglected.

Genomics is essentially a descriptive science and as we are in the early stages of this technological development, we do not have a full understanding of the malleability of plant biological processes. Obtaining a thorough understanding of plant genomes, both transcribed regions and structural components, will suggest applications that are not conceivable currently. The knowledge gained through the systematic functional analysis of arabidopsis and rice genes will provide a technological base that can be tapped for future decades in the quest of developing plant-derived products to benefit the world and its ever-expanding population.

## REFERENCES

1.  The *Arabidopsis* Initiative. Analysis of the genome sequence of the flowering plant *Arabidopsis thaliana*. Nature 408:796–815, 2000.
2.  D Dickson, D Cyranoski. Commercial sector scores success with whole rice genome. Nature 409:551, 2001.
3.  SJ Clough, AF Bent. Floral dip: A simplified method for *Agrobacterium*-mediated transformation of *Arabidopsis thaliana*. Plant J 16:735–743, 1998.
4.  C Alonso-Blanco, M Koornneef. Naturally occurring variation in *Arabidopsis*: An underexploited resource for plant genetics. Trends Plant Sci 5:22–29, 2000.
5.  DR Cook. *Medicago truncatula*—a model in the making. Curr Opin Plant Biol 2: 301–304, 1999.
6.  K Handberg, J Stougaard. *Lotus japonicus*, an autogamous, diploid legume species for classical and molecular genetics. Plant J 2:487–496, 1992.
7.  AT Trieu, SH Burleigh, IV Kardailsky, IE Maldonado-Mendoza, WK Versaw, LA Blaylock, H Shin, T-J Chiou, H Katagi, GR Dewbre, D Weigel, MJ Harrison. Transformation of *Medicago truncatula* via infiltration of seedlings or flowering plants with *Agrobacterium*. Plant J 22:531–542, 2000.
8.  KL Kindle. High-frequency nuclear transformation of *Chlamydomonas reinhardtii*. Proc Natl Acad Sci USA 87:1228–1232, 1990.
9.  K Shimogawara, S Fujiwara, A Grossman, H Usuda. High-efficiency transformation of *Chlamydomonas reinhardtii* by electroporation. Genetics 148:1821–1828, 1998.
10. JA Nelson, PA Lefebvre. Targeted disruption of the *NIT8* gene in *Chlamydomonas reinhardtii*. Mol Cell Biol 15:5762–5769, 1995.
11. DG Schaefer, JP Zryd. Efficient gene targeting in the moss *Physcomitrella patens*. Plant J 11:1195–206, 1997.
12. H Puchta, B Hohn. From centimorgans to base pairs: Homologous recombination in plants. Trends Plant Sci 1:340–348, 1996.
13. SA Kempin, SJ Liljegren, LM Block, SD Rounsley, MF Yanofsky, E Lam. Targeted disruption in *Arabidopsis*. Nature 389:802–803, 1997.
14. S Wattler, M Kelly, M Nehls. Construction of gene targeting vectors from λ KOS genomic libraries. BioTechniques 26:1150–1160, 1999.
15. P Ng, MD Baker. The molecular basis of multiple vector insertion by gene targeting in mammalian cells. Genetics 151:1143–1155, 1999.
16. GP Redei, C Koncz. Classical mutagenesis. In: C Koncz, NH Chua, J Schell, eds. Methods in *Arabidopsis* Research. River Edge, New Jersey: World Scientific Publishers, 1992, pp 16–83.
17. CM McCallum, L Comai, EA Greene, S Henikoff. Targeted screening for induced mutations. Nat Biotechnol 18:455–457, 2000.
18. T Maes, PD Keukeleire, T Gerats. Plant tagnology. Trends Plant Sci 4:90–96, 1999.
19. RA Martienssen. Functional genomics: Probing plant gene function and expression with transposons. Proc Natl Acad Sci USA 95:2021–2026, 1998.
20. H Enoki, T Izawa, M Kawahara, M Komatsu, S Koh, J Kyozuka, K Shimamoto. *Ac* as a tool for the functional genomics in rice. Plant J 19:605–613, 1999.
21. AF Tissier, S Marillonnet, V Klimyuk, K Patel, MA Torres, G Murphy, JDG Jones.

Multiple independent defective *suppressor-mutator* transposon insertions in *Arabdopsis*: A tool for functional genomics. Plant Cell 11:1841–1852, 1999.

22. V Sundaresan, P Springer, T Volpe, S Haward, JD Jones, C Dean, H Ma, R Martienssen. Patterns of gene action in plant development revealed by enhancer trap and gene trap transposable elements. Genes Dev 9:1797–1810, 1995.

23. R Azpiroz-Leehan R, KA Feldmann. T-DNA insertion mutagenesis in *Arabidopsis*: Going back and forth. Trends Genet 13:152–156, 1997.

24. PJ Krysan, JC Young, MR Sussman. T-DNA as an insertional mutagen in *Arabidopsis*. Plant Cell 11:2283–2290, 1999.

25. JS Jeon, S Lee, KH Jung, SH Jun, DH Jeong, J Lee, C Kim, S Jang, S Lee, K Yang, J Nam, K An, MJ Han, RJ Sung, HS Choi, JH Yu, JH Choi, SY Cho, SS Cha, SI Kim, G An. T-DNA insertional mutagenesis for functional genomics in rice. Plant J 22:561–570, 2000.

26. D Weigel, JH Ahn, MA Blazquez, JO Borevitz, SK Christensen, C Fankhauser, C Ferrandiz, I Kardailsky, EJ Malancharuvil, MM Neff, JT Nguyen, S Seto, ZY Wang, Y Xia, RA Dixon, MJ Harrison, CJ Lamb, MF Yanofsky, J Chory. Activation tagging in *Arabidopsis*. Plant Physiol 122:1003–1013, 2000.

27. MM Neff, SM Nguyen, EL Malancharuvil, S Fujioka, T Noguchi, H Seto, M Tsubuki, T Honda, S Takatsuto, S Yoshida, J Chory. *BAS1*: A gene regulating brassinosteroid levels and light responsiveness in *Arabidopsis*. Proc Natl Acad Sci USA 96:15316–15323, 1999.

28. D Bauchez, H Hofte. Functional genomics in plants. Plant Physiol 118:725–732, 1998.

29. S Parinov, V Sundaresan. Functional genomics in *Arabidopsis*: Large-scale insertional mutagenesis complements the genome sequencing project. Curr Opin Biotechnol 11:157–161, 2000.

30. JG Izant, H Weintraub. Inhibition of thymidine kinase gene expression by antisense RNA: A molecular approach to genetic analysis. Cell 36:1007–1013, 1984.

31. SR Rodermel, MS Abbott, L Bogorad. Nuclear-organelle interactions: Nuclear antisense gene inhibits ribulose bisphosphate carboxylase enzyme levels in transformed tobacco plants. Cell 55:673–679, 1988.

32. T Hjalt, EGH Wagner. The effect of loop size in antisense and target RNAs on the efficiency of antisense RNA control. Nucleic Acids Res 20:6723–6732, 1992.

33. DC Baulcombe. Fast forward genetics based on virus-induced gene silencing. Curr Opin Plant Biol 2:109–113, 1999.

34. SW Ding SW. RNA silencing. Curr Opin Biotechnol 11:152–156, 2000.

35. P Meyer, H Saedler. Homology-dependent gene silencing in plants. Annu Rev Plant Physiol Plant Mol Biol 47:23–48, 1996.

36. JM Kooter, MA Matzke, P Meyer. Listening to the silent genes: Transgene silencing gene regulation and pathogen control. Trends Plant Sci 4:340–347, 1999.

37. A Cole-Strauss, K Yoon, Y Xiang, BC Byrne, MC Rice, J Gryn, WK Holloman, EB Kmiec. Correction of the mutation responsible for sickle cell anemia by an RNA-DNA oligonucleotide. Science 273:1386–1389, 1996.

38. V Alexeev, K Yoon. Stable and inheritable changes in genotype and phenotype of albino melanocytes induced by RNA-DNA oligonucleotides. Nat Biotechnol 16: 1343–1346, 1998.

39. PR Beetham, PB Kipp, XL Sawycky, CJ Arntzen, GD May. A tool for functional plant genomics: Chimeric RNA/DNA oligonucleotides cause *in vivo* gene-specific mutations. Proc Natl Acad Sci USA 96:8774–8778, 1999.

40. A Cole-Strauss, H Gamper, WK Holloman, M Munoz, N Cheng, EB Kmiec. Targeted gene repair directed by the chimeric RNA/DNA oligonucleotide in a mammalian cell-free extract. Nucleic Acids Res 27:1323–1330, 1999.

41. C Alonso-Blanco, SE El-Assal, G Coupland, M Koorneef. Analysis of natural allelic variation at flowering time loci in the Landsberg erecta and Cape Verde Islands ecotypes of *Arabidopsis thaliana*. Genetics 149:749–764, 1998.

42. V Hartenstein. Atlas of *Dropsophila* morphology and development. Plainview, New York: Cold Spring Harbor Laboratory Press, 1993.

43. AS Wilkins. Genetic analysis of animal development. New York: Wiley-Liss, Inc., 1993.

44. PD Lancashire, H Bleiholder, TVD Boom, P Langeluddeke, R Stauss, E Weber, A Witzenberger. A uniform decimal code for growth stages of crops and weeds. Ann Appl Biol 119:561–601, 1991.

45. JM Diamond. Guns, germs, and steel. New York: WW Norton and Company, 1997, pp 85–92.

46. N Smith. Seeds of opportunity: An assessment of the benefits, safety, and oversight of plant genomics and agricultural biotechnology: Report transmitted to the Committee on Science for the 106th Congress, 2nd session. Washington: US Government Printing Office, 2000, pp 27–28.

47. D Guyer, A Tutle, S Rouse, S Volrath, M Johnson, S Potter, J Gorlach, S Goff, L Crossland, E Ward. Activation of latent transgenes in *Arabidopsis* using hybrid transcription factor. Genetics 149:633–639, 1998.

48. D Berg, K Tietjen, D Wollweber, R Hain. From genes to target: Impact of functional genomics on herbicide discovery. In: The 1999 Brighton Crop Protection Conference: Weeds, 1999, pp 491–500.

49. JS Boyer. Plant productivity and the environment. Science 228:443–448, 1982.

50. M Kasuga, Q Liu, M Setsuka, K Yamaguchi-Shinozaki, K Shinozaki. Improving plant drought, salt, and freezing tolerance by gene transfer of a single stress-inducible transcription factor. Nat Biotechnol 17:287–291, 1999.

51. MP Apse, GS Aharon, WA Sneedon, E Blumwald. Salt tolerance conferred by overexpression of vaculoar $Na^+/H^+$ antiport in *Arabidopsis*. Science 285:1256–1258, 1999.

52. J Doebley, L Lukens. Transcriptional regulators and the evolution of plant form. Plant Cell 10:1075–1082, 1998.

53. RA Raff. The shape of life: Genes development, and the evolution of animal form. Chicago: University of Chicago Press, 1996.

54. K Maleck, A Levine, T Euglem, A Morgan, J Schmid, KA Lawton, JL Dangl, RA Dietrich. The transcriptome of *Arabidopsis thaliana* during systemic acquired resistance. Nat Genet 26:403–410, 2000.

55. RA Narang, A Bruene, T Altmann. Analysis of phosphate acquisition efficiency in different *Arabidopsis* accessions. Plant Physiol 124:1786–1799, 2000.

# 11

# Functional Genomics of Plant Abiotic Stress Tolerance

**John C. Cushman**
*University of Nevada–Reno, Reno, Nevada, U.S.A.*

## I. INTRODUCTION

Archeological investigations of past agrarian-based civilizations in Mesopotamia recount the devastating impact of drought and salinity build-up caused by irrigation (Jacobson and Adams 1958; Kerr 1998; Ashraf, 1994). Today, physiochemical factors that impose water-deficit stress, such as drought, salinity, and temperature extremes, remain the dominant factors that suppress agricultural productivity (Boyer 1982; Kramer and Boyer 1985). Salt build-up from irrigation water forces the abandonment of approximately 10 million hectares each year, especially in semi-arid and arid regions of the world. Between 20% and 50% of the areas under irrigation are estimated to suffer from some salinization-related losses in productivity (Seckler et al. 1998; Flowers and Yeo 1995). Scarce water resources in such areas and an ever-increasing demand for water by growing urban populations threaten agricultural production systems that must supply food, feed, and fiber for a human population that is predicted to reach 9 billion within the next 30 to 50 years (Khush 1999). During the 1990s, the rate of population growth exceeded the rate of growth of food-grain production. To offset this trend, new cultivars with higher yield potential and stability must be developed to avoid serious food shortages in the 21st century. Such improvements in crop productivity will increasingly rely on a detailed understanding of the complexity of plant responses to environmental stress (Miflin 2000). Recent advances in genome mapping and sequencing technologies and the development of high throughput, multiparallel functional genomics approaches such as comprehensive gene expression profiling and systematic gene disruption strategies will play key roles in improving our knowledge of how plants respond and adapt to abiotic stress. More importantly, keeping track of the massive amounts of biological data using robust bioinformatic systems will become more important in the development of systems approaches and corresponding models of biological responses to environmental perturbations.

## II.  MOLECULAR BREEDING STRATEGIES FOR ABIOTIC STRESS TOLERANCE

Evolution in environmental niches exposed to differing magnitudes of environmental stress has resulted in an extraordinary genetic and phenotypic diversity with the greatest diversity in the most stressful environments (Nevo 1997; Nevo et al., 1997; Nevo 2001). Tapping into the diverse genetic resources of naturally occurring wild relatives of crop species is likely to play a central role in improving modern varieties (Hoisington et al. 1999; Ellis et al. 2000; Zamir 2001). Recently, considerable attention has been paid to wild relatives of barley (Ivandic et al. 2000; Forster et al. 2000) and tomato (Foolad et al. 1998a, b; Chen and Foolad 1999; Foolad and Chen, 1999). Traditional breeding attempts to generate plant varieties with improved stress tolerance by capturing genetic variation arising from exotic or varietal germplasm, interspecific or intergeneric hybridization, induced mutations, and somaclonal variation of cell and tissue cultures have proved largely unsuccessful (Flowers and Yeo 1995). Few, if any, new introductions of plants with improved stress resistance under field conditions have become commercialized (Price and Courtois 1999; Flowers et al. 2000). In addition to an historical lack of urgency to improve such traits, selection-based breeding approaches are limited by the complexity of stress tolerance mechanisms (e.g., escape, avoidance, tolerance, and recovery), the interaction of resistance or tolerance determinants, the differential sensitivity to stress depending on developmental stage, the interaction of environmental conditions, the low genetic variance of yield components under stress conditions, the lack of efficient selection techniques, and the logistical constraints of physiological screening of large breeding populations on a field scale (Ribaut et al. 1996, 1997; Frova et al. 1999a, b; Price and Courtois 1999; Quarrie et al. 1999; Flowers et al. 2000). Furthermore, cross-breeding of elite varieties with exotic species may require extensive back-crossing (typically six generations) to restore desirable traits along with the introgressed tolerance traits (Zamir 2001). The efficiency of traditional breeding approaches may be improved by the identification of discrete chromosomal regions identified by molecular markers that have a major effect on the enhancement of stress tolerance through quantitative trait loci (QTLs) mapping (Quarrie 1996). The availability of genetic linkage maps and physical maps with a high density of phenotypic and molecular markers and improved mapping techniques and software is expected to greatly enhance the identification and mapping of QTLs for stress tolerance traits (Koyama et al. 2001; Childs et al. 2001; Chen et al. 2002a; Wu et al. 2002). Once identified, breeders may use the molecular markers as selection criteria in place of complex physiological or morphological traits such as osmotic adjustment or root structure. In addition to marker-assisted selection programs (Stuber et al. 1999), molecular markers closely associated with QTLs can be used for QTL pyramiding and positional cloning of candidate genes (Pfieger et al. 2001).

Considerable progress has been made in the identification of QTLs associated with abiotic stress tolerance traits, and in a few cases, in the colocalization with candidate genes associated with one or more of these different traits (Table 1). Controlled experiments performed under greenhouse and reasonably controlled field conditions have led to the identification of QTLs associated with abiotic stress resistance or tolerance related traits. In rice, for example, QTLs have been identified for cold tolerance (Saito et al. 1995; 2001; Misawa et al. 2000), root volume and thickness (Price and Tomos 1997), root penetration ability (Ray et al. 1996; Zheng et al. 2000; Ali et al. 2000; Price et al. 2000; 2002a),

**Table 1**  QTL or Candidate Genes Associated with Abiotic Stress Tolerance Traits

| Trait | Species | Reference |
|---|---|---|
| Na$^+$/K$^+$ uptake/discrimination | *Oryza sativa* | Koyama et al., 2001 |
| Root depth | *Oryza sativa* | Shen et al, 2001 |
| Root penetration | *Oryza sativa* | Ray et al., 1996 |
| Root penetration/morphology/ distribution | *Oryza sativa* | Price et al., 2000 |
| Root penetration/morphology/ distribution | *Oryza sativa* | Zheng et al., 2000 |
| Root volume /thickness | *Oryza sativa* | Price and Tomos, 1997b |
| Root morphology/distribution | *Oryza sativa* | Yadav et al., 1997 |
| Cold tolerance | *Oryza sativa* | Saito et al., 1995 |
| Cold tolerance | *Oryza sativa* | Misawa et al., 2000 |
| Cold tolerance | *Oryza sativa* | Saito et al., 2001 |
| Drought avoidance/root morphology | *Oryza sativa* | Hemamalini et al., 2000 |
| Drought avoidance/root morphology | *Oryza sativa* | Champoux et al., 1995 |
| Drought avoidance | *Oryza sativa* | Courtois et al., 2000 |
| Salt tolerance | *Oryza sativa* | Gong et al, 1999 |

**Table 1** Continued

| Trait | Species | Reference |
| --- | --- | --- |
| Salt tolerance | *Oryza sativa* | Zhang et al, 1995 |
| Membrane stability | *Oryza sativa* | Tripathy et al., 2000 |
| Stomatal frequencies | *Oryza sativa* | Ishimaru et al., 2001 |
| Stomatal conductance/leaf rolling | *Oryza sativa* | Price et al., 1997 |
| Leaf rolling/drying/water content | *Oryza sativa* | Price et al., 2002b |
| Osmotic adjustment/ion homeostasis | *Oryza sativa* | Zhang et al., 1999a |
| Osmotic adjustment | *Oryza sativa* | Lilley et al., 1996 |
| Stay-green | *Oryza sativa* | Cha et al., 2002 |
| Water status | *Hordeum vulgare* | Teulat et al., 1997 |
| ABA accumulation | *Zea mays* | Tuberosa et al., 1998 |
| ABA accumulation | *Zea mays* | Sanguineti et al., 1999 |
| ABA accumulation | *Hordeum vulgare* | Sanguineti et al., 1994 |
| ABA accumulation/stomatal conductance/water potential/root number/pulling resistance | *Zea mays* | Lebreton et al., 1995 |
| Stomatal conductance /water status | *Helianthus annum* | Hervé et al., 2001 |
| $Na^+/Cl^-$ accumulation | *Citrus/Poncirus* | Tozlu et al., 1999 |
| ABA accumulation | *Triticum aestivum/Zea mays* | Quarrie et al., 1994 |
| ABA accumulation | *Oryza sativa* | Quarrie et al., 1997 |
| Salt tolerance | *Lycopersicon esculentum* | Foolad et al., 1998 |
| Salt tolerance | *Hordeum vulgare* | Mano and Takeda, 1997 |
| Freezing tolerance | *Hordeum vulgare* | Tuberosa et al., 1997 |
| Salt tolerance | *Hordeum vulgare* | Ellis et al., 1997 |
| Osmotic adjustment/dehydration tolerance | *Hordeum vulgare* | Teulat et al., 1998 |
| Osmotic adjustment/water status | *Hordeum vulgare* | Teulat et al., 2001 |
| Osmotic adjustment | *Triticum aestivum* | Morgan and Tan, 1996 |
| Drought tolerance | *Stylosanthes scabra* | Thumma et al., 2001 |

**Table 1** Continued

| Trait | Species | Reference |
| --- | --- | --- |
| Water use efficiency | *Lycopersicon esculentum* | Martin et al, 1989 |
| Water use efficiency | *Glycine max* | Mian et al., 1996 |
| Freezing tolerance | *Hordeum vulgare* | Ismail et al., 1999 |
| Salt tolerance: seed germination/vegetative growth | *Lycopersicon esculentum x L. pimpinellifolium* | Foolad 1999 |
| Salt tolerance: seed germination | *Lycopersicon esculentum x L. pimpinellifolium* | Foolad et al., 1998 |
| Salt tolerance: seed germination | *Lycopersicon esculentum x L. pennellii* | Foolad et al., 1997 |
| Salt tolerance: vegetative growth | *Lycopersicon esculentum x L. pimpinellifolium* | Foolad and Chen, 1999 |
| Salt tolerance: vegetative growth | *Lycopersicon esculentum x L. pimpinellifolium* | Foolad et al., 2001 |
| Drought tolerance | *Zea mays* | Frova et al., 1999 |
| Anthesis-silking interval | *Zea mays* | Jiang et al., 1999 |
| Drought tolerance: anthesis-silking interval | *Zea mays* | Sari-Gorla et al., 1999 |
| Root length/diameter/weight | *Zea mays* | Tuberosa et al., 2002 |
| Drought tolerance: anthesis-silking interval | *Zea mays* | Veldboom and Lee, 1996 |
| Drought tolerance: anthesis-silking interval | *Zea mays* | Ribaut et al., 1996; 1997 |
| Drought tolerance: anthesis-silking interval | *Zea mays* | Sari-Gorla et al., 1999 |
| Thermotolerance | *Zea mays* | Frova and Sari-Gorli, 1994 |
| Heat shock proteins | *Zea mays* | Jorgensen and Nguyen, 1995 |
| Thermotolerance/membrane stability | *Zea mays* | Ottaviano et al., 1991 |
| Salt tolerance | *Hordeum vulgare* | Ellis et al., 1997 |
| Drought tolerance | *Hordeum vulgare* | Forster et al., 2000 |
| Drought tolerance | *Sorghum bicolor* | Tuinstra et al., 1998 |
| Preflowering drought tolerance/stay-green | *Sorghum bicolor* | Kebede et al., 2001 |
| Stay-green | *Sorghum bicolor* | Wu et al., 2000 |
| Stay-green | *Sorghum bicolor* | Crasta et al., 1999 |
| Stay-green | *Sorghum bicolor* | Wu et al., 2000 |
| Stay-green | *Sorghum bicolor* | Sanchez et al., 2002 |

root morphology (Champoux et al. 1995; Yadav et al. 1997; Price et al., 1997a; Shen et al. 2001; Price et al., 2002c), stomatal conductance, leaf rolling, leaf drying, and leaf water content (Price et al., 1997b; Price et al. 2002c), drought avoidance (Courtois et al., 2000) and its relationship to root growth (Hemamalini et al., 2000), osmotic adjustment and dehydration tolerance (Lilley et al., 1996; Zhang et al., 1999a), membrane stability under drought (Tripathy et al., 2000), salt tolerance (Zhang et al., 1995), and quantitative sodium and potassium uptake (Koyama et al. 2001). Because of their importance in abiotic stress adaptation, QTLs have also been identified for ABA accumulation in rice (Quarrie et al. 1997) as well as barley, wheat, and maize (Quarrie et al. 1994; Tuberosa et al. 1998; Sanguineti et al. 1994, 1999). Interestingly, QTLs for dehydration tolerance (Lilley et al. 1996), stomatal conductance (Price et al. 1997b), and ABA accumulation (Quarrie et al. 1997) were found to colocalize, suggesting these may participate in a common mechanism. In addition, QTLs for root penetration ability or rooting depth, a putative measure of drought resistance, have been colocalized with QTLs for root length, weight, and thickness in different mapping populations of rice on multiple chromosomes (Price et al. 2000; 2002a, b, c). Multiple QTLs for drought avoidance have been repeatedly mapped to the same location in different mapping populations of rice with individual QTLs for leaf rolling, leaf drying, and relative leaf water content (Price et al. 2002a, c). However, few such QTLs colocalize with QTLs for root-growth characteristics, suggesting that selection for such traits may not be an efficient strategy for improving drought resistance. The lack of colocalization of QTLs for drought avoidance and root morphology may be caused by the difference in root growth responses measured under control and field conditions and the differences in specific environmental conditions, such as the physical or biological properties of soils at different sites among individual trials, and shoot-related traits (Price et al., 2002a). Multiple QTLs for stomatal frequencies have also been identified in rice, but none overlap with QTLs for leaf rolling, suggesting that osmotic tolerance mechanisms are independent of stomatal behavior (Ishimaru et al. 2001).

Although fewer studies have been performed involving root characteristics in maize compared with rice, QTLs for seminal root traits (e.g., root length, primary root diameter, primary root weight, and adventitious root weight) have been identified in young (3-week-old), hydroponically grown maize plants that consistently overlap with QTLs for grain yield under field conditions (Tuberosa et al. 2002). Furthermore, some root trait QTLs were found to overlap with QTLs controlling leaf ABA content (Tuberosa et al. 1998; Sanguineti et al. 1999), root pulling resistance under greenhouse (Lebreton et al. 1995) and grain yield under dry field conditions (Ribaut et al. 1997).

Other species have also received considerable attention in the identification of QTLs for drought stress tolerance traits. In barley, for example, QTLs have been identified for water status or osmotic adjustment (Teulat et al. 1997, 1998,

2001), and are under development for drought tolerance (Forster et al. 2000). Quantitative trait loci identified for drought tolerance in the pasture legume, *Stylosanthes scabra*, were found to be correlated with low specific leaf area and negatively correlated with high biomass production and transpiration efficiency (Thumma et al. 2001). Related studies have found QTLs for water use efficiency (WUE) in tomato (Martin, et al., 1989) and soybean (Mian et al., 1996). Quantitative trait loci for water status traits, such as stomatal conductance, transpiration, leaf water potential, and relative water content, have also been identified in sunflower (Hervé et al., 2001). However, measurement of WUE relies on carbon isotope discrimination techniques and has not always been found to correlate positively with productivity under stress conditions; therefore, the use of WUE alone as a selection trait may be unadvisable (Price and Courtois, 1999). Furthermore, measurement of WUE relies on carbon isotope discrimination techniques, results from which may be confounded by differences in water availability owing to root architecture rather than simply stomatal conductance.

Quantitative Trait loci mapping and marker-assisted selection strategies for secondary traits that have a major effect on stress tolerance or are adversely affected by abiotic stresses have been extensively studied in sorghum and maize (Bruce et al. 2002). Grain sorghum has been extensively used to identify QTLs associated with both preflowering and postflowering drought tolerance (stay-green) (Tuinstra et al., 1998; Kebede et al., 2001; Sanchez et al. 2002). The stay-green character is a post-flowering drought tolerance trait associated with premature senescence under drought stress during the grainfilling stage. The majority of QTLs identified for the stay-green trait were independent of maturity QTLs demonstrating the possibility of pyramiding desirable QTLs for each trait, resulting in improved drought tolerance (Crasta et al. 1999). More recently, candidate genes such as key photosynthetic enyzmes, heat shock proteins, and ABA-responsive genes were found to colocalize with two consistent stay-green QTLs (Xu et al. 2000). A stay-green locus has also been mapped in rice (Cha et al. 2002). However, it remains difficult to separate the mechanisms of green leaf retention from those associated with drought resistance and recovery. Environmental stresses such as drought can also have profound effects on secondary yield traits associated with reproductive processes. Male gametogenesis, anthesis, ovary development, and the early stage of embryo development are particularly susceptible to water deficit stress. One of the best example of this type of trait is anthesis-silking interval. Drought stress causes a delay in silking, resulting in an increase in the anthesis-silking interval, which in turn causes significant yield loses (Bolaños and Edmeades 1996). Quantitative Trait loci for anthesis-silking interval have been identified in several studies, some associated with similar chromosome map positions (Ribaut et al. 1996; 1997; Veldboom and Lee 1996; Jiang et al. 1999; Sari-Gorla et al. 1999). Quantitative Trait loci for yield components under drought stress (Frova et al. 1999a), and thermotolerance and membrane stability (Ottaviano et al. 1991; Frova and Sari-Gorla 1994; Frova et al. 1999b) have also been identified in maize.

Quantitative Trait loci mapping of salinity tolerance traits have been conducted in several different species. In barley, for example, QTLs have been identified for salinity tolerance (Ellis et al. 1997; Mano and Takeda 1997). Quantitative Trait loci mapped at germination show no overlap with those mapped at the seedling stage (Mano and Takeda 1997). Whole plant and tissue-specific QTLs for $Na^+$ and $Cl^-$ accumulation were identified in an intergeneric cross between *Citrus grandis* and a related genus of extremely salt-sensitive trifoliate orange, *Poncirus trifoliata* (Tozlu et al. 1999). In addition, interspecific hybrids of salt-sensitive and salt-tolerance tomato have been used to develop QTLs for salinity tolerance (Foolad et al. 1997; Foolad et al. 1998a; Foolad 1999; Foolad and Chen 1999; Foolad et al. 2001). Similar to studies performed in barley, mapping of QTLs at different stages of development identified QTLs that differed depending on whether tolerance was assayed during seed germination or vegetative growth, although some QTL locations from these independent analyses did coincide (Foolad 1999) and were validated in later studies (Foolad et al. 2001). Quantitative Trait Loci associated with cold tolerance during seed germination have also been described for tomato (Foolad et al. 1998b; Foolad et al. 1999). Some QTLs were associated with only cold or salt stress tolerance, whereas others were correlated with both stress conditions as well as nonstress conditions and generally accounted for a larger portion of phenotypic variation than did stress-specific QTLs (Foolad et al. 1999). These results are consistent with gene expression studies that demonstrate that some genes exhibit only salt- or cold-responsive changes in expression, whereas others respond to multiple stresses (Shinozaki and Yamaguchi-Shinozaki 1997). Overall, QTL studies in tomato indicate that salinity and cold tolerance are controlled by a few major QTLs, with additional smaller effect QTLs with little or no epistatic interactions. Thus, MAS may facilitate breeding efforts to identify cultivars with improved tolerance at different developmental stages or to more than one abiotic stress (Foolad 2000).

Despite its potential utility, QTL mapping still has several drawbacks. First, QTLs may only be identified at one stage in development or may differ from those identified at other stages (Price and Tomos 1997; Foolad et al. 1998a; Foolad 1999; 2000). Second, the likelihood of success and accuracy of QTL mapping is, improved by having as large a population as possible, but the need for large populations sometimes restricts the sophistication of the analytical techniques that can be used for screening. Third, QTLs for certain traits identified under tightly controlled greenhouse conditions, may not be possible or may give different results when conducted under less tightly controlled field conditions, as tolerance mechanisms may be expressed differently, depending on growth conditions and additional environmental variability, particularly soil conditions, associated with field conditions. Fourth, relatively few major QTLs for environmental stress tolerance have been identified, despite the fact that much larger numbers of genes are most likely involved in tolerance traits. Finally, the low resolution of mapping in many cases limits the precise localization of candidate genes. However, improvements in the resolution of genetic and physical maps (Yano et al. 2002) have permitted the identification of candidate plant genes in several cases (Fridman et al. 2000; Yano et al. 2000). These results confirm that the molecular basis of a QTL can be linked directly to a single gene encoding an enzyme affecting a biochemical pathway or a transcription factor controlling the expression of many genes.

For abiotic stress-related traits, loci encoding dehydrin proteins and other drought-induced genes have also been found to cosegregate with major QTLs for cold hardiness, salt and drought tolerance, and osmotic adjustment in barley (Teulet et al. 1998; Ismail et al. 1999; Zhu et al. 2000; Cattivelli et al. 2002). Similarly, QTLs for the role of heat shock proteins (HSPs) in thermotolerance have been identified in maize (Ottaviano et al. 1991; Jorgensen and Nguyen 1995). Quantitative trait loci for ion homeostasis have been shown to colocalize with candidate genes, such as the plasma membrane $H^+$-ATPase (Zhang et al. 1999b). Still, there remain many examples of stress-related genes located outside any stress tolerance QTLs, a situation that serves to illustrate the limitation of QTL mapping approaches.

One approach to improve the reproducibility and precision of QTL identification, known as "advanced back-cross QTL analysis," is a combination of marker-assisted selection and the development of recombinant inbred or isogenic lines (Tansley and Nelson 1996). Using this approach it should be possible to efficiently select for QTLs of interest using molecular markers flanking the QTL and for alleles of the recurrent parental at all other markers, thereby obtaining near isogenic lines for all QTLs within a few years (Price and Courtois 1999). The use of such recombinant inbred lines should streamline breeding trials at many different conditions or locations and at multiple developmental stages. Smaller population sizes will also suffice, allowing more sophisticated trait analyses to be performed. Improved precision of QTL mapping should facilitate the identification of candidate genes that can be readily isolated using chromosome landing or positional cloning approaches (Tanksley et al. 1995). Advanced multivariate methodologies may improve the sensitivity and efficiency of QTL localization when considering several traits simultaneously (Calinski et al. 2000; Korol et al. 2001). New physiological assay methodologies to screen large numbers of plants in field settings, such as infrared thermometry and carbon or oxygen isotope discrimination measurements, would streamline and improve the accuracy of plant water status, stomatal function, and water use efficiency determinations (Price et al. 2002a). The identification of multiple QTLs for different stress tolerance traits presents the possibility that improvements can be made by pyramiding these favorable QTLs or consecutive selection for and accumulation of component physiological traits, such as the transfer of QTLs for root length and thickness to improve drought tolerance traits (Flowers and Yeo 1995; Price and Courtois 1999) or QTLs for ion uptake and selectivity to improve salt tolerance traits (Koyama et al. 2001) into popular, but stress-sensitive cultivars. Such strategies should prove increasingly useful as the resolution of genetic and physical chromosome maps of major crops improves.

## III.  DEFINING THE OSMOME, XEROME, AND THERMOME

In contrast to the slow but steady progress in QTL identification and positional cloning of the candidate genes, more rapid gene discovery progress has been made through high throughput genome sequencing of model organism, such as cyanobacteria, yeast, *Arabidopsis*, and rice, and large-scale, partial sequencing of randomly selected cDNA clones or expressed sequence tags (ESTs) from these models, agronomically important crops, and less well studied halophytic and poikyolohydric species that are adapted to extremely saline or dry conditions, respectively. Such gene discovery programs can rapidly identify candidate genes that can be used to map QTLs efficiently (Wang et al. 2001). More importantly, genome and EST sequence information has brought about the development of accurate and efficient methods for large-scale or genome-wide analysis of genetic variation and gene transcript profiling (Donson et al. 2002; Aharoni and Vorst 2002). With these and other tools it has become possible for the first time to define the full complement of genes that contribute to abiotic stress adaptation (Hasegawa et al. 2000a,b). We have coined the terms osmome, xerome, and thermome to describe those portions of the genome involved in ionic or osmotic, water deficit or drought, and cold or heat stress perception and adaptive responses, respectively (Cushman and Bohnert 2000).

## A.   Gene Discovery Programs

The complete sequence of the *Synechocystis* (Kaneko et al. 1996a, b), *Saccho-myces cereviseae* (Goffeau et al. 1997), *Arabidopsis* genomes (The Arabidopsis Genome Initiative 2000), and recent draft genome sequences for indica (Yu et al. 2002) and japonica (Goff et al. 2002) varieties of rice have provided the first step toward cataloging and categorizing the true genetic complexity of abiotic stress responses. These genome sequencing programs have been complemented by rapid gene discovery from large-scale EST sequencing in *Arabidopsis* (White et al. 2000; Asamizu et al. 2000b; Seki et al. 2001b; Seki et al. 2002) and rice (Goff 1999; Ewing et al. 1999). To improve the accuracy of genomic sequence annotation of predicted transcription units and gene products, current efforts are focused on the characterization of full-length complementary DNAs (cDNAs) in

*Arabidopsis* (Seki et al. 2002). Genome scale collections of full-length cDNAs are essential for gene product structure-function studies. In addition to transcript profiling, full-length cDNA sequences can be used for the creation of promoter sequence databases by mapping 5′ ends of cDNA to genomic DNA. Finally, EST projects in organisms for which incomplete or no genome sequence is available provide invaluable resources for proteomic studies using peptide mass finger-printing (Lisacek et al. 2001; Porubleva et al. 2001). For those species for which the complete genome has not been attempted or is in progress, large scale EST sequencing initiatives represent cost-effective, economical means of obtaining information about the transcribed genes of individual species (Mayer and Mewes 2001). Large-scale EST sequencing efforts are well underway for various crop model species (Walbot 1999), including maize (Gai et al. 2000; Fernandes et al. 2002), barley (Druka et al. 2002), *Medicago truncatula* (Bell et al. 2001), *Lotus japonicus* (Asamizu et al. 2000a), potato (Crookshanks et al. 2001), *Sorghum bicolor* (Childs et al. 2001; Mullet et al. 2002), loblolly pine (Allona et al. 1998), poplar (Sterky et al. 1998; Vander Mijnsbrugge et al. 2000), and Japanese cedar (Ujino-Ihara et al. 2000). The number of publicly available ESTs available is growing rapidly and this growth can be monitored at the dbEST section of Gen-Bank (http://www.ncbi.nlm.nih.gov/dbEST/dbEST_summary.html).

Gene indices or integrated data from international EST sequencing and analysis of the transcribed sequences represented in the world's public EST data, have been created for model glycophytes (*Arabidopsis, Chlamydomonas rein-hardtii, Oryza sativa* (rice), *Lotus japonicus, Medicago truncatula*), halophytes (*Mesembryanthemum crystallinum* [ice plant]), and major crop species (*Gos-sipium hirsutem* [cotton], *Zea mays* [maize], *Solanum tuberosum* [potato], *Sor-ghum bicolor, Glycine max* [soybean], *Hordeum vulgare* [barley], *Lycopersicon esculentum* [tomato], *Secale cereale* [rye], *Triticum aestivum* [wheat]) (Quacken-bush et al. 2001; http://www.tigr.org/tdb/tgi/).

## 1. Gene Discovery in Stress Sensitive Models

Although the vast majority of EST data is derived from cDNAs prepared from different tissues, organs or cells, developmental states, and various external stim-uli and plant growth regulator treatments, relatively few studies have focused specifically on EST production from plants exposed to environmental stress con-ditions. Early attempts to identify stress-specific transcripts in suspension cell cultures of rice exposed to salinity or nitrogen-starvation stress, using random sequencing of 780 ESTs, revealed that salinity stress induced the expression of several glycolysis- and TCA cycle-related enzymes that contribute to ATP pro-duction (Umeda et al. 1994). Sequencing of 220 randomly selected ESTs from a subtracted *Arabidopsis* cDNA library included 15 osmotic stress-induced genes with early, late, and continuous patterns of expression with a range of 2- to 50-

fold induction (Pih et al. 1997). Other EST sequencing projects have discovered a large proportion of genes associated with abiotic stress or defense responses. For example, stress or defense-related transcripts were found to be abundantly represented in EST collections from Petunia protoplasts (Yu et al. 1999) and rice suspension cell cultures (Umeda et al. 1994). Genes involved in abiotic stress responses, and pathogen wounding and drought responses, in particular, are also abundant in the inner bark of Japanese cedar where they comprise about 12% of sampled ESTs (Ujino-Ihara et al. 2000). More recently, concerted efforts to conduct large-scale EST sequencing using tissue and developmental stage-specific cDNA libraries generated exclusively from RNA of salinity (or drought) stressed *Arabidopsis*, barley, maize, rice, and tobacco (Bohnert et al. 2001; Kawasaki et al. 2001). This approach has permitted the rapid characterization of randomly selected ESTs diagnostic of abiotic stress responses that have been historically underrepresented by EST sequencing projects. Expressed Sequence Tag collections derived from stressed plants are clearly enriched for genes with functions involved with transport facilitation, cell rescue and defense, energy metabolism, proteome restructuring (e.g., proteases, ubiquitinases), and osmotic and dehydration stress adaptation with the increased appearance of gene-encoding distinct, presumably stress-induced, enzyme or protein isoforms. Sampling differences between unstressed and stressed plants also revealed pronounced down-regulation in transcript abundance for components of the photosynthetic apparatus. Interestingly, EST collections from abiotically stressed plants also show significantly increased numbers of functionally unknown ESTs, presumably due to the greater diversity of these previously underrepresented transcripts in tissues collected under unstressed conditions. Selected EST datasets can be browsed and searched on-line at the Stress Functional Genomics Consortium website (http://stress-genomics.org/).

Using a more targeted approach, screening of a subtracted cDNA library from wild-type *Arabidopsis* seedlings treated for 4 hours with or without 160 mM NaCl by successive rounds of differential subtraction resulted in the identification of 84 nonredundant, salt-regulated genes, most of which had not been previously described as salt-responsive (Gong et al. 2001). Comparison of the salt-regulated expression profiles of these salt-regulated ESTs (SREs) in wild-type and the sos3 mutant, which carries a defect in a $Ca^{2+}$-binding protein, revealed that the SOS pathway functions to reduce salinity-induced signals to genes that are induced coincidentally by related stresses and that function in cold adaptation or during plant pathogen defense responses (Gong et al. 2001).

Standard and normalized leaf and root libraries have been used to enrich for drought stress-responsive genes from drought-stressed *Oryza saltiva* (cv. indica) seedlings (Reddy et al. 2002). Normalization reduced the redundancy from 10% to 3.5% for the first 200 ESTs analyzed, and increased the recovery of novel ESTs greater than five-fold relative to standard libraries. Approximately 10% of the ESTs recovered from the normalized library prepared from drought-stressed rice seedlings showed significant homology to genes having abiotic stressed-related functions. Thus, the use of normalized libraries provides a cost-effective strategy for gene discovery. Standard, normalized, and subtracted cDNA libraries have also been created and sequenced from various tissues of *Arabidopsis* exposed to various abiotic stress treatments including cold, dehydration, NACl, heat, ultraviolet (UV) light, and abscisic acid (ABA) (Seki et al. 2002). Many large scale EST sequencing efforts from, for example, wheat and barley (Cattivelli et al. 2002; Druka et al. 2002; Ozturk et al. 2002), rice (Kawasaki et al. 2001), maize (Bohnert et al. 2001), and *M. truncatula* (Bell et al. 2001) now include cDNA libraries derived from plant tissues exposed to abiotic stresses.

## 2. Gene Discovery in Stress Tolerant Models

Although the complement of genes present in glycophytes and in salt or desiccation-tolerant organisms is likely to be largely conserved among all land plants, there exists the possibility that plants adapted to extreme environments have evolved unique genetic determinants or unique combinations of such determinants. To test this possibility, direct functional comparisons between genes from sensitive versus tolerance organisms need to be conducted. Although glycophytes, halophytes, or desiccation-tolerant plants appear to use similar tolerance effectors and regulatory pathways, it now appears that these organisms have evolved subtle regulatory differences in the timing, abundance, and location of gene expression that account for their ability to withstand severe osmotic, ionic, or dehydration stress (Hasegawa et al. 2000a, b; Zhu 2000). For example, moderate overexpression of a homologous cDNA encoding a sodium/proton antiporter can confer improved salinity tolerance to *Arabidopsis* (Apse et al. 1999). This observation suggests that salt-sensitive species, like *Arabidopsis*, possess the genetic potential and biochemical machinery for efficient stress tolerance, but have lost the regulatory or signaling controls required for up-regulation of key stress-adaptive components to confer stress protection. The halophyte, *Aster tripolium* possesses a $Na^+$-sensing system that down regulates $K^+$ uptake by guard cells in response to high salt concentrations. The resulting stomatal closure prevents excessive accumulation of $Na^+$ uptake by the transpiration stream. Nonhalophytic relatives appear to lack this specialized sensing ability and may actually respond to $Na^+$ ions by increasing stomatal apertures (Véry et al. 1998). One of the key mechanisms of salinity tolerance is the ability of halophytes to sequester $Na^+$ into plant vacuoles. The halophyte *Mesembryanthemum crystallinum* has evolved the ability to induce increases in vacuolar $Na^+/H^+$ exchange in parallel with V-ATPase $H^+$-transport activities in leaf tissues with the highest activities being measured in epidermal bladder cells. These specialized structures, which are modified trichomes, are thought to serve as an important tissue in which the plant can sequester Na$^+$ to concentrations nearing 500 mM (Barkla et al. 2002). Vegetative desiccation tolerance, a evolutionary adaptation thought to have been critical to land colonization by primitive plant, is widespread, but uncommon in the modern plant kingdom (Bohnert 2000; Oliver et al. 2000b). The mechanisms of desiccation tolerance involve constitutive cellular protection coupled with active cellular repair. These traits were apparently lost during the evolution in lieu of increased growth rates, structural and morphological complexity, and mechanisms for increased water conservation and efficient carbon fixation. However, many independent cases of re-evolution of vegetative desiccation tolerance have occurred in bryophytes, club-mosses (*Selaginella*), ferns, and a few angiosperms (Oliver et al. 2000b). Constitutive protection mechanisms involve maintaining high concentrations of sucrose or sucrose-related carbohydrates and dehydrin (e.g., hydrophilins or late embryogenesis abundant) proteins. The sequestration of mRNA in mRNP particles also serves to hasten recovery of damage induced by desiccation or rehydration by minimizing the time required for growth reinitiation following rehydration.

To identify the unique genetic or regulatory differences between glycophytes and halophytes, large ESTs sequencing efforts have been initiated for several halophytic species. Expressed sequence tags collections have been reported for *Suaeda salsa* (Zhang et. al. 2001) and *Mesembryanthemum crystallinum* (Bohnert and Cushman 2000; Bohnert et al. 2001). Both species are leaf succulent euhalophytes that accumulate and sequester high concentrations of salt within foliar tissues and remove sodium from their roots and thus can survive and complete their lifecycles when irrigated with seawater-strength saline solutions. *Mesembryanthemum crystallinum* is also a popular model for the study of Crassulacean acid metabolism, as this alternative mode of photosynthetic carbon metabolism that improves water use efficiency, is induced by salinity or water-deficit stress (Cushman 2001). Recently, a salt tolerant relative of *A. thaliana*, *Thellungiella halophila*, or salt cress, equally amenable to molecular genetic analysis as *A. thaliana*, has been promoted as an ideal model for the study of salinity tolerance mechanisms (Bressan et al. 2001; Zhu 2001a). Other halophytes for which large EST collections are available include sugar beet (*Beta vulgaris*) (Schneider et al.1999) and the wild salt-tolerant relative of tomato, *Lycopersicon pennellii*. Expressed sequence tags sequencing in the unicellular, halotolerant green alga, *Dunaliella salina* (Gokhman et al 1998), is also underway (Bohnert et al. 2001).

In resurrection plants, several major models have emerged that illustrate distinct adaptive mechanisms for vegetative desiccation tolerance. Bryophytes, such as *Tortula ruralis* rely on a primitive, constitutive protection system of sucrose and dehydrins (or rehydrins) (Velten and Oliver 2001), coupled with an active rehydration-induced recovery mechanism to limit damage during rehydration and, therefore, to survive rapid drying. During slow drying, large (>150 kDa) messenger ribonucleoprotein (mRNP) particles form, which allows for the rapid restoration of protein synthesis after rehydration and facilitates survival of the desiccated vegetative cells (Wood and Oliver 1999; Oliver et al. 2000a). Initial gene discovery efforts in this desiccation-tolerant moss from EST sequencing (152 ESTs) from a cDNA library from polysomal mRNP fractions of desiccated tissue, showed that the majority (70%) of the ESTs represented novel sequences. Several ESTs showed homology with unidentified desiccation-tolerance genes identified in *C. plantagineum* (see below), suggesting common mechanistic ties to the more primitive desiccation-tolerance syndrome found in bryophytes. Such EST collections should help to define the range of gene products essential for cellular repair and recovery after vegetative desiccation tolerance (Wood et al. 1999). The large proportion of novel genes isolated suggests that many new insights into cellular mechanisms of vegetative desiccation tolerance lie ahead. Expressed sequence tags collections have also been initiated in *Physcomitrella patens*, a moss model with efficient gene targeting (Machuka et al. 1999). To identify genes associated with desiccation tolerance, 169 ESTs were characterized from *P. patens* protonema after ABA treatment. Most ESTs (69%) shared homology with known sequences, although many of the clones encoded proteins induced during heat shock, cold acclimation, oxidative stress adaptation and xenobiotic detoxification (Machuka et al. 1999).

In contrast to bryophytes, tracheophytes, such as *Selaginella lepidophylla* (Iturriaga et al. 2000), and angiosperms such as the dicotyledonous South African *Craterostigma plantagineum* (Bockel et al. 1998) and the monocotyledonous African desert grass *Sporobolus stapfianus* (Blomstedt et al. 1998; Neale et al. 2000) require slow drying over several days to survive desiccation and rely predominantly on ABA and/or drying-induced cellular protection mechanisms to accumulate gene products such as LEA proteins and sugars (e.g., sucrose, raffinose, trehalose) and establish cellular protection before desiccation. (Oliver and Bewley 1997; Oliver et al. 2000b; Kirch et al. 2001; Bartels and Salamini 2002). Expressed sequence tags sequencing from *S. lepidophylla* cDNA libraries prepared from plants undergoing desiccation is in progress (Bohnert et al. 2001). Differential, subtractive, or cold plaque screening was used for 200 cDNA clones from *C. plantagineum* leaves dried for 1 hour or totally dried down (Bockel et al. 1998). One half of the sequences showed no significant similarity to those in public databases, and of those with a predicted function, 6% and 58% were upregulated or transiently up-regulated, respectively, whereas 35.8% were down-regulated after dehydration (Bockel et al. 1998). The main functional groups of genes encoded proteins involved in cellular protection, transport processes, transcription factors, and signal transduction. Genes encoding abundant drought-induced genes correlated with desiccation tolerance or low abundance transcripts encoding gene products not previously associated with drought stress in *S. stapfianus* were isolated by differential screening (Blomstedt et al. 1998) or by "cold-plaque" hybridization procedures (Neale et al. 2000), respectively, suggesting that resurrection plants may possess unique genes or regulatory processes that confer desiccation tolerance. Comparisons of *S. stapfianus* protein expression profiles with those in a closely related desiccation-sensitive species, *S.* pyramidalis revealed a set of 12 novel proteins associated with desiccation tolerance in support this hypothesis (Gaff et al. 1997).

## B. High-Throughout Stress-Specific Gene Expression Analysis

Although gene discovery programs in model single-celled organisms and abiotic stress sensitive or tolerant plants are an important first step in defining the osmome, xerome, or thermones, knowledge of primary sequence information alone cannot provide precise information about gene function. Less than 70% of predicted protein coding regions within the *Arabidopsis* genome share some sequence similarity with genes with known function in other organisms, and the function of only about 8% to 9% of *Arabidopsis* genes have been characterized experimentally (The Arabidoposis Genome Initiative 2000; Cho and Walbot 2001). Genome sequencing or EST clone collections and associated sequence databases do, however, provide the necessary prerequisite for high-throughput, large-scale gene expression profiling. In the absence of other information, an important approach for exploiting genomic or EST resources is the use of large-scale transcript profiling revealing differential expression patterns that can often provide clues to gene function (Eisen et al. 1998). Functionally unknown sequences can be grouped with known transcripts having similar expression patterns, possibly because of functional relatedness, which can then be tested. Measuring changes in transcript abundance has been carried out using a variety of direct or indirect methods (for review see Donson et al. 2002). Table 2 summarizes transcript-profiling studies related to abiotic stress responses.

### 1. Direct Analysis of Transcript Profiling

Direct technologies such as EST sequencing measure the frequency of occurrence of ESTs in a cDNA library that reflects abundance variations of corresponding mRNAs. This so called "digital northern" or "electronic RNA" approach favors identification of more abundant, significantly up- or down-regulated genes and may also reflect bias of cloning or cDNA synthesis. Confidence in the significance of EST frequency differences, particularly for rare transcripts, can only be gained by increasing the sampling size of the EST collection and validation of results through statistical models (Audic and Claverie 1997). Nonetheless, with adequate EST sampling from nonnormalized libraries, this approach can provide reliable quantitative comparions of transcript profiles among different conditions, cDNA libraries, or different organisms (Ewing et al. 1999; Ewing et al. 1999-2000;

**Table 2** Large-Scale Abiotic Stress Responsive Gene Expression Profiling

| Treatment | Species | Reference |
|---|---|---|
| NaCl (0.5 M, 30 min.) stress/osmotic (0.5 M sorbitol, 30 min) stress | *Synechocystis sp.* | Kanesaki et al., 2002 |
| High light stress | *Synechocystis sp.* | Hihara et al., 2001 |
| Cold stress | *Synechocystis sp.* | Suzuki et al., 2001 |
| NaCl (0.7 M) 60 min. | *Saccharomyces cerevisiae* | Planta et al., 1999; Brown et al., 2001 |
| NaCl stress (0.4 M and 0.8 M);10 min, 20 min. | *Saccharomyces cerevisiae* | Posas et al., 2000 |
| NaCl stress (0.5 M, 0.7 M) /osmotic stress (0.95 M sorbitol); 30, 45 min. | *Saccharomyces cerevisiae* | Rep et al., 2000 |
| NaCl (1 M) 0, 10, 30, 90 min. | *Saccharomyces cerevisiae* | Yale and Bohnert, 2001 |
| Asexual development (conidiation) | *Aspergillus nidulans* | Prade et al., 2001 |
| Dehydration (60 min. or complete) | *Craterostigma plantagineum* | Bockel et al., 1998 |
| Dehydration (mixed stages) | *Sporobolus stapfianus* | Neale et al., 2000 |
| NaCl stress (160 mM, 4 h) | *Arabidopsis thaliana* | Gong et al., 2001 |
| NaCl stress | *Arabidopsis thaliana* | Bohnert et al., 2001 |
| Mechanical wounding/ insect feeding | *Arabidopsis thaliana* | Reymond et al., 2000 |
| Mechanical wounding | *Arabidopsis thaliana* | Cheong et al., 2002 |
| Drought and cold | *Arabidopsis thaliana* | Seki et al., 2001 |
| Cold (4°C), salt (100mM NaCl), Osmotic (200mM mannitol) | *Arabidopsis thaliana* | Chen et al., 2002 |
| Oxidative ($H_2O_2$) stress, harpin, drought stress | *Arabidopsis thaliana* | Desikan et al., 2001 |
| Ozone stress | *Arabidopsis thaliana* | Matsuyama et al., 2002 |
| NaCl stress (tissue specific) | *Mesembryanthemum crystallinum* | Bohnert et al., 2001 |
| NaCl (150 mM, 0, 15 min., 1, 3, 6, 24, 168 h.) | *Oryza sativa* | Kawasaki et al., 2001 |
| Drought (6, 10 h.) and NaCl (150 mM, 24 h)1 | *Hordeum vulgare* | Ozturk et al., 2002 |

Ewing and Claverie 2000; Mekhedov et al. 2000; Stekel et al. 2000; Prade et al. 2001). For example, comparison of the accumulation of stress-related mRNAs between conidiating *A. nidulans* and vegetative tissues of *Neuspora crassa* revealed pronounced differential expression of transcripts encoding heat shock proteins, DNA repair components, trehalose and sorbitol recycling enzymes, starvation response proteins, proton transporters, reactive oxygen scavenging enzymes, and metal homeostasis (Prade et al. 2001). A direct comparison of EST sampling and microarray hybridization techniques in maize suggested that while both reliably demonstrate differences in expression profiles, microarrays analyses can detect much higher qualitative overlap in gene expression among different tissues relative to EST sequencing (Fernandes et al. 2002). One key component of successful EST sequencing programs is the availability of dedicated bioinformatics tools for the automatic cleansing, editing, clustering, functional assignment or gene ontology annotation, and submission of polished ESTs to public databases (Bell et al. 2001; Ayoubi et al. 2002)

Alternative approaches, such as serial analysis of gene expression (SAGE), have been developed for rapidly quantitating the occurrence of large numbers of transcripts in a particular cell population, tissue type, or under different treatment conditions. With a 9-14-base size for each tag, SAGE unambiguously identifies individual transcripts, yet improves the efficiency (up to 40-fold) of generating extremely large EST databases by sequencing multiple tags within a single clone (Bertelsen and Velculescu 1999). Modifications of this procedure have been developed to accommodate limited starting amounts of tissue (Datson et al. 1999; Peters et al. 1999). Although there have been few reports of this approach in plants, SAGE has been used to compare gene expression profiles in rice seedlings exposed to control and anaerobic conditions (Matsumura et al. 1999) and to examine gene expression in lignifying xylem of loblolly pine (*Pinus taeda* L.) (Lorenz and Dean 2000). Massively parallel signature sequencing (MPSS), based on in vitro cloning of 16-20 base tags on microbeads, retains the digital sample features of EST sequencing or SAGE, but with vastly superior throughput (Brenner et al. 2000a, b). However, there have been no reports of applying MPSS to plant gene expression research to date. Another alternative, called nuclear expressed sequence tag (NEST), combines fluorescence-assisted nucleus sorting, based on nucleus-targeted GFP expression, and cDNA generation from RNA of isolated nuclei (Macas et al. 1998). By avoiding the influence of posttranscriptional RNA turnover in the cytosol, this approach more accurately reflects nuclear transcript abundance. Differential display reverse-transcription polymerase chain reaction (DDRTPCR) (Liang and Pardee 1992) and the related RNA arbitrarily primed PCR (RAP-PCR) (Welsh et al. 1992) provide simple and cost-effective approaches for the discovery of differentially expressed genes, but have been less widely used as a means of large-scale expression profiling in plants. These techniques suffer from inaccurate quantitation when conducting large-scale expression profiling and the generation of false-positives. Nonetheless, DDRT-PCR has been used to isolate genes from barley roots (Ueda et al. 2001; Shi et al. 2001) or leaves (Muramoto et al. 1999) and tomato roots (Wei et al., 2000) exposed to high salinity as well as UV-B irradiation exposure in pea (Brosche and Strid 1999) and light stress in *Arabidopsis* leaves (Dunaeva and Adamska 2001).

In contrast to differential display, cDNA-amplified fragment length polymorphism analysis (cDNA-AFLP) (Bachem et al. 1996) and related methods (Broude et al. 1999) offer a more reproducible approach to systematic surveying of transcript expression. Systematic sequencing of cDNA-AFLP fragments allows differential expression results to be linked to sequence information without the need for further sequencing (Donson et al. 2002). Also, cDNA-AFLP has been used to identify and isolate desiccation- or abscisic acid (ABA)-responsive genes in almond (Campalans et al. 2001) and heat stress responsive genes in cowpea nodules (Simoes-Araujo et al. 2002). Fluorescent labeling, multiplexing strategies, and capillary-based electrophoresis detection methods promise greater automation and throughput for cDNA-AFLP systems (Cho et al. 2001). Finally, a novel technique for the characterization of expression profiles of up to 100 genes at a time, called Beads Array for the Detection of Gene Expression (BADGE), has been reported (Yang et al. 2001). This technique uses commercially available, fluorescent, color-coded microspheres with an attached gene-specific oligonucleotide probe to selectively hybridize to labeled cRNA, which are then passed through a flow cytometer to identify and quantify the hybridization associates with each class of beads.

## 2. Indirect Analysis of Transcript Profiling

To obtain novel insights into gene function and regulatory control of biological processes associated with drought, salinity, or freezing stress responses, oligonucleotide-based or cDNA-based microarrays offer a high-throughput approach to obtaining comprehensive gene expression profiles (for reviews, see Lemieux et al. 1998; Baldwin et al. 1999; Kehoe et al. 1999; van Hal et al. 2000; Schaffer et al. 2000; Donson et al. 2002; Aharoni and Vorst 2002). High-throughput, parallel gene expression monitoring using microarray-based methods has been used to examine gene expression patterns in different tissues and under an ever-expanding set of treatment conditions. Detailed surveys of gene expression profiles should allow new insight into how plants integrate abiotic stress responses in different developmental contexts and a complex assortment of tissues having differential sensitivity or susceptibility to different environmental stresses. Microarray analysis will also permit comparisons among evolutionarily diverse glycophytic, halophytic, and desiccation-tolerant models to identify differences and commonalties in gene expression patterns or gene complements that contribute to tolerance (Bohnert et al. 2001). High throughput expression profiling should also allow the delineation of discrete responses, and the identification of common adaptive responses to different stresses such as salinity, drought, and temperature extremes. Finally, microarrays offer a rapid and comprehensive way to identify stress tolerance determinants by detecting transcripts exhibiting differential expression patterns in wild type versus mutant backgrounds dysfunctional in biochemical endpoint or signaling pathway components under stress conditions.

Comprehensive genome-wide surveys of stress-responsive gene expression using DNA microarrays have been conducted in single-celled model organisms whose entire genome sequence is known, including *Synechocystis* sp. PCC 6803 and *S. cerevisiae*. A DNA microarray containing 3,079 of 3,168 possible open reading frames (ORFs) in the cyanobacterium *Synechocystis* sp. PCC 6803 genome were used to examine the differential effects on gene expression of ionic stress and osmotic stress by the short-term (e.g., 30 minutes) treatments of 0.5 M NaCl and sorbitol, respectively (Kanesaki et al. 2002.). The expression of approximately 12% of genes were significantly ($>$ twofold change) altered by salt stress, whereas the expression of only 8% of genes was altered by sorbitol treatment. In each treatment, fewer genes were induced than were repressed. About half the genes showing enhanced or repressed expression patterns encoded proteins of unknown function. Of the ORFs for which functional assignments could be made, salt stress strongly and specifically induced the expression of certain ribosomal proteins and proteins involved in degradation of photosystem II reaction center complex protein D1. In contrast, hyperosmotic stress specifically enhanced the expression of genes for fatty acid biosynthesis and rare lipoproteins. Both stresses enhanced the expression of heat shock proteins, superoxide dismutase, osmoprotectant (e.g., glucosylglycerol) biosynthesis, and sigma 70 factors, the latter being possibly involved in the regulation of the expression of stress-inducible genes. Salt stress specifically repressed genes for fatty acid desaturation and acyl carrier protein, enzymes that play important roles in the maintenance of biological membrane integrity. In contrast, hyperosmotic stress decreased the expression of a repressor (e.g., homologue of the lexA gene) of the SOS response to stimulate the repair of DNA damage. Both stresses repressed the expression of photosystem I and phycobilisome components. DNA microarrays have also been used to examine the effects of high light (Hihara et al. 2001) and cold stress (Suzuki et al. 2001) in *Synechocystis*.

Yeast provides an excellent model in which to study osmotic stress signaling and adaptation pathways (Hohmann 2002a, b). Furthermore, many of the salinity and osmotic stress signaling pathways and biochemical determinants for stress adaptation found in yeast are also conserved in higher plants. Posas et al. (2000) used DNA microarray analysis to survey the global transcriptional response of the yeast genome under salinity stress. A mild saline stress treatment (0.4 M NaCl or 10 minutes) of yeast cells results in more than 1,300 genes being induced more than threefold (Posas et al. 2000). However, the majority of genes were transiently induced with a longer salinity stress (20 minutes), resulting in a tenfold reduction in the number of salt-responsive genes. Interestingly, more severe stress (0.8 M NaCl for 10 minutes) resulted in only about 400 genes being induced more than threefold; however, this number more than doubled after a longer duration of stress (20 minutes) (Posas et al. 2000). Thus, the transcriptional response appears to be organized into early and largely transient responses to osmotic shock and a delayed and more sustained response initiated at higher salinity. These results underscore the importance of examining the timing of stress responses. Specifically, examination of later time points (1 hour) revealed far fewer changes in gene expression than earlier time points (Planta et al. 1999; Brown et al. 2001). Major salt-responsive genes included those encoding proteins involved in carbohydrate metabolism, including glycerol and trehalose biosynthesis and metabolism, protein biosynthesis, ion homeostasis including P-type and vacuolar $H+$-ATPases, and signal transduction. Comparison of induction responses to salt stress in wild-type and a mutant strain lacking the Hog1 mitogen-activated protein (MAP) kinase gene revealed that more than 200 yeast genes depend exclusively on the high osmolarity glycerol (HOG) signaling pathway for transcriptional activation after stress (Posas et al. 2000). Other signaling pathways, however, participate in salinity stress responsiveness.

In a related study, genome-wide transcriptional responses to salinity (0.5 or 0.7 M NaCl) and osmotic (0.95 M sorbitol) were analyzed using filter microarrays (Rep et al. 2000). Transcript abundance of 186 genes was greater than three fold higher 30 or 45 minutes after osmotic and salinity stress, respectively. The majority of induced genes encoded proteins or enzymes involved in carbohydrate metabolism and transport, protection against oxidative damage, cell surface assembly, cell wall formation, vacuolar water and ion homeostasis, transcription, and signaling. The expression of more than 100 genes was also repressed under these same conditions. Many down-regulated genes encoded either ribosomal or translational proteins reflecting the well-documented temporary growth arrest after stress. Surprisingly, both salinity and osmotic (sorbitol) stress-specific gene expression responses were detected. Nearly 50 genes were identified whose expression is controlled, at least in part, by the HOG pathway, with a similar number of genes being identified whose expression was moderately or not affected by deletion of *Hog1*. In addition, the expression of a small subset of HOG-dependent genes was found to be dependent on various transcription factors that control HOG-dependent responses (Rep et al. 2000). A more detailed time course analysis of yeast cells exposed to hyperosmotic shock (1 M NaCl for 0–90 minutes) revealed that approximately 10% of all genes showed a greater than twofold increase or decrease in transcript abundance with the number of genes showing salinity-induced increases in transcript abundance growing larger over time (Yale and Bohnert 2001). Genes with early expression patterns (10 and 30 minutes) shared more coincident expression patterns (67% identical at both early time points) compared with those expressed later (90 minutes), which revealed different up-regulated transcripts (only 13% and 22% identity, respectively). This study discovered that a succession of genes with different biochemical functions become progressively up-regulated over time. For example, genes involved in early stress responses had functions related to nucleotide and amino acid metabolism and protein biosynthetic pathways. Transcripts of genes associated with energy production were up-regulated throughout the time-course with respiration-associated gene expression peaking at 30 minutes. Genes strongly up-regulated at 90 minutes included known salinity stress-induced genes (e.g., glycerol and trehalose metabolism), intracellular protein and metabolite transport detoxification-related responses, transport facilitation and ion homeostasis, metabolism of energy reserves, nitrogen and sulfur compounds, and lipid and fatty acid or isoprene biosynthesis (Yale and Bohnert 2001). Although this study confirmed the expression behavior of many transcripts with known biochemical stress responses, the temporal exploration of stress responses revealed a set of about 200 salt-regulated genes of unknown function and provided the first information about the functional roles of these unknown ORFs in cellular stress adaptation processes. Furthermore, many open reading frames with unknown function that exhibited salt-responsive expression patterns shared significant homology with ESTs from the halophyte, *M. crystallinum*, rice, and maize (Bohnert et al. 2001). Further comparisons revealed that more than 50 yeast genes with known functions are similarly up-regulated in salinity-stressed rice (Kawasaki et al. 2001). Large-scale EST sequencing in *Aspergillus nidulans* (Prade et al. 2001) has also permitted the in silico reconstruction of the HOG pathway (Han and Prade 2002). Deletion of the *Hog1* homologue resulted in a reduction in polar growth, alterations in hyphal branching patterns, and growth expansion rates under high-salinity conditions documenting the crucial role of this gene in *A. nidulans* (Han and Prade 2002).

The utility of cDNA microarrays to not only quantitate gene expression profiles, but also to identify novel genes that are expressed under specific abiotic stress conditions has recently been demonstrated in a number of higher plant studies. To monitor the expression patterns of *Arabidopsis* genes in response to drought and cold stress, Seki et al. (2001a) constructed a cDNA microarray containing about 1300 full-length *Arabidopsis* cDNAs. Of the 44 drought-inducible genes identified, 30 had not been reported previously to be drought-inducible. These new drought-inducible transcripts shared homology with putative cold-acclimation, LEA, nonspecific lipid transfer, and water channel proteins. Likewise, of the 19 cold-inducible cDNAs characterized, 10 were reported as novel. These new cold-inducible cDNAs shared homology with putative cold-acclimation and LEA proteins, ferritin, β-amylase, nodulin-like proteins and glyoxalase among others. This full-length cDNA microarray was also useful in identifying target genes of the DREB1A transcription factor by comparing the expression profiles of wild-type plants with transgenic *Arabidopsis* plants that overexpress

DREB1A under the control of CaMV 35S promoter. A total of 12 DREB1A targets were identified. Six were previously identified and six were found to be novel DREB1A gene targets. A majority of these DREB1A targets displayed both drought-and cold-inducible expression patterns. One drawback of full-length cDNA microarrays is that cross-hybridization may occur between closely related gene family members sharing more than 70% to 80% sequence identity (Richmond and Somerville 2000; Donson et al. 2002). This disadvantage may be overcome through the use of probe sets representing 3′-UTR regions (Yazaki et al. 2000) or oligonucleotide-based microarrays which can discriminate between nucleotide sequences sharing less than 93% identity (Lipshutz et al. 1999). However, cDNA-based microarrays have been used successfully to distinguish tissue-specific expression profiles of different cytochrome P450 (P450s) gene family members within this very large superfamily in *Arabidopsis* (Xu et al. 2001).

To gain a more refined analysis of gene responses than single time point analyses, Kawasaki et al. (2001) performed a comparative analysis of the effects of salt shock on gene expression in a salt-sensitive variety (e.g., IR29) and moderately salt tolerant (e.g., Pokkali) lines of rice over a time-course of 15 minutes to 1 week (Kawasaki et al. 2001). Transcript profiles in roots of stressed plants were investigated using a cDNA microarray containing 1,728 probes derived from roots of unstressed or salt stress (12–72 hours duration). The two cultivars of rice displayed very distinct expression patterns after stress, reflecting their relative sensitivities to salinity stress. Pokkali (salt-tolerant) showed immediate and significant increases and decreases in approximately 10% of the transcripts surveyed, with these initial differences becoming less pronounced as adaptation progressed and eventually returning to prestress expression levels. In contrast, IR29 (salt-sensitive) exhibited a delay in the immediate response and then a general decrease in transcript abundance after about 3 hours of stress, leading ultimately to plant death. The profile of gene expression changes in Pokkali was dynamic over time with different categories of transcripts encoding stress- and defense-related gene products. Transcripts were organized into clusters according to temporal response timeframes. An "instantaneous" cluster (15 minutes) was represented by transcripts encoding several GRP isoforms. An "early response" (1 hour) was exemplified by transcripts encoding a calcium-dependent protein kinase previously reported to be involved in cold, high salt, and drought adaptive signaling (Saijo et al. 2000), as well as ribosomal proteins and translational elongation factors that may direct retasking of the translational machinery toward the synthesis of adaptive stress determinants. Transcripts abundant in the "early recovery" phase (3–6 hours) were functionally diverse and included those thought to be associated with maintenance of cell wall integrity, protease inhibitors, polyamine biosynthesis, and pathogen defense reactions. Genes encoding reactive oxygen scavenging functions dominated transcripts induced during the "stress compensation" cluster (24 hours). Overall, Pokkali appeared to overcome salt stress, in part, through its ability to rapidly induce transcripts that initiate signaling circuits and protein synthesis, whereas IR29 failed to initiate these immediate responses and, through its delay, mobilized fewer responses overall, which eventually resulted in a general decrease in transcription and death within 24 hours.

Gene expression charges in response to drought and salinity stress were monitored by hybridization to a microarray containing 1,463 DNA elements derived from cDNA libraries constructed from drought-stressed barley (Ozturk et al. 2002). About 15% of all transcripts showed significant increases or decreases in abundance after drought stress, whereas only about 5% of transcripts responded significantly to NaCl stress. Drought-induced transcripts were represented by jasmonate- and ABA-responsive genes and metallothionein-like, pathogenesis-related, and late embryogenesis abundant (LEA) proteins. These latter two gene products were induced by both types of stress treatments. Salt-induced transcripts were typified by ubiquitin-related genes. In contrast, transcripts encoding photosynthetic apparatus components and associated proteins, and enzymes concerned with photorespiration, amino acid, and carbohydrate metabolism, showed consistent declines in abundance after salinity and drought stress treatments. Microarray hybridization results were validated in this study by real-time PCR.

High-density oligonucleotide-based microarrays (e.g., GeneChips®) have been developed that contain probes corresponding to 8,300 unique *Arabidopsis* genes (Zhu et al. 2001a; Zhu and Wang 2000). Such GeneChips provided the platform for in-depth expression profiling of 402 transcription factors in plants exposed to cold, salt, and osmotic stress, in addition to infection by various bacterial pathogens, wounding, and treatment with jasmonic acid (Chen et al. (2002b). Most transcription factors analyzed showed some degree of differential expression after these various stress treatments, allowing them to be clustered according to the type of treatments or coregulation in expression observed. One group of 12 factors induced preferentially by abiotic stress (e.g., cold, high salt, osmoticum) included the genes encoding DREB1/CBF1, DREBIC/CBF2, and DREB1A/ CBF3 DRE/CRT binding factors as well as various genes encoding zinc finger proteins, Myb proteins, bZIP/HD-ZIPS, and AP2/EREBP domain-containing proteins. A second group of five genes was found to be activated preferentially by both abiotic stress and bacterial infection and included genes encoding leucine zipper DNA-binding proteins (e.g., bZIPs, HD-ZIPs) that play roles in light and ABA or water stress-responsive gene expression. Another bZIP and Myb transcription factor was also included in this group. The expression of the remaining groups of factors was activated mainly by infection by bacteria or other types of pathogens including fungi, oomycetes, and viruses. Gene expression profiles were also gathered after pathogen infection in a collection of different mutant backgrounds to dissect which factors function in which signaling pathways. This work also confirmed that substantial overlap exists in the signaling pathways activated by stress and senescence as well as plant hormones. Of the 43 transcription factor genes induced during senescence, 28 also displayed stress-responsive expression patterns. Additional support for crosstalk among signal transduction pathways comes from large-scale expression analysis of herbivory- and jasmonate-responsive genes (Arimura et al. 2000; Sasaki et al., 2000; Sasaki et al., 2001). Interestingly, a greater percentage of transcription factors ($\sim15\%$) were expressed at relatively high levels in roots, whereas lower percentages, 6% and 3%, were highly expressed in leaves and flowers/siliques, respectively. Finally, statistical analysis of promoter regions of cold-regulated genes indicated an unambiguous enrichment for the occurrence of known conserved transcription factor-binding sites associated with the cold response.

Interestingly, cDNA microarray analysis of gene expression changes of 150 genes in response to mechanical wounding, dehydration, and insect feeding of *Arabidopsis* leaves revealed that wounding and dehydration treatments displayed coordinate activation of a much larger number of genes than did wounding and feeding (Reymond et al. 2000). These results support the interaction of signaling pathways (e.g., crosstalk) between wounding and dehydration responses and suggest that insect larvae may feed in a manner designed to minimize the induction of host defense genes. A more recent study using an oligonucleotide-based microarray to profile wounding responses has confirmed that significant crosstalk exists among signaling pathways for wounding, pathogen infection, hormal responses, and abiotic stresses (e.g., cold, salt, ABA, heat shock) (Cheong et al. 2002). In plants, various biotic and abiotic stresses result in the accumulation of reactive oxygen species, such as hydrogen peroxide ($H_2O_2$), leading to oxidative stress. Hydrogen peroxide functions as a stress signal that could be a common mediator in the adaptive response pathways to various stresses. Comparison of the gene expression changes in response to $H_2O_2$-treated *Arabidopsis* leaves using cDNA microarray analysis clearly indicated many of the genes induced by oxidative stress are also induced by other stresses, such as drought, UV irradiation, and fungal elicitor treatment (Desikan et al. 2001). In a related study, gene expression changes of 12 genes were monitored after exposure to ozone, drought, and wounding treatments using a custom *Arabidopsis* cDNA microarray (Matsuyama et al. 2002).

## IV. CONCLUSIONS AND FUTURE PERSPECTIVES

With the identification of candidate genes using QTL mapping and the comprehensive gene discovery and transcript profiling studies summarized above, it should be clear that the discovery phase of functional genomics, in which all the genetic parts required for adaptation to abiotic stresses will be inventoried and catalogued, is rapidly coming to a close for glycophytic model organisms. However, this process is just beginning for stress tolerance halophytes and resurrection plants. Thus, it is expected that gene discovery efforts in these stress tolerance models will continue well into the future. Defining all of the components or formulating a "genetic parts list" essential for stress adaptation is the first step in the creation of a systems biology framework (Ideker et al. 2001). For organisms in which the complete genome sequence is in hand, one important challenge will be to dissect the complex regulatory circuitry associated with abiotic stress signaling-response pathways. Although high-throughput gene expression profiling is now in wide use, this approach cannot directly identify sets of genes that belong to the same regulatory group. Even though statistical algorithms based on microarray expression data and the identification of common *cis*-regulatory elements can lead to hierarchical clustering of gene sets, these are indirect approaches that can only identify putatively coregulated genes. One useful approach for directly determining promoter regions that are bound by or controlled by the same transcription factor or group of related factors combines chromatin immunoprecipitation with DNA microarrays to obtain a direct assessment of transcription factor-*cis* element interactions, rather than interactions that are merely inferred from *cis* element surveys (Nal et al. 2001; Weinmann et al. 2002). Such an approach should be useful for dissecting global regulatory networks controlled by single or groups of related transcription factors and to elucidate the intricate genetic networks that coordinate the transcriptional activation of different regulatory circuits within a cell, a tissue, or an entire organism after abiotic stress exposure.

Research on the protein complement expressed by the genome (e.g., proteomics) of plants under abiotic stress is still in its infancy. The majority of global gene expression studies has focused on transcript abundance changes. Various studies have shown that stress treatments result in profound changes in transcript abundance of genes encoding ribosomal proteins and translational factors. Such changes are expected to have complex effects on various aspects of the translational apparatus that may not necessarily be reflected at the transcript level. Recent studies in yeast and mammalian systems have demonstrated that abiotic stresses and associated oxidative stresses exert multiple and complex inhibitory effects on the translational machinery (Uesono and Toh 2002; Patel et al. 2002). Another great challenge will be to integrate the information gathered from transcript profiling programs with data on comprehensive protein expression studies, posttranslational protein modification networks (e.g., phosphorylome analysis), protein-ligand interactions, and protein-protein interaction networks (e.g., protein linkage mapping) (Kersten et al. 2002; Roberts 2002) to determine how abiotic stresses influences these processes and to define which are relevant to stress tolerance phenotypes. The first step in protein linkage mapping is identifying the constituent protein complexes. Recent advances in affinity purification of protein complexes coupled with mass spectrometry, and bioinformatic analysis have resulted in the first proteome-wide analysis of protein complexes in yeast (Gavin

et al 2002). However, changes in protein abundance, modification, or their interactions may not necessarily indicate changes in metabolic activities associated
with proteins or protein assemblies (Trethewey 2001). Therefore, monitoring
changes in metabolite concentrations (e.g., metabolic profiling) will provide another powerful approach for assessing gene function and relationships to different
phenotypes (Fiehn 2002; Phelps et al., 2002). Thus, integrated studies using combinations of these multiparallel approaches (Fiehn et al. 2001) should greatly
improve our understanding of the intricate biochemical regulatory networks that
control metabolic responses to abiotic stress. Ultimately, a systematic effort to
mutagenize all components of the "osmome," "xerome," or "thermone" is required to complement gene discovery and integrative profiling information. Once
all the candidates have been identified and analyzed, systematic perturbation of
the system using forward or reverse genetic strategies, complementation, and
overexpression tests will be necessary to determine the function of these components (Zhu 2000; Zhu 2001b). The generation and screening of large T-DNA or
transposon insertional mutant collections of *Arabidopsis* and rice are well advanced (Bohnert et al. 2001). Comparative analyses of osmotic-, desiccation-, or
temperature-specific stress responses across a set of glycophytic, halophytic, and
resurrection plant models will assist in resolving those cellular tolerance mechanisms that are evolutionarily conserved. Ultimately, such systematic analyses are
expected to provide an integrated understanding of the biochemical and physiological basis of stress responses in plants. With a detailed knowledge about signaling hierarchies and the metabolic impact of alterations involved in stress responses, it will be possible to rationally engineer and optimize tolerance traits
for more effective crop protection to meet the agricultural demands of the 21st
century.

## ACKNOWLEDGMENTS

The author gratefully acknowledges support from the National Science Foundation Plant Genome Program (NSF DBI 9813360) and the Nevada Agricultural
Experiment Station.

## REFERENCES

Aharoni A, Vorst O. DNA microarrays for functional plant genomics. Plant Mol Biol 48:
    99–118, 2002.
Akhtar N, Blomberg A, Adler L. Osmoregulation and protein expression in a pbs2delta
    mutant of *Saccharomyces cerevisiae* during adaptation to hypersaline stress. FEBS
    Lett 403:173–180, 1997.

Ali ML, Pathan MS, Zhang J, Bai G, Sarkarung S, Guyen HT. Mapping QTL for root traits in a recombinant inbred population from two indica ecotypes in rice. Theoret Appl Genet 101:756–766, 2000.

Allona I, Quinn M, Shoop E, Swope K, St Cyr S, Carlis J, Riedl J, Retzel E, Campbell MM, Sederoff R, Whetten RW. Analysis of xylem formation in pine by cDNA sequencing. Proc Natl Acad Sci USA 95:9693–9698, 1998.

Apse MP, Aharon GS, Snedden WA, Blumwald E. Salt tolerance conferred by over-expression of a vacuolar Na+/H+ antiport in *Arabidopsis*. Science 285:1256–1258, 1999.

Arimura G, Tashiro K, Kuhara S, Nishioka T, Ozawa R, Takabayashi J. Gene responses in bean leaves induced by herbivory and by herbivore-induced volatiles. Biochem Biophys Res Commun 277:305–310, 2000.

Asamizu E, Nakamura Y, Sato S, Tabata S. Generation of 7137 non-redundant expressed sequence tags from a legume, *Lotus japonicus*. DNA Res 7:127–130, 2000a.

Asamizu E, Nakamura Y, Sato S, Tabata S. A large scale analysis of cDNA in *Arabidopsis thaliana*: Generation of 12,028 non-redundant expressed sequence tags from nor-malized and size-selected cDNA libraries. DNA Res 7:175–180, 2000b.

Ashraf M. Breeding for salinity tolerance in plants. Crit Rev Plant Sci 13:17–42, 1994.

Audic S, Claverie J-M. The significance of digital gene expression profiles. Genome Res 7:986–995, 1997.

Ayoubi P, Jin X, Leite S, Liu X, Martajaja J, Abduraham A, Wan Q, Yan W, Misawa E, Prade RA. PipeOnline 2.0: automated EST processing and functional data sorting. Nucleic Acids Res 30:4761–4769, 2002.

Bachem CW, van der Hoeven RS, de Bruijn SM, Vreugdenhil D, Zabeau M, Visser RG. Visualization of differential gene expression using a novel method of RNA finger-printing based on AFLP: Analysis of gene expression during potato tuber develop-ment. Plant J 9:745–753, 1996.

Baldwin D, Crane V, Rice D. A comparison of gel-based, nylon filter and microarray techniques to detect differential RNA expression in plants. Cuff Opin Plant Biol 2:96–103, 1999.

Barkla BJ, Vera-Estrella R, Camacho-Emiterio J, Pantoja D. Na$^+$/H$^+$ exchange in the halophyte *Mesembryanthemum crystallinum* is associated with cellular sites of Na$^+$ storage. Funct Plant Biol 29:1017–1024, 2002.

Bartels D, Salamini F. Desiccation tolerance in the resurrection plant *Craterostigma plan-tagineum*. A contribution to the study of drought tolerance at the molecular level. Plant Physiol 127:1346–1353, 2001.

Bell CJ, Dixon RA, Farmer AD, Flores R, Inman J, Gonzales RA, Harrison MJ, Paiva NL, Scott AD, Weller JW, May GD. The Medicago Genome initiative: A model legume database. Nucleic Acids Res 29:114–117, 2001.

Bertelsen AH, Valculescu VE. High-throughput gene expression analysis using SAGE. DDT 3:152–159, 1998.

Blomstedt CK, Gianello RD, Gaff DF, Hamill JD, Neale AD. Differential gene expression in desiccation-tolerant and desiccation-sensitive tissue of the resurrection grass, *Sporobolus stapfianus*. Aust J Plant Physiol 25:937–946, 1998.

Bockel C, Salamini F, Bartels D. Isolation and characterization of genes expressed during early events of the dehydration process in the resurrection plant *Craterostigama plantagineum*. J Plant Physiol 152:158–166, 1998.

Bohnert HJ. What makes desiccation tolerable? Genome Biol 1:1010.1–1010.4, 2000.

Bohnert HJ, Cushman JC. The ice plant cometh: Lessons in abiotic stress tolerance. J Plant Growth Reg 19:334–346, 2000.

Bohnert HJ, Ayoubi P, Borchert C, Bressan RA, Burnap RL, Cushman JC, Cushman MA, Deyholos M, Galbraith DW, Hasegawa PM, Jenks M, Kawasaki S, Koiwa H, Koreeda S, Lee B-H, Michalowski, CB, Misawa E, Nomura M, Ozturk M, Postier B, Prade R, Song C-P, Tanaka Y, Wang H, Zhu J-K. A genomics approach towards salt stress tolerance. Plant Physiol Biochem 39:295–311, 2001.

Bolaños J, Edmeades GO. The importance of the anthesis-silking interval in breeding for drought tolerance in tropical maize. Field Crops Res 48:65–80, 1996.

Boyer JS Plant productivity and environment. Science 218:443–448, 1982,

Brenner S, Johnson M, Bridgham J, Golda G, Lloyd DH, Johnson D, Luo S, McCurdy S, Foy M, Ewan M, Roth R, George D, Eletr S, Albrecht G, Vermaas E, Williams SR, Moon K, Burcham T, Pallas M, DuBridge RB, Kirchner J, Fearon K, Mao J, Corcoran K. Gene expression analysis by massively parallel signature sequencing (MPSS) on microbead arrays. Nat Biotechnol 18:630–634, 2000a.

Brenner S, Williams SR, Vermaas EH, Storck T, Moon K, McCollum C, Mao JI, Luo S, Kirchner JJ, Eletr S, DuBridge RB, Burcham T, Albrecht G. In vitro cloning of complex mixtures of DNA on microbeads: Physical separation of differentially expressed cDNAS. Proc Natl Acad Sci USA 97:1665–16670, 2000b.

Bressan RA, Zhang C, Zhang H, Hasegawa PM, Bohnert HJ, Zhu JK. Learning from the Arabidopsis experience. The next gene search paradigm. Plant Physiol 127:135–1360, 2001.

Brosche M, Strid A. Cloning, expression, and molecular characterization of a small pea gene family regulated by low levels of ultraviolet B radiation and other stresses. Plant Physiol 121:479–487, 1999.

Broude NE, Storm N, Malpel S, Graber JH Lukyanov S, Sverdlov E, Smith CL. PCR based targeted genomic and cDNA differential display. Genet Anal 15:51–63, 1999.

Brown AJ, Planta RJ, Restuhadi F, Bailey DA, Butler PR, Cadahia JL, Cerdan ME, De Jonge M, Gardner DC, Gent ME, Hayes A, Kolen CP, Lombardia LJ, Murad AM, Oliver RA, Sefton M, Thevelein JM, Tournu H, van Delft YJ, Verbart DJ, Winderickx J, Oliver SG. Transcript analysis of 1003 novel yeast genes using high-throughput northern hybridizations. EMBO J 20:3177–3186, 2001.

Bruce WB, Edmeades GO, Barker TC. Molecular and physiological approaches to maize improvement for drought tolerance. J Exp Bot 53:13–25, 2002.

Calinski T, Kaczmarek Z, Krajewski P, Frova C, Sari-Gorla M. A multivariate approach to the problem of QTL localization. Heredity 84:303–310, 2000.

Campalans A, Pages M, Messeguer R. Identification of differentially expressed genes by the cDNA-AFLP technique during dehydration of almond (*Prunus amygdalus*). Tree Physiol 21:633–643, 2001.

Cattivelli L, Baldi P, Crosatti C, Di Fonzo N, Faccioli P, Grossi M, Mastrangelo AM, Pecchioni N, Stanca AM. Chromosomal regions and stress-related sequences involved in resistance to abiotic stress in *Triticaeae*. Plant Mol Biol 48:649–665, 2002.

Cha K-W, Lee Y-J, Koh H-J, Lee B-M, Nam Y-W, Paek N-C. Isolation, characterization, and mapping of the stay green mutant in rice. Theor Appl Genet 104:526-532, 2002.

Champoux MC, Wang G, Sarkarung S, Mackill DJ, O'Toole JC, Huang N, McCouch SR Locating genes associated with root morphology and drought avoidance in rice via linkage to molecular markers. Theor Appl Genet 90:969–981, 1995.

Chen FQ Foolad MR. A molecular linkage map of tomato based on a cross between *Lycopersicon esculentum* and *L. pimpinellifolium* and its comparison with other molecular maps of tomato. Genome 42:94–103, 1999.

Chen M, Presting G, Barbazuk WB, Goicoechea JL, Blackmon B, Fang G, Kim H, Frisch D, Yu Y, Sun S, Higingbottom S, Phimphilai J, Phimphilai D, Thurmond S, Gaudette B, Li P, Liu J, Hatfield J, Main D, Farrar K, Henderson C, Barnett L, Costa R, Williams B, Walser S, Atkins M, Hall C, Budiman MA, Tomkins JP, Luo M Bancroft I, Salse J, Regad F, Mohapatra T, Singh NK, Tyagi AK, Soderlund C, Dean RA, Wing RA. An integrated physical and genetic map of the rice genome. Plant Cell 14:537–545, 2002a.

Chen W, Provart NJ, Glazebrook J, Katagiri F, Chang HS, Eulgem T, Mauch F, Luan S, Zou G, Whitham SA, Budworth PR, Tao Y, Xie Z, Chen X, Lam S, Kreps JA, Harper JF, Si-Ammour A, Mauch Mani B, Heinlein M, Kobayashi K, Hohn T, Dangl JL, Wang X Zhu T. Expression profile matrix of Arabidopsis transcription factor genes suggests their putative functions in response to environmental stresses. Plant Cell 14:559–574, 2002b.

Cheong YH, Chang HS, Gupta R, Wang X, Zhu T, Luan S. Transcriptional profiling reveals novel interactions between wounding, pathogen, abiotic stress, and hormonal responses in Arabidopsis. Plant Physiol 129:661–677, 2002.

Childs KL, Klein RR, Klein PE, Morishige DT, Mullet JE. Mapping genes on an integrated sorghum genetic and physical map using cDNA selection technology. Plant J 27: 243–255, 2001.

Cho Y, Walbot V. Computational methods for gene annotation: The Arabidopsis genome. Curr Opin Biotech 12:126–130, 2001.

Cho YJ, Meade JD, Walden JC, Chen X, Guo Z, Liang P. Multicolor fluorescent differential display. Biotechniques 30:562–568, 570, 572, 2001.

Courtois B, McLaren G, Sinha PK, Prasad K, Yadav R, Shen L. Mapping QTLs associated with drought avoidance in upland rice. Molecular Breeding 6:55-66, 2000.

Crasta OR, Xu WW, Rosenow DT, Mullet J, Nguyen HT. Mapping of post-flowering drought resistance traits in grain sorghum: association between QTLs influencing premature senescence and maturity. Mol Gen Genet 262:579–588, 1999.

Crookshanks M, Emmersen J, Welinder KG, Nielsen KL. The potato tuber transcriptome: Analysis of 6077 expressed sequence tags. FEES Lett 506:123–126, 2001.

Cushman JC. Crassulacean acid metabolism (CAM): A plastic photosynthetic adaptation to arid environments. Plant Physiol 127:1439–1448, 2001.

Cushman JC, Bohnert HJ. Genomic approaches to plant stress tolerance. Curr Opin Plant Biol 3:117–124, 2000.

Datson NA, van der Perk-de Jong J, van den Berg MP, de Kloet ER, Vreugdenhil E. MicroSAGE: A modified procedure for serial analysis of gene expression in limited amounts of tissue. Nucleic Acids Res 27:1300–1307, 1999.

Desikan R, A-H-Mackerness S, Hancock JT, Neill SJ. Regulation of the Arabidopsis transcriptome by oxidative stress. Plant Physiol 127:159–172, 2001.

Donson J, Fang Y, Espiritu-Santo G, Xing W, Salazar A, Miyamoto S, Armendarez V, Volkmuth W. Comprehensive gene expression analysis by transcript profiling. Plant Mol Biol 48:75-97, 2002.

Druka A, Kudrna D, Kannangara CG, von Wettstein D, Kleinhofs A. Physical and genetic mapping of barley (*Hordeum vulgare*) germin-like cDNAs. Proc Nail Acad Sci USA 99:850–8505, 2002.

Dunaeva M, Adamska I. Identification of genes expressed in response to light stress in leaves of *Arabidopsis thaliana* using RNA differential display. Eur J Biochem) 268: 5521–5529, 2001.

Eisen MB, Spellman PT, Brown PO, Botstein D. Cluster analysis and display of genome-wide expression patterns. Proc Natl Acad Sci USA 95:14863–14868, 1998.

Ellis RP, Forster BP, Robinson D, Handley LL, Gordon DC, Russell JR, Powell W. Mapping physiological traits in barley. New Phytol 137:149–157, 1997.

Ellis RP, Forster BP, Robinson D, Handley LL, Gordon DC, Russell JR, Powell W. Wild barley: A source of genes for crop improvement in the 21st century? J Exp Bot 51: 9–17, 2000.

Ewing RM, Kahla AB, Poirot O, Lopez F, Audic S, Claverie JM. Large-scale statistical analyses of rice ESTs reveal correlated patterns of gene expression. Genome Res 9:950–959, 1999.

Ewing R, Poirot O, Claverie JM. Comparative analysis of the Arabidopsis and rice expressed sequence tag (EST) sets. In Silico Biol 1:197–213, 1999–2000.

Ewing RM, Claverie JM. EST databases as multi-conditional gene expression datasets. Pac Symp Biocomput 430–442, 2000.

Fernandes J, Brendel V, Gai X, Lal S, Chandler VL, Elumalai RP, Galbraith DW, Pierson EA, Walbot V. Comparison of RNA expression profiles based on maize expressed sequence tag frequency analysis and micro-array hybridization. Plant Physiol 128: 896–910, 2002.

Fiehn O. Metabolomics—the link between genotypes and phenotypes. Plant Mol Biol 48: 155–171, 2002.

Fiehn O, Kloska S, Altmann T. Integrated studies on plant biology using multiparallel techniques. Curr Opin Biotechnol 12:82–86, 2001.

Flowers TJ, Yeo AR. Breeding for salinity tolerance in crop plants: Where next? Aust J Plant Physiol 22:875–884, 1995.

Flowers TJ, Koyama ML, Flowers SA, Sudhakar C, Singh KP, Yeo AR. QTL: Their place in engineering tolerance of rice to salinity. J Exp Bot 51:99–106, 2000.

Foolad MR. Comparison of salt tolerance during seed germination and vegetative growth in tomato by QTL mapping. Genome 42:727–734, 1999.

Foolad MR. Genetic bases of salt and cold tolerance in tomato and the prospects for developing tolerant cultivars. Curr Top Plant Biol 2:35–49, 2000.

Foolad MR, Chen FQ. RFLP mapping of QTLs conferring salt tolerance during vegetative stage in tomato. Theor Appl Genet 99:235–243, 1999.

Foolad MR, Chen FQ, Lin GY. RFLP mapping of QTLs conferring salt tolerance during germination in an intraspecific cross of tomato. Theor Appl Genet 97:1133–1144, 1998a.

Foolad MR, Chen FQ, Lin GY. RFLP mapping of QTLs conferring cold tolerance during germination in an intraspecific cross of tomato. Mol Breed 4:519–529, 1998b.

Foolad MR, Lin GY, Chen FQ. Comparison of QTLs for seed germination under non-stress, cold stress and salt stress in tomato. Plant Breed 118:167–173, 1999.

Foolad MR, Stoltz T, Dervinis C, Rodriquez RL, Jones RA. Mapping of QTLs conferring salt tolerance during germination in tomato by selective genotyping. Mol Breed 3: 269–277, 1997.

Foolad MR, Zhang LP, Lin GY. Identification and validation of QTLs for salt tolerance during vegetative growth in tomato by selective genotyping. Genome 44:444–454, 2001.

Forster BP, Ellis RP, Thomas WT, Newton AC, Tuberosa R, This D, el-Enein RA, Bahri MH, Ben Salem M. The development and application of molecular markers for abiotic stress tolerance in barley. J Exp Bot 51:19–27, 2000.

Fridman E, Pleban T, Zamir D. A recombinant hotspot delimits a wild-species quantitative trait locus for tomato sugar content to 484 bp within an invertase gene. Proc Natl Acad Sci USA 97:4718–4723, 2000.

Frova C, Sari-Gorla M. Quantitative trait loci (QTLs) for pollen thermotolerance detected in maize. Mol Gen Genet 245:424–430, 1994.

Frova C, Krajewski P, di Fonzo N, Villa M, Sari-Gorla M. Genetic analysis of drought tolerance in maize by molecular markers. 1. Yield components. Theor Appl Genet 99:280–288, 1999a.

Frova C, Caffulli A, Pallavera E. Mapping quantitative trait loci for tolerance to abiotic stresses in maize. J Exp Zool 282:164–170, 1999b.

Gaff DF, Bartels D, Gaff JL. Changes in gene expression during drying in a desiccation-tolerant grass *Sporobolus stapfianus* and a desiccation-sensitive grass *Sporobolus pyramidalis*. Aust J Plant Physiol 24:617–622, 1997.

Gai X, Lal S, Xing L, Brendel V, Walbot V. Gene discovery using the maize genome database ZmDB. Nucleic Acids Res 28:94–96, 2000.

Gavin AC, Bosche M, Krause R, Grandi P, Marzioch M, Bauer A, Schultz J, Rick JM, Michon AM, Cruciat CM, Remor M, Hofert C, Schelder M, Brajenovic M, Ruffner H, Merino A, Klein K, Hudak M, Dickson D, Rudi T, Gnau V, Bauch A, Bastuck S, Huhse B, Leutwein C, Heurtier MA, Copley RR, Edelmann Al, Querfurth E, Rybin V, Drewes G, Raida M, Bouwmeester T, Bork P, Seraphin B, Kuster B, Neubauer G, Superti-Furga G. Functional organization of the yeast proteome by systematic analysis of protein complexes. Nature 415:141–147, 2002.

Goff SA. Rice as a model for cereal genomics. Curr Opin Plant Biol 2:86–89, 1999.

Goff SA, Ricke D, Lan TH, Presting G, Wang R, Dunn M, Glazebrook J, Sessions A, Oeller P, Varma H, Hadley D, Hutchison D, Martin C, Katagiri F, Lange BM, Moughamer T, Xia Y, Budworth P, Zhong J, Miguel T, Paszkowski U, Zhang S, Colbert M, Sun WL, Chen L, Cooper B, Park S, Wood TC, Mao L, Quail P, Wing R, Dean R, Yu Y, Zharkikh A, Shen R, Sahasrabudhe S, Thomas A, Cannings R, Gutin A, Pruss D, Reid J, Tavtigian S, Mitchell J, Eldredge G, Scholl T, Miller RM, Bhatnagar S, Adey N, Rubano T, Tusneem N, Robinson R, Feldhaus J, Macalma T, Oliphant A, Briggs S. A draft sequence of the rice genome (*Oryza sativa* L. ssp. *japonica*). Science 296:92–100, 2002.

Goffeau A, Barrell BG, Bussey H, Davis RW, Dujon B, Feldmann H, Galibert F, Hoheisel

JD, Jacq C, Johnston M, Louis EJ, Mewes HW, Murakami Y, Philippsen P, Tettelin H,Oliver SG. Life with 6000 genes. Science 274:563–567, 1996.

Gokhman I, Fisher M, Pick U, Zamir A. New insights into the extreme salt tolerance of the unicellular green alga *Dunaliella*. Micro Biogeochem Hypersaline Environ. CRC Press Inc, Boca Raton, FL. pp. 203–213, 1998.

Gong Z, Koiwa H, Cushman MA, Ray A, Bufford D, Kore-eda S, Matsumoto T, Zhu J, Cushman J, Bressan T, Hasegawa M. Genes that are uniquely stress regulated in salt overly sensitive (sos) mutants. Plant Physiol 126:363–375, 2001.

Han KH, Prade RA. Osmotic stress-coupled maintenance of polar growth in *Aspergillus nidulans*. Mol Microbiol. 43:1065-1078, 2002.

Hasegawa PM, Bressan RA, Pardo JM. The dawn of plant salt tolerance genetics. Trends Plant Sci 5:317–319, 2000a.

Hasegawa PM, Bressan RA, Zhu J-K, Bohnert HJ. Plant cellular and molecular responses to high salinity. Annu Rev Plant Physiol Plant Mol Biol 51:463–499, 2000b.

Hemamalini GS, Shashidar HE, Hittalmani S. Molecular marker assisted tagging of mophological and physiological traits under two contrasting moisture regimes at peak vegetative stage in rice (*Oryza sativa* L.). Euphytica 112:69–78, 2000.

Hervé D, Fabre F, Berrios EF, Leroux N, Chaarani GA, Planchon C, Sarrafi A, Gentzbittel L. QTL analysis of photosynthesis and water status traits in sunflower (*Helianthus annum* L.) under greenhouse conditions. J Exp Bot 52:1857–1864, 2001.

Hihara Y, Kamei A, Kanehisa M, Kaplan A, Ikeuchi M. DNA microarray analysis of cyanobacterial gene expression during acclimation to high light. Plant Cell 13:793–806, 2001.

Hohmann S. Osmotic stress signaling and osmoadaptation in yeasts. Microbiol Mol Biol Rev 66:300–372, 2002a.

Hohmann S. Osmotic adaptation in yeast—control of the yeast osmolyte system. Int Rev Cytol 215:149–187, 2002b.

Hoisington D, Khairallah M, Reeves T, Ribaut JM, Skovmand B, Taba S, Warburton M. Plant genetic resources: What can they contribute toward increased crop productivity? Proc Natl Acad Sci USA 96:5937–5943, 1999.

Ideker T, Galitski T, Hood L. A new approach to decoding life: Systems biology. Annu Rev Genomics Hum Genet 2:343–372, 2001a.

Ideker T, Thorsson V, Ranish JA, Christmas R, Buhler J, Eng JK, Bumgarner R, Goodlett DR, Aebersold R, Hood L. Integrated genomic and proteomic analyses of a systematically perturbed metabolic network. Science 292:929–934, 2001b.

Ishimaru K, Shirota K, Higa M, Kawamitsu Y. Identification of quantitative trait loci for adaxial and abaxial stomatal frequencies in *Oryza sativa*. Plant Physiol Biochem 39:173–177, 2001.

Ismail AM, Hall AE, Close TJ. Allelic variation of a dehydrin gene cosegregates with chilling tolerance during seedling emergence. Proc Natl Acad Sci USA 96:13566–13570, 1999.

Iturriaga G, Gaff DF, Zentella R. New desiccation-tolerant plants, including a grass, in the central highlands of Mexico, accumulate trehalose. Aust J Bot 48:153–158, 2000.

Ivandic V, Hackett CA, Zhang ZJ, Staub JE, Nevo E, Thomas WT, Forster BP. Phenotypic responses of wild barley to experimentally imposed water stress. J Exp Bot 51:2021–2029, 2000.

Jacobson T, Adams RM. Salt and silt in ancient mesopotamian agriculture. Science 128: 1251–1258, 1958.

Jiang C, Edmeades GO, Armstead I, Lafitte HR, Hayward MD, Hoisington D. Genetic analysis of adaptation differences between highland and lowland tropical maize using molecular markers. Theor Appl Genet 99:1106–1119, 1999.

Jorgensen JA, Nguyen HT. Genetic analysis of heat shock proteins in maize. Theor Appl Genet 91:38–46, 1995.

Kaneko T, Sato S, Kotani H, Tanaka A, Asamizu E, Nakamura Y, Miyajima N, Hirosawa M, Sugiura M, Sasamoto S, Kimura T, Hosouchi T, Matsuno A, Muraki A, Nakazaki N, Naruo K, Okumura S, Shimpo S, Takeuchi C, Wada T, Watanabe A, Yamada M, Yasuda M, Tabata S. Sequence analysis of the genome of the unicellular cyanobacterium Synechocystis sp. strain PCC6803. II. Sequence determination of the entire genome and assignment of potential protein-coding regions. DNA Res 3:109–136, 1996a.

Kaneko T, Sato S, Kotani H, Tanaka A, Asamizu E, Nakamura Y, Miyajima N, Hirosawa M, Sugiura M, Sasamoto S, Kimura T, Hosouchi T, Matsuno A, Muraki A, Nakazaki N, Naruo K, Okumura S, Shimpo S, Takeuchi C, Wada T, Watanabe A, Yamada M, Yasuda M, Tabata S. Sequence analysis of the genome of the unicellular cyanobacterium Synechocystis sp. strain PCC6803. II. Sequence determination of the entire genome and assignment of potential protein-coding regions (supplement). DNA Res 3:185–209, (1996b).

Kanesaki Y, Suzuki I, Allakhverdiev SI, Mikami K, Murata N. Salt stress and hyperosmotic stress regulate the expression of different sets of genes in Synechocystis sp. PCC 6803. Biochem Biophys Res Commun 290:339–348, 2002.

Kawasaki S, Borchert C, Deyholos M, Wang H, Brazille S, Kawai K, Galbraith D, Bohnert HJ. Gene expression profiles during the initial phase of salt stress in rice. Plant Cell 13:889–905, 2001.

Kebede H, Subudhi PK, Rosenow DT, Hguyen HT. Quantitative trait loci influencing drought tolerance in grain sorghum (Sorghum bicolor L. Moench). Theor Appl Genet 103:266–276, 2001.

Kehoe DM, Villand P, Somerville S. DNA microarrays for studies of higher plants and other photosynthetic organisms. Trends Plant Sci 4:38–41, 1999.

Kerr RA. Sea-floor dust shows drought felled Akkadian empire. Science 279:325–326, 1998.

Kersten B, Burkle L, Kuhn EJ, Giavalisco P, Konthur Z, Lueking A, Walter G, Eickhoff H, Schneider U. Large-scale plant proteomics. Plant Mol Biol 48:133–141, 2002.

Khush GS. Green revolution: Preparing for the 21st century. Genome 42:646–655, 1999.

Kirch HH, Nair A, Bartels D. Novel ABA- and dehydration-inducible aldehyde dehydrogenase genes isolated from the resurrection plant Craterostigma plantagineum and Arabidopsis thaliana. Plant J 28:555–567, 2001.

Korol AB, Ronin YI, Itskovich AM, Peng J, Nevo E. Enhanced efficiency of quantitative trait loci mapping analysis based on multivariate complexes of quantitative traits. Genetics 157:1789–1803, 2001.

Koyama ML, Levesley A, Koebner RM, Flowers TJ, Yeo AR. Quantitative trait loci for component physiological traits determining salt tolerance in rice. Plant Physiol 125: 406–422, 2001.

Kramer PJ, Boyer JS. Evolution and agricultural water use. In: Water Relations of Plants and Soils. Academic Press, San Diego, pp. 377–404, 1985.

Lebreton C, Lazic-Jancic V, Steed A, Pekic S, Quarrie SA. Identification of QTL for drought responses in maize and their use in testing causal relationships between traits. J Exp Bot 46:853–865, 1995.

Lemieux B, Aharoni A, Schena M. Overview of DNA chip technology. Mol Breed 4: 277–289, 1998.

Liang P, Pardee AB. Differential display of eukaryotic messenger RNA by means of the polymerase chain reaction. Science 257:967–971, 1992.

Lilley JM, Ludlow MM, McCrouch SR, O'Toole JC. Locating QTLs for osmotic adjustment and dehydration tolerance in rice. J Exp Bot 47:1427–1436, 1996.

Lipshutz RJ, Fodor SP, Gingeras TR, Lockhart DJ. High density synthetic oligonucleotide arrays. Nat Genet 21:20–24, 1999.

Lisacek FC, Traini MD, Sexton D, Harry JL, Wilkins MR. Strategy for protein isoform identification from expressed sequence tags and its application to peptide mass fingerprinting. Proteomics 1:186–193, 2001.

Lorenz WW, Dean JF. SAGE profiling and demonstration of differential gene expression along the axial developmental gradient of lignifying xylem in loblolly pine (*Pinus taeda*). Tree Physiol 22:301–310, 2000.

Macas J, Lambert GM, Dolezel D, Galbraith DW. Nuclear expressed sequence Tag (NEST) analysis: A novel means to study transcription through amplification of nuclear RNA. Cytometry 33:460–468, 1998.

Machuka J, Bashiardes S, Ruben E, Spooner K, Cuming A, Knight C, Cover D. Sequence analysis of expressed sequence tags from an ABA-treated cDNA library identifies stress response genes in the moss *Physcomitrella patens*. Plant Cell Physiol 40: 378–387, 1999.

Mano Y, Takeda K. Mapping quantitative trait loci for salt tolerance at germination and the seedling stage in barley (*Hordeum vulgare* L.). Euphytica 94:263–272, 1997.

Martin B, Nienhuis J, King G, Schaefer A. Restriction fragment length polymorphisms associated with water use efficiency in tomato. Science 243:1725–1728, 1989.

Matsumura H, Nirasawa S, Terauchi R. Transcript profiling in rice (*Oryza sativa* L.) seedlings using serial analysis of gene expression (SAGE). Plant J 20:719–726, 1999.

Matsuyama T, Tamaoki M, Nakajima N, Aono M, Kubo A, Moriya S, Ichihara T, Suzuki O, Saji H. cDNA microarray assessment for ozone-stressed Arabidopsis thaliana. Environ Pollut 117:191–194, 2002.

Mayer K, Mewes H-W. How can we deliver the large plant genomes? Strategies and perspectives. Curr Opin Plant Biol 5:173–177, 2001.

Mekhedov S, de Ilarduya OM, Ohlrogge J. Toward a functional catalog of the plant genome. A survey of genes for lipid biosynthesis. Plant Physiol 122:389–402, 2000.

Mian MAR, Bailey MA, Ashley DA, Wells R, Carter TE, Parrot WA, Boerma HR. Molecular markers associated with water use efficiency and leaf ash in soybean. Crop Sci 36:1252–1257, 1996.

Miflin B. Crop improvement in the 21st century. J Exp Bot 51:1–8, 2000.

Misawa S, Mori N, Takumi S, Yoshida S, Makamura C. Mapping of QTLs for low temperature response in seedling of rice (*Oryza sativa* L.). Cereal Res Comm 28:33–41, 2000.

Morgan JM, Tan MK. Chromosomal location of a wheat osmoregulation gene using RFLP analysis. Aust J Plant Physiol 23:803–806, 1996.

Mullet JE, Klein RR, Klein PE. *Sorghum bicolor*–an important species for comparative grass genomics and a source of beneficial genes for agriculture. Curr Opin Plant Biol 5:118–121, 2002.

Muramoto Y, Watanabe A, Nakamura T, Takabe T. Enhanced expression of a nuclease gene in leaves of barley plants under salt stress. Gene 234:315–321, 1999.

Nal B, Mohr E, Ferrier P. Location analysis of DNA-bound proteins at the whole-genome level: Untangling transcriptional regulatory networks. Bioessays 23:473–476, 2001.

Neale AD, Blomstedt CK, Bronson P, Le T-N, Guthridge K, Evans J, Gaff DF, Hamill JD. The isolation of genes from the resurrection grass *Sporobolus stapfianus* which are induced during severe drought stress. Plant Cell Enivon 23:265–277, 2000.

Nevo E. Evolution in action across phylogeny caused by microclimatic stresses at "Evolution Canyon." Theor Popul Biol 52:231–243, 1997.

Nevo E, Apelbaum-Elkaher I, Garty J, Beiles A. Natural selection causes microscale allozyme diversity in wild barley and a lichen at 'Evolution canyon' Mt. Carmel Israel. Heredity 78:373–382, 1997.

Nevo E. Evolution of genome-phenome diversity under environmental stress. Proc Natl Acad Sci USA 98:6233–6240, 2001.

Oliver MJ, Bewley JD. Desiccation tolerance of plant tissues: A mechanistic overview. Hort Rev 18:171–214, 1997.

Oliver MJ, Velten J, Wood AJ. Bryophytes as experimental models for the study of environmental stress tolerance: *Tortula ruralis* and desicccation-tolerance in mosses. Plant Ecol 151:73–84, 2000a.

Oliver MJ, Tuba Z, Mishler BD. The evolution of vegetative desicccation tolerance in land plants. Plant Ecol 151:85–100, 2000b.

Ottaviano E, Sari Gorla M, Pé E, Frova C. Molecular makers (RFLPs and HSPs) for the genetic dissection of thermotolerance in maize. Theor Appl Genet 81:713–719, 1991.

Ozturk ZN, Talamé V, Deyholos M, Michalowski CB, Galbraith DW, Gozukirmizi N, Tuberosa R, Bohnert HJ. Monitoring large-scale changes in transcript abundance in drought- and salt-stressed barley. Plant Mol Biol 48:551–573, 2002.

Patel J, McLeod LE, Vries RG, Flynn A, Wang X, Proud CG. Cellular stresses profoundly inhibit protein synthesis and modulate the states of phosphorylation of multiple translation factors. Eur J Biochem 269:3076–3085, 2002.

Peters DG, Kassam AB, Yonas H, O'Hare EH, Ferrell RE, Brufsky AM. Comprehensive transcript analysis in small quantities of mRNA by SAGE-lite. Nucleic Acids Res 27:e39, 1999.

Pflieger S, Lefebvre V, Causse M. The candidate gene approach in plant genetics: A review. Mol Breed 7:275–291, 2001.

Phelps TJ, Palumbo AV, Beliaev AS. Metabolomics and microarrays for improved understanding of phenotypic characteristics controlled by both genomics and environmental constraints. Curr Opin Biotechnol 13:20–24, 2002.

Pih KY, Jang HJ, Kang SG, Piao HL, Hwang I. Isolation of molecular markers for salt stress responses in *Arabidopsis thaliana*. Mol Cells 7:567–571, 1997.

Planta RJ, Brown AJ, Cadahia JL, Cerdan ME, de Jonge M, Gent ME, Hayes A, Kolen

CP, Lombardia LJ, Sefton M, Oliver SG, Thevelein J, Tournu H, van Delft YJ, Verbart DJ, Winderickx J. Transcript analysis of 250 novel yeast genes from chromosome XIV. Yeast 15:329–350, 1999.

Porubleva L, Vander Velden K, Kothari S, Oliver DJ, Chitnis PR. The proteome of maize leaves: Use of gene sequences and expressed sequence tag data for identification of proteins with peptide mass fingerprints. Electrophoresis 22:1724–1738, 2001.

Posas F, Chambers JR, Heyman JA, Hoeffler JP, de Nadal E, Arino J. The transcriptional response of yeast to saline stress. J Biol Chem 275:17249–17255, 2000.

Prade RA, Ayoubi P, Krishnan S, Macwana S, Russell H. Accumulation of stress and inducer-dependent plant-cell-wall-degrading enzymes during asexual development in *Aspergillus nidulans*. Genetics 157:957–967, 2001.

Price AH, Tomos AD. Genetic dissection of root growth in rice (*Oryza sativa* L.). II. Mapping quantitative trait loci using molecular markers. Theor Appl Genet 95:143–152, 1997.

Price AH, Virk DS, Tomos AD. Genetic dissection of root growth in rice (*Oryza sativa* L.). I. A hydroponic screen. Theor Appl Genet 95:132–142, 1997a.

Price AH, Young EM, Tomos AD. Quantitative trait loci associated with stomatal conductance, leaf rolling, and heading date in upland rice (*Oryza sativa* L.). New Phytol 137:83–91, 1997b.

Price A, Courtois B. Mapping QTLs associated with drought resistance in rice: Progress, problems and prospects. Plant Growth Regulation 29:123–133, 1999.

Price AH, Steele KA, Moore BJ, Barraclough PB, Clark LJ. A combined RFLP and AFLP linkage map of upland rice (*Oryza sativa* L.) used to identify QTLs for root-penetration ability. Theor Appl Genet 100:49–56, 2000.

Price AH, Cairns JE, Horton P, Jones HG, Griffiths H. Linking drought-resistance mechanisms to drought avoidance in upland rice using a QTL approach: Progress and new opportunities to integrate stomatal and mesophyll responses. J Exp Bot 53:989–1004, 2002a.

Price AH, Steele KA, Moore BJ, Jones RGW. Upland rice grown in soil-filled chambers and exposed to contrasting water-deficit regimes: II. Mapping QTL for root morphology and distribution. Field Crops Research. In press, 2002b.

Price AH, Townsend J, Jones MP, Audebert A, Courtois B. Mapping QTLs associated with drought avoidance in upland rice grown in the Philippines and West Africa. Plant Mol Biol 48:683–695, 2002c.

Quackenbush J, Cho J, Lee D, Liang F, Holt I, Karamycheva S, Parvizi B, Pertea G, Sultana R, White J. The TIGR Gene Indices: Analysis of gene transcript sequences in highly sampled eukaryotic species. Nucleic Acids Res 29:159–164, 2001.

Quarrie SA, Lebreton C, Gulli M, Calestani C, Marmiroli N. QTL analysis of ABA production in wheat and maize and associated physiological traits. Russian J Plant Physiol 41:565–571, 1994.

Quarrie SA, Laurie DA, Zhu J, Lebreton C, Semikhodskii A, Steed A, Witsenboer H, Calestani C. QTL analysis to study the association between leaf size and abscisic acid accumulation in droughted rice leaves and comparisons across cereals. Plant Mol Biol 35:155–165, 1997.

Quarrie SA. New molecular tools to improve the efficiency of breeding for increased drought tolerance. J Plant Growth Regul 20:167–178, 1996.

Quarrie SA, Stojanovic J, Pekic S. Improving drought resistance in small-grained cereals: A case study, progress and prospects. J Plant Growth Regul 29:1–21, 1999.

Ray JD, Yu L, McCouch SR, Champoux MC, Wang G, Nguyen HT. Mapping quantitative traits loci associated with root penetration ability in rice (*Oryza sativa* L.) Theor Appl Genet 92:627–636, 1996.

Rep M, Krantz M, Thevelein JM, Hohmann S. The transcriptional response of *Saccharomyces cerevisiae* to osmotic shock. Hot1p and Msn2p/Msn4p are required for the induction of subsets of high osmolarity glycerol pathway-dependent genes. J Biol Chem 275:8290–8300, 2000.

Reddy AR, Ramakrishna W, Sekhar AC, Ithal N, Babu PR, Bonaldo MF, Soares MB, Bennetzen JL. Novel genes are enriched in normalized cDNA libraries from drought-stressed seedlings of rice (*Oryza sativa* L. subsp. *indica* cv. Nagina 22). Genome. 45:204–211, 2002.

Reymond P, Weber H, Damond M, Farmer EE. Differential gene expression in response to mechanical wounding and insect feeding in Arabidopsis. Plant Cell 12:707–720, 2000.

Ribaut JM, Hosington DA, Deitsch JA Jiang C, Gonzalez-de-Leon D. Identification of quantitative trait loci under drought conditions in tropical maize. 1. Flowering parameters and the anthesis-silking interval. Theor Appl Genet 92:905–914, 1996.

Ribaut JM, Jiang C, Gonzalez-de-Leon D, Edmeades GO, Hosington DA. Identification of quantitative trait loci under drought conditions in tropical maize. 2. Yield components and marker-assisted selection strategies. Theor Appl 94:887–896, 1997.

Richmond T, Somerville S. Chasing the dream: Plant EST microarrays. Curr Opin Plant Biol 3:108–116, 2000.

Roberts JK. Proteomics and a future generation of plant molecular biologists. Plant Mol Biol 48:143–154. 2002.

Saijo Y, S Hata, J Kyozuka, K Shimamoto, K Izui. Over-expression of a single $Ca^{2+}$-dependent protein kinase confers both cold and salt/drought tolerance on rice plants. Plant J 23:319–327, 2000.

Saito K, Miura K, Nagano K, Hayano-Saito Y, Saito A, Araki H, Kato A. Chromosomal location of quantitative trait loci for cool tolerance at the booting stage in rice variety 'Norin-PL8'. Breed Sci 45:337–340, 1995.

Saito K, Miura K, Nagano K, Hayano-Saito Y, Araki H, Kato A. Identification of two closely linked quantitiative trait loci for cold tolerance on chromosome 4 of rice and their association with anther length. Theor Appl Genet 103:862–868, 2001.

Sanchez AC, Subudhi PK, Rosenow DT, Nguyen HT. Mapping QTLs associated with drought resistance in sorghum (*Sorghum bicolor* L. Moench). Plant Mol Biol 48: 713–726, 2002.

Sanguineti MC, Tubersoa R, Stefanelli, Noli E, Blake TK, Hayes PM. Utilization of a recombinant inbred population to localize QTLs for absicisic-acid content in leaves of drought-stressed barley (*Hordeum vulgare* L.) Russian J Plant Physiol 41:572–576, 1994.

Sanguineti MC, Tubersoa R, Landi P, Salvi S, Maccaferri M, Casarinin E, Conti S. QTL analysis of drought-related traits and grain yield in relation to genetic variation for leaf abscisic acid concentration in field-grown maize. J Exp Bot 50:1289–1297, 1999.

Sari-Gorla M, Krajewsi P, Di Fonzo N, Villa M, Frova C. Genetic anaysis of drought tolerance in maize by molecular markers. II. Plant height and flowering. Theor Appl Genet 99:289–295, 1999.

Sasaki Y, Asamizu E, Shibata D, Nakamura Y, Kaneko T, Awai K, Masuda T, Shimada H, Takamiya K, Tabata S, Ohta H. Genome-wide expression-monitoring of jasmonate-responsive genes of Arabidopsis using cDNA arrays. Biochem Soc Trans 28:863–864, 2000.

Sasaki Y, Asamizu E, Shibata D, Nakamura Y, Kaneko T, Awai K, Amagai M, Kuwata C, Tsugane T, Masuda T, Shimada H, Takamiya K, Ohta H, Tabata S. Monitoring of methyl jasmonate-responsive genes in Arabidopsis by cDNA macroarray: Self-activation of jasmonic acid biosynthesis and crosstalk with other phytohormone signaling pathways. DNA Res 8:153–161, 2001.

Schaffer R, Landgraf J, Perez-Amador M, Wisman E. Monitoring genome-wide expression in plants. Curr Opin Biotechnol 11:162–167, 2000.

Schneider K, Borchardt DC, Schafer-Pregl R, Nagl N, Glass C, Jeppsson A, Gebhardt C, Salamini F. PCR-based cloning and segregation analysis of functional gene homologues in *Beta vulgaris*. Mol Gen Genet 262:515–524, 1999.

Seckler D, Amarasinghe U, Molden D, da Silva R, Barker R. World water demand and supply, 1990 to 2025: Scenarios and issues. IWMI Research Report 19, Colombo, Sri Lanka: IWMA 1998.

Seki M, Narusaka M, Abe H, Kasuga M, Yamaguchi-Shinozaki K, Carninci P, Hayashizaki Y, Shinozaki K. Monitoring the expression pattern of 1300 Arabidopsis genes under drought and cold stresses by ssing a full-Length cDNA microarray. Plant Cell 13:61–72, 2001a.

Seki M, Narusaka M, Yamaguchi-Shinozaki K, Carninci P, Kawai J, Hayashizaki Y, Shizaki K. Arabidopsis encyclopedia using full-length cDNAs and its application. Plant Physiol Biochem 39:211–220, 2001b.

Seki M, Narusaka M, Kamiya A, Ishida J, Satou M, Sakurai T, Nakajima M, Enju A, Akiyama K, Oono Y, Muramatsu M, Hayashizaki Y, Kawai J, Carninci P, Itoh M, Ishii Y, Arakawa T, Shibata K, Shinagawa A, Shinozaki K. Functional annotation of a full-length Arabidopsis cDNA collection. Science 296:141–145, 2002.

Shen L, Courtois B, McNally KL, Robin S, Li Z. Evaluation of near-isogenic lines of rice introgressed with QTLs for root depth through marker-aided selection. Theor Appl Genet 103:75–83, 2001.

Shi WM, Muramoto Y, Ueda A, Takabe T. Cloning of peroxisomal ascorbate peroxidase gene from barley and enhanced thermotolerance by overexpressing in Arabidopsis thaliana. Gene 273:23–27, 2001.

Shinozaki K, Yamaguchi-Shinozaki K. Gene expression and signal transduction in water-stress response. Plant Physiol 115:327–334, 1997.

Simoes-Araujo JL, Rodrigues RL, de A Gerhardt LB, Mondego JM, Alves-Ferreira M, Rumjanek NG, Margis-Pinheiro M. Identification of differentially expressed genes by cDNA-AFLP technique during heat stress in cowpea nodules. FEBS Lett 515:44–50, 2002.

Stekel DJ, Git Y, Falciani F. The comparison of gene expression from multiple cDNA libraries. Genome Res 10:2055–2061, 2000.

Sterky F, Regan S, Karlsson J, Hertzberg M, Rohde A, Holmberg A, Amini B, Bhalerao

R, Larsson M, Villarroel R, Van Montagu M, Sandberg G, Olsson O, Teeri TT, Boerjan W, Gustafsson P, Uhlen M, Sundberg B, Lundeberg J. Gene discovery in the wood-forming tissues of poplar. Analysis of 5,692 expressed sequence tags. Proc Natl Acad Sci USA 95:13330–13335, 1998.

Stuber CW, Polacco M, Senior ML. Synergy of empirical breeding, marker-assisted selection, and genomics to increase crop yield potential. Crop Science 39:1571-1583, 1999.

Suzuki I, Kanesaki Y, Mikami K, Kanehisa M, Murata N. Cold-regulated genes under control of the cold sensor Hik33 in *Synechocystis*. Mol Microbiol 40:235–244, 2001.

Tanksley SD, Ganal MW, Martin GB. Chromosome landing—a paradigm for map-based gene cloning in plants with large genomes. Trends Genet 11:63–68, 1995.

Tanksley SD, Nelson JC. Advanced back-cross QTL-analysis: A method for the simultaneous discovery and transfer of valuable QTLs from unadapted germplasm into elite breeding lines. Theor Appl Genet 92:191–203, 1996.

Teulat B, Monneveux P, Werry J, Borries C, Souyris I, Charrier A, This D. Relationship between relative water content and growth parameters under water stress in barley: A QTLs study. New Phytol 137:99–107, 1997.

Teulat B, This D, Khairallah M, Borries C, Ragot C, Sourdille P, Leroy P, Monneveux P, Charrier A. Several QTLs involve in osmotic-adjustment trait variation in barley (*Hordeum vulgare* L.) Theor Appl Genet 96:688–698, 1998.

Teulat B, Borries C, This D. 2001. New QTLs for plant water status, water-soluble carbohydrates and osmotic adjustment in a barley population grown in a growth-chamber under two water regimes. Theor Appl Genet. 103:161–170, 2001.

The Arabidopsis Genome Initiative. Analysis of the genome sequence of the flowering plant *Arabidopsis thaliana*. Nature 408:796–815, 2000.

Thumma BR, Naidu BP, Chandra A, Cameron DF, Bahnisch LM, Liu C. Identification of causal relationships among traits related to drought resistance in *Stylosanthes scabra* using QTL analysis. J Exp Bot 52:203–214, 2001.

Tozlu I, Guy CL, Moore GA. QTL analysis of Na+ and Cl− accumulation related traits in an intergenic BC1 progeny of *Citrus* and *Poncirus* under saline and non-saline environments. Genome 42:692–705, 1999.

Trethewey RN. Gene discovery via metabolic profiling. Curr Opin Biotechnol 12:135–138, 2001.

Tripathy N, Zhang J, Robin S, Nguyen TT, Nguyen HT. QTLs for cell-membrane stability mapped in rice (*Oryza sativa* L.) under drought stress. Theor Appl Genet 100:1197–1202, 2000.

Tuberosa R, Galiba G, Sanguineti MC, Noli E, Sutka J. Identification of QTL influencing freezing tolerance in barley. Acta Agric Hung 45:413–417, 1997.

Tuberosa R, Sanguineti MC, Landi P, Salvi S, Casarin E, Conti S. RFLP mapping of quantitative trait loci controlling abscisic acid concentrations in leaves of drought-stressed maize (*Zea mays* L.). Theor Appl Genet 97:744–755, 1998.

Tuberosa R, Sanguineti MC, Landi P, Giuliani MM, Salvi S, Conti S. Identification of QTLs for root characteristics in maize grown in hydroponics and analysis of their overlap with QTLs for grain yield in the field at two water regimes. Plant Molec Biol 48:697–712, 2002.

Tuinstra MR, Eheta G, Goldsbrough P. Evaluation of near-isogenic Sorghum lines contrasting for QTL markers associated with drought tolerance. Crop Sci 38:835–842, 1998.

Ueda A, Shi W, Sanmiya K, Shono M, Takabe T. Functional analysis of salt-inducible proline transporter of barley roots. Plant Cell Physiol 42:1282–1289, 2001.

Uesono Y, Toh-E A. Transient inhibition of translation initiation by osmotic stress. J Biol Chem 277:13848–13855, 2002.

Ujino-Ihara T, Yoshimura K, Ugawa Y, Yoshimaru H, Nagasaka K, Tsumura Y. Expression analysis of ESTs derived from the inner bark of *Cryptomeria japonica*. Plant Mol Biol 43:451–457, 2000.

Umeda M, Hara C, Matsubayashi Y, Li H-H, Liu Q, Tadokoro F, Aotsuka S, Uchimiya H Expressed sequence tags from cultured cells of rice (*Oryza sativa* L.) under stressed conditions: Analysis of transcripts of genes engaged in ATP generating pathways. Plant Mol Biol 25:469–478, 1994.

Vander Mijnsbrugge K, Meyermans H, Van Montagu M, Bauw G, Boerjan W. Wood formation in poplar: Identification, characterization, and seasonal variation of xylem proteins. Planta 210:589–598, 2000.

van Hal NL, Vorst O, van Houwelingen AM, Kok EJ, Peijnenburg A, Aharoni A, van Tunen AJ, Keijer J. The application of DNA microarrays in gene expression analysis. J Biotechnol 78:271–280, 2000.

Veldboom LR, Lee M. Genetic mapping of quantitative trait loci in maize in stress and non-stress environments. II. Plant height and flowering. Crop Sci 36:1320–1327, 1996.

Velten J, Oliver MJ. Tr288, a rehydrin with a dehydrin twist. Plant Mol Biol 45:713–722, 2001.

Véry AA, Robinson MF, Mansfield TA, Sanders D. Guard cell cation channels are involved in Na+-induced stomatal closure in a halophyte. Plant J 14:509–521, 1998.

Walbot V. Genes, genomes, genomics. What can plant biologists expect from the 1998 national science foundation plant genome research program? Plant Physiol 119:1151–1155, 1999.

Wang Z, Taramino G, Yang D, Liu G, Tingey SV, Miao GH, Wang GL. Rice ESTs with disease-resistance gene- or defense-response gene-like sequences mapped to regions containing major resistance genes or QTLs. Mol Genet Genomics 265:302–310, 2001.

Wei J, Tirajoh A, Effendy J, Plant AL. Characterization of saltinduced changes in gene expression in tomato (*Lycopersicon esculentum*) roots and the role played by abscisic acid. Plant Sci 159:135–148, 2000.

Weinmann AS, Yan PS, Oberley MJ, Huang T H-M, Farnham PJ. Isolating human transcription factor targets by coupling chromatin immunoprecipitation and CpG island microarray analysis. Genes Dev 16:235–244, 2002.

Welsh J, Chada K, Dalal SS, Cheng R, Ralph D, McClelland M. Arbitrarily primed PCR fingerprinting of RNA. Nucleic Acids Res 20:4965–4970, 1992.

White JA, Todd J, Newman T, Focks N, Girke T, de Ilarduya OM, Jaworski JG, Ohlrogge JB, Benning C. A new set of Arabidopsis expressed sequence tags from developing seeds. The metabolic pathway from carbohydrates to seed oil. Plant Physiol 124:1582–1594, 2000.

Wood AJ, Oliver MJ. Translational control in plant stress: The formation of messenger ribonucleoprotein particles (mRNPs) in response to desiccation of *Tortula ruralis* gametophytes. Plant J 18:359–370, 1999.

Wood AJ, Duff RJ, Oliver MJ. Expressed sequence tags (ESTs) from desiccated *Tortula ruralis* identify a large number of novel plant genes. Plant Cell Physiol 40:361–368, 1999.

Wu J, Maehara T, Shimokawa T, Yamamoto S, Harada C, Takazaki Y, Ono N, Mukai Y, Koike K, Yazaki J, Fujii F, Shomura A, Ando T, Kono I, Waki K, Yamamoto K, Yano M, Matsumoto T, Sasaki T. A comprehensive rice transcript map containing 6591 expressed sequence tag sites. Plant Cell 14:525–535, 2002.

Xu W, Subudhi PK, Crasta OR, Rosenow DT, Mullet JE, Nguyen HT. Molecular mapping of QTLs conferring stay-green in grain. sorghum (*Sorghum bicolor* L. Moench). Genome 43:461–469, 2000.

Xu W, Bak S, Decker A, Paquette SM, Feyereisen R, Galbraith DW. Microarray-based analysis of gene expression in very large gene families: The cytochrome P450 gene superfamily of *Arabidopsis thaliana*. Gene 272:61–74, 2001.

Yadav R, Courtois B, Huang N, McLaren G. Mapping genes controlling root morphology and root distribution in a double-haploid population of rice. Theor Appl Genet 94:619–632, 1997.

Yale J, Bohnert HJ. Transcript expression in *Saccharomyces cerevisiae* at high salinity. J Biol Chem 276:15996–16007, 2001.

Yang L, Tran DK, Wang X. BADGE, Beads Array for the Detection of Gene Expression, a high-throughput diagnostic bioassay. Genome Res 11:1888–1898, 2001.

Yano M, Katayose Y, Ashikari M, Yamanouchi U, Monna L, Fuse T, Baba T, Yamamoto K, Umehara Y, Nagamura Y, Sasaki T. Hd1, a major photoperiod-sensitivity quantitatitve trait locus in rice, is closely related to the Arabidopsis flowering time gene CONSTANT. Plant Cell 12:2473–2484, 2000.

Yano M, Matsumoto T, Sasaki T. A comprehensive rice transcript map containing 6591 expressed sequence tag sites. Plant Cell 14:525–535, 2002.

Yazaki J, Kishimoto N, Nakamura K, Fujii F, Shimbo K, Otsuka Y, Wu J, Yamamoto K, Sakata K Sasaki T, Kikuchi S. Embarking on rice functional genomics via cDNA microarray: Use of 3' UTR probes for specific gene expression analysis. DNA Res 7:367–370, 2000.

Yu H-J, Moon M-S, Lee H-S, Mun J-H, Kwon YM, Kim S-G. Analysis of cDNAs expressed during first cell division of petunia petal protoplast cultures using expressed sequence tags. Mol Cells 9:258–264, 1999.

Yu J, Hu S, Wang J, Wong GK, Li S, Liu B, Deng Y, Dai L, Zhou Y, Zhang X, Cao M, Liu J, Sun J, Tang J, Chen Y, Huang X, Lin W, Ye C, Tong W, Cong L, Geng J, Han Y, Li L, Li W, Hu G, Huang X, Li W, Li J, Liu Z, Li L, Liu J, Qi Q, Liu J, Li L, Li T, Wang X, Lu H, Wu T, Zhu M, Ni P, Han H, Dong W, Ren X, Feng X, Cui P, Li X, Wang H, Xu X, Zhai W, Xu Z, Zhang J, He S, Zhang J, Xu J, Zhang K, Zheng X, Dong J, Zeng W, Tao L, Ye J, Tan J, Ren X, Chen X, He J, Liu D, Tian W, Tian C, Xia H, Bao Q, Li G, Gao H, Cao T, Wang J, Zhao W, Li P, Chen W, Wang X, Zhang Y, Hu J, Wang J, Liu S, Yang J, Zhang G, Xiong Y, Li Z, Mao L, Zhou C, Zhu Z, Chen R, Hao B, Zheng W, Chen S, Guo W, Li G, Liu S, Tao M, Wang J, Zhu L, Yuan L, Yang H. A draft sequence of the rice genome (*Oryza sativa* L. *ssp. indica*) Science 296:79–92, 2002.

Zamir D. Improving plant breeding with exotic genetic libraries. Nat Rev Genet 2:983–989, 2001.

Zhang GY, Guo Y, Chen SL, Chen SY. RFLP tagging of a salt tolerant gene in rice. Plant Sci 110:227–234, 1995.

Zhang J-S, Xie C, Li Z-Y, Chen S-Y. Expression of the plasma membrane H+-ATPase gene in response to salt stress in a rice salt-tolerant mutant and its original variety. Theor Appl Genet 99:1006–1011, 1999a.

Zhang L, Ma X-L, Zhang Q, Ma C-L, Wang P-P, Sun Y-F, Zhao Y-X, Zhang H. Expressed sequence tags from a NaCl-treated *Suaeda salsa* cDNA library. Gene 267:193–200, 2001.

Zhang JX, Nguyen HT, Blum A. Genetic analysis of osmotic adjustment in crop plants. J Exp Bot 50:291–302, 1999b.

Zheng HG, Babu RC, Pathan MS, Ali L, Huang N, Courtois B, Nguyen HT. Quantitative trait loci for root penetration ability and root thickness in rice: Comparison of genetic backgrounds. Genome 43:53–61, 2000.

Zhu B, Choi DW, Fenton R, Close TJ. Expression of the barley dehydrin multigene family and the development of freezing tolerance. Mol Gen Genet 264:145–153, 2000.

Zhu J-K. Genetic analysis of plant salt tolerance using Arabidopsis. Plant Physiol 124:941–948, 2000.

Zhu J-K. Plant salt tolerance. Trends Plant Sci 6:66–71, 2001a.

Zhu J-K. Cell signaling under salt, water and cold stresses. Curr Opin Plant Biol 4:401–406, 2001b.

Zhu T, Wang X. Large-scale profiling of the Arabidopsis transcriptome. Plant Physiol 124:1472–1476, 2000.

Zhu T, Budworth P, Han B, Brown D, Chang H-S, Zou G, Wang X. Toward elucidating the global gene expression patterns of developing Arabidopsis: Parallel analysis of 8300 genes by a high-density oligonucleotide probe array. Plant Physiol Biochem 39:221–242, 2001.

# 12

# Using Workflow to Build an Information Management System for a Geographically Distributed Genome Sequencing Initiative

**David Hall**
*Research Triangle Institute, Research Triangle Park, North Carolina, U.S.A.*

**John A. Miller, Jonathan Arnold, Krzysztof J. Kochut, and Amit P. Sheth**
*University of Georgia, Athens, Georgia, U.S.A.*

**Michael J. Weise**
*Accelys, Inc., San Diego, California, U.S.A.*

## I. INTRODUCTION

Genomics is a discipline that investigates biological problems by looking at entire genomes or large numbers of genes at one time. Immediate goals of a typical genome project include the creation of high-resolution physical and genetic maps, determining the complete DNA sequence of the genome, and identifying, mapping, and determining the function of all genes (Collins et al. 1998). Numerous laboratory and analytical processes are carried out on materials and data in prescribed orderings. Examples of these processes include preparation of template DNA for sequencing, DNA sequencing, sequence assembly, gene prediction, and synthesis of oligonucleotide sequencing primers. We will call these processes *tasks*. The amount of material and data being processed at any given time is typically very large. This necessitates the creation of automated systems to help manage efficiently the processing of materials and data (Smith et al. 1997). Laboratory information management systems (LIMS) have been used in different types

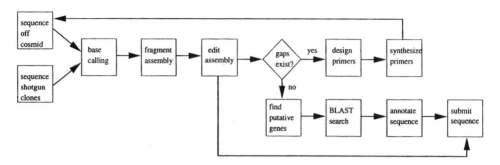

**Figure 1**  Flow chart description of simplified workflow to produce contiguous genomic sequence of a cosmid clone with annotation of putative genes.

of analytical laboratories to automate many aspects of data analysis, sample tracking, and scheduling of experiments (Steel et al. 1999), and are used extensively in genome laboratories (Goodman et al. 2001).

Major genome research activities are organized into workflows (Goodman et al. 1998b). Each workflow is composed of a series of individual tasks. Data and materials "flow" from one task to the next in a prescribed order. An individual task may create, consume, or modify data. Figure 1 shows a flow chart illustrating the order of tasks in a workflow that carries out sequencing and the annotation of genes for a cosmid clone. Tasks may make use of existing applications, or they may be specifically created for the workflow.

Because of high cost and the need for a large number of research personnel, genome projects are often carried out collaboratively. Within a genome project, different research centers or individual researchers may specialize in different analytical or production tasks. There may be specialized hardware at one of the centers or a person who is an expert at a particular analytical task. For example, one center may provide custom DNA sequencing primers for all research centers. Thus, the tasks in a single workflow may be distributed across multiple research centers. Most genome projects have a unique organization, so custom information management systems must be developed for each project. Research methodologies often change over time, which may make modifications to the supporting software necessary.

Information management systems for supporting workflows are called workflow applications. Workflow applications in genomics have been created ad hoc, which requires a significant amount of coding by developers. Developing a workflow application to support a collaborative genome project presents developers with many challenges. Mechanisms for scheduling and invoking applications running on multiple systems and for delivering data among those applications must be implemented. These applications may be running on different computing platforms as well. Recovery from system failures should be addressed to protect valuable data. User interfaces must be created. Tools that allow running workflows to be monitored are desirable. Finally, the workflow application should be designed in such a way that modifications can be easily made.

Workflow management systems (WfMSs) are widely used in many application domains where there is a need for the computational support of complex organizational processes that may involve large volumes of data and the participation of multiple, perhaps geographically distributed personnel and processing entities (Cichocki et al. 1998). They offer an environment for systematically integrating applications into workflows. These systems can significantly decrease development time and the amount of coding that must be done to create robust, distributed workflow applications. Workflow management systems provide tools for rapidly developing workflow applications, an execution environment for running the applications, and tools for monitoring their execution (Anyanwu et al. 1998; Miller et al. 1998).

Workflow management is an active area of research and a number of general purpose WfMSs that have been developed by research groups including Mentor (Wodtke et al. 1997), JEDI (Cugola et al. 1998), WIDE (Casati et al. 1998), CodAlf (Schill and Mittasch 1997), and METEOR (Sheth et al. 1997). There are also a number of popular general purpose commercial WfMSs available from vendors such as FileNet (http://www.filenet.com), IBM (http://www.ibm.com), and InConcert (http://www.inconcert.com). General purpose WfMSs have been widely used in many application domains including healthcare (Anyanwu et al. 1998) and manufacturing (Kovacs et al. 1998). Goodman et al. (1998b) have developed a specialized WfMS for analytical laboratories call LabFlow. This system consists of a class library written in Perl 5 and is designed to run on top of the LabBase data management system (Goodman et al. 1998c). The library contains classes that application programmers can use to build workflow applications. The classes take care of many aspects of task scheduling and invocation, data delivery, data storage, and recovery. However, at this time there is no high-level design tool or code generator for developing LabFlow applications.

The Fungal Genome Initiative is a multi-institution project to create high-resolution physical maps and determine the complete genome sequences of important fungal organisms (Bennett 1997). In suport of this initiative, an informatics infrastructure is being developed at the University of Georgia. This infrastructure consists of database, data sharing, and workflow tools. The database for the project was described in Kochut et al. (1993). A suite of applications for data entry, visualization, manipulation, and analysis were implemented on top of this database. These applications run on the SunOS-SPARC platfrom. Java-based remote access tools for the database have also been developed (Natarajan 1998), which allow access from clients running on other platforms. The database and associated applications were used in the physical mapping of *Aspergillus nidulans* (Prade et al. 1997) and are currently being used in physical mapping projects for *Neurospora crassa, Aspergillus flavus*, and *Nectria haematococca*. (For information about these projects, please see http://gene.genetics.uga.edu.)

A workflow system is also being developed to serve the initiative. A prototype has been built using the METEOR WfMS (Sheth et al. 1997) and is currently being evaluated with sequencing data from the *Pneumocystis carinii* (Arnold and Cushion 1997) and *N. crassa* genome projects. The METEOR system was chosen for a number of reasons, including its graphical workflow design tool, automatic code generation, support for multiple platforms, exception handling, and the capability for creating fully distributed workflow applications. The METEOR system has significantly shortened development time by providing much of the workflow infrastructure. This chapter discusses how METEOR is being used to create a workflow application for genomics and describes the prototype system that is in testing.

## II.  SYSTEM AND METHODS

### A.  Creating Workflow Applications with METEOR

The METEOR WFMS provides an automated framework for managing intra- and interenterprise organizational processes (Krishnakumar and Sheth 1995; Sheth et al. 1996). METEOR workflow applications have been tested and deployed by numerous organizations, including the Medical College of Georgia (MCG), Connecticut Healthcare Research and Education Foundation (CHREF), Advanced Technology Institute (ATI), Microelectronics and Computer Technology Corporation (MCC), Bellcore, Boeing, and Visigenic. This paper describes the first use of METEOR in a high-throughput research environment. The METEOR system has four main components. These are a graphical user interface (GUI) based workflow design tool, workflow enactment services, a workflow repository, and a workflow manager service.

The design tool, called METEOR Builder, is a high-level graphical tool that allows developers to design workflow applications by "dragging and dropping" icons and filling out electronic forms. Developers can define tasks and associate these tasks with existing applications. Some simple types of tasks, such as the entering or viewing of data, or the updating of a database can be created entirely by the system. Using the designer, the developer also creates data objects and specifies the order in which these objects flow among the tasks. In a typical workflow application there are different types of users that will access the system. For example, in a healthcare workflow, these users may include nurses, doctors, and receptionists. Each type of user will interact with only part of the system. In METEOR, developers create *roles* for different classes of users using METEOR

Builder. Roles specify which parts of the system different classes of users have access to.

Workflow enactment services provide command, communication, and control ($C^3$) for individual tasks. The main components of a METEOR enactment service are workflow schedulers, task managers, and (application) tasks. *Task managers*, as the name suggests, are used to control the execution of tasks (e.g., when they execute, where they get their input, what they do when they fail, and where to send their output). To establish global control as well as facilitate recovery and monitoring, the task managers communicate with *workflow schedulers*. It is possible for scheduling to be either centralized or distributed, or even some hybrid between the two (Miller et al. 1996). Developers can choose to deploy METEOR applications using one of two enactment services. The first is a fully distributed CORBA-based enactment system, called ORBWork, that is designed for maximum scalability and reliability (Kochut et al. 1999). The second is a web-based system, called WebWork, that is designed for ease of installation and use (Miller et al., 1998). Web servers are the only required infrastructure for WebWork. The repository provides tools for storage and retrieval of workflow process definitions. The workflow builder can use the repository for rapid application development. The enactment services can use the repository for controlling the workflow. The workflow manager provides for the monitoring of workflow instances, as well as the configuration and installation of the enactment services.

The prototype genomics workflow application we have implemented is based on the WebWork enactment services. After a workflow application has been designed with METEOR Builder, WebWork generates and installs a set of CGI applications that implement the workflow. Basically each task is bundled together with a task manager and scheduler into a single CGI application. Code for interfacing with users and databases is also generated by WebWork. Outside of the individual tasks, very little additional code must be written to develop a WebWork application.

## B. Implementation of the Prototype

The prototype workflow application that we have developed encapsulates a set of sequence analysis tasks. This application will ultimately be expanded as the activities of the Fungal Genome Initiative expand. The prototype is based on a series of tasks for producing the complete annotated sequence of a cosmid clone. A flow chart describing the laboratory and analytical steps for such a workflow is shown in Figure 1. Admittedly, this workflow likely omits some steps that should be present in a production system, such as quality assurance. However, we believe it is sufficient to test the suitability of the overall software design and to gain early feedback from users.

The first step in the workflow is the sequencing of shotgunned clones from a cosmid. Sequencing is assumed to be done using automated sequencers. Base calling of trace files generated by the sequencers is then carried out. The resulting sequences are assembled automatically. Human inspection and editing of the assembled sequences is performed, and if physical or quality gaps remain, primers are designed for direct sequencing of the cosmid. After direct cosmid sequencing, all sequences are again assembled. This cycle repeats until a single contiguous sequence remains. At this point annotation begins. Prospective genes are identified and a search for similar sequences is then carried out using the BLAST program (Altschul et al. 1997). Human annotation using the output from the gene finding and BLAST programs is then performed.

Workflow applications are designed around the structure of the organization that they support. The following organizational structure was the basis for the prototype design. Note that this is a scaled down version of the structure of the actual sequencing initiative, which we feel is complex enough for realistic software testing. Sequencing takes place at two separate research centers. Sequence assembly is done at each center on one of their servers. Each center has a *finisher*, who is responsible for editing assembled sequences and designing primers to close gaps in the cosmid sequences. Synthesis of new primers is done by a support staff at a single location that serves the entire enterprise. There is a person called the *submitter* whose role is to inspect all newly assembled and annotated sequences and send them to a high-throughput sequencing public database. Likewise there is an *annotator*, whose role is to annotate sequences using the results of the gene finding and BLAST programs. The gene finding and BLAST programs are run on special parallel hardware at a central location.

Figure 2 is a screen shot of METEOR Builder displaying the top level hierarchical design of the application. Each icon in the figure represents either a single task, or a compound task containing its own subworkflow. Figure 3 shows the design of each subworkflow. A legend that explains some of the METEOR Builder icons is given in Table 1. Table 2 lists the individual legacy applications encapsulated in each of the main subworkflows in Figure 2. The individual tasks in this application are spread across several UNIX servers at the University of Georgia. ASSEMBLY_1 and ASSEMBLY_2 subworkflows are executed on separate servers. The ANNOTATION subworkflow is executed on an SGI Origin 2000 server. The remaining tasks run on another server. Roles were created for the finisher, submitter, and annotater. WebWork creates URLs for each role with links to all tasks associated that role.

Table 3 shows the logical design of the data object that flows through the application. This object consists of all data from a single cosmid clone. Initially the object contains only the trace files from the sequencers. As the object flows among the different tasks, the remaining fields are filled in. Note that it may take more than one pass through the application to completely fill in a data object.

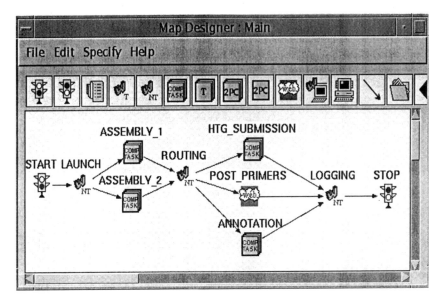

**Figure 2**   Screen shot of METEOR Builder displaying the top level hierarchical design of the prototype genome workflow application.

(As a simplification for the prototype, the back arrows in Figure 1 do not appear in the METEOR design tool. Instead, a new data object is initialized for each cycle through the application.) Some of the data fields in Table 3 may contain very large amounts of data. For instance, each individual trace file is on the order of 500 kB. Because of this, pointers to files are used for large data elements. In the prototype, this design suffices, as all applications that need a given trace file run on the same server. However, in future versions of the application, it may be necessary to pass the trace data on to different servers, depending on the organizational structure of the project. This will be a design challenge that may require increasing network bandwidth, as a 40 kilobase cosmid with five fold coverage will have around 200 megabytes of associated trace data.

Only a small amount of code had to be written for the prototype. Four small Java programs for performing database updates were created. These correspond to the tasks *assembly_to_db, submission_record_to_db, genes_to_db*, and *annotation_to_db* in Figure 3. In the current version of WebWork, database access code generated by the system uses call level interfaces. At this time, only Oracle and MiniSQL DBMSs are supported. The database update programs we wrote use JDBC for database access, which gives us more flexibility in choosing a DBMS for deployment. The only other code that had to be written for the

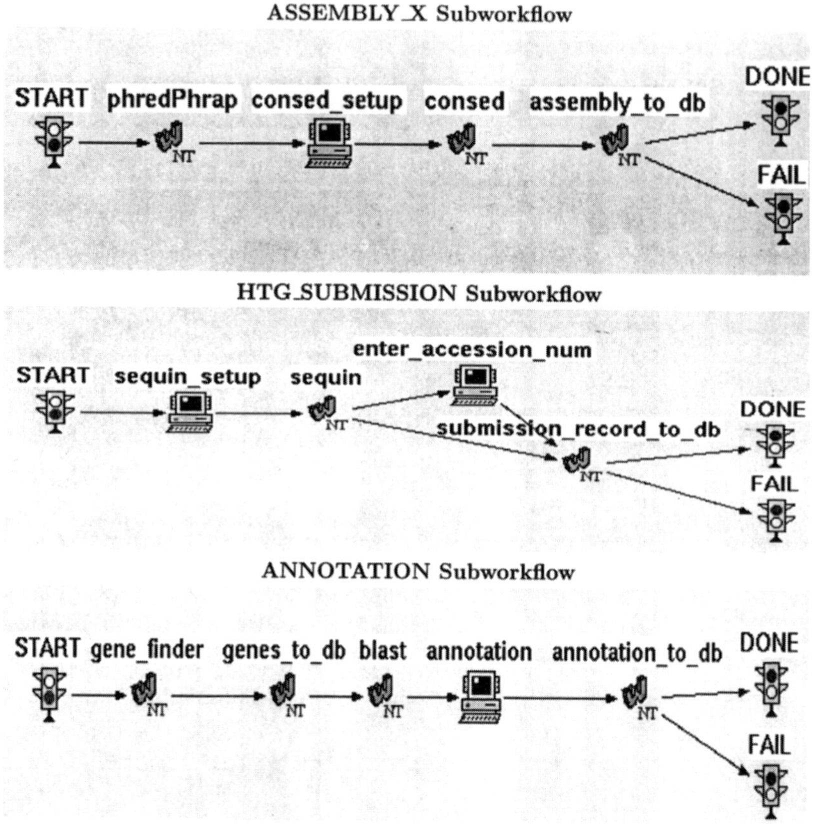

ASSEMBLY_X Subworkflow

HTG_SUBMISSION Subworkflow

ANNOTATION Subworkflow

**Figure 3**   Screen shot of METEOR Builder showing the ASSEMBLE subworkflow.

prototype consisted of a set of small "wrapper" programs. Most of the existing applications, such as the gene finding program, *ffg*, read input data from files and generated files containing output. The wrappers created the required input files for the applications using a workflow data object, and parsed the output from the application creating or modifying a workflow data object. Several applications in the workflow, such as *Consed*, require human interaction via an X Windows interface. These applications themselves run on a central server, but the workflow allows their graphical output to be redirected to any machine with a routable network address and an X server. We found little difference in the performance of these applications when the client and server were on the same or different machines within our networks. This may not be true for other networks. Network

**Table 1** Legend of Some of the METEOR Builder Icons

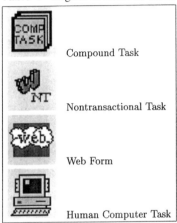

Compound Task

Nontransactional Task

Web Form

Human Computer Task

latency may need to be addressed in the design of the production system as the workflow is expanded to other institutions.

The application is designed to be automatically launched at specified time intervals. It first checks special directories for the existence of new trace files. If new trace files are found, then a new data object is initialized for each cosmid containing new sequences. This data object then enters the workflow. In the compound task ASSEMBLER_X (Fig. 2), base calling is carried out on new trace

**Table 2** Legacy Applications Embedded in Subworkflows

| Subworkflow | Applications | Purpose |
|---|---|---|
| ASSEMBLY_X | Phred (Ewing and Green 1998) | Base calling |
| | | Sequence assembly |
| | Phrap | Editing assembled sequence |
| | Consed (Gordon et al. 1998) | Primer design |
| HTG_SUBMISSION | Sequin | Submission of high throughput sequence to NCBI |
| ANNOTATION | ffg (E. Kraemer, unpublished) | Searching for predicted genes |
| | BLAST (Altschul et al. 1997) | Sequence similarity search |

**Table 3** GeneFlow Data Object

| Field name | Type |
| --- | --- |
| Cosmid name | String |
| Accession number | String |
| Traces | Array of binary objects |
| Sequence fragments | Array of strings |
| Consensus sequence | String |
| Predicted genes | Array of gene objects |
| Annotation | Array of strings |

files, then all sequences from the cosmid are assembled. The data object is then placed on the worklist for the finisher, which is accessed through *consed_setup*. The task *consed_setup* in Figure 3 is a Web-based task in which the finisher can retrieve his or her worklist. In this task, the network address of another machine can also be specified for redirecting the output of *Consed*. Worklists are displayed in a browser as hyperlinked cosmid names. After clicking on a cosmid name, Consed will be launched with the corresponding data and will appear on the user's screen. After termination of Consed, the data object will be sent to the next task, which writes all new trace reads and assembly data to the project database. The task *routing* routes the data to subsequent tasks. If quality or physical gaps remained after the assembled sequences are edited, the user may have designed primers in Consed for sequencing of the cosmid. If any primers were designed, these are placed on the worklist of the oligonucleotide synthesis staff. This worklist can be accessed through the Web task *post_primers* in Figure 2. The data object is always sent to the task *submission*, which is a compound task responsible for sequence submission to the high-throughput sequencing database at NCBI. If no quality or physical gaps remain in the cosmid consensus sequence, then the data object is sent to the *annotation* subworkflow. The *submission* and *annotation* subworkflows will not be described in detail, as the meaning of the tasks should be clear from Figure 2 and the description of the *assemble* subworkflow.

An online demonstration of this application is available at http://gene.genetics.uga.edu/workflow. It implements a slightly modified version of this prototype that has been made suitable for public access.

## III. DISCUSSION

Many of the activities associated with a genome project contain workflows. For example, Figure 1 shows a workflow for generating annotated sequences of clones. Likewise, other activities, such as constructing physical maps, generating ESTs, and carrying out functional studies on genes, contain workflows. The high cost and high labor requirements of a genome project often require the collaboration of multiple research centers. Many of the challenges of building an information management system to manage a geographically distributed research project can be more easily addressed using a WfMS than in creating a system ad hoc. The integration of applications into workflows is systematic. Workflow management systems, such as METEOR, provide easy to use tools for workflow application design and implementation. They also provide run-time enactment systems that handle the invocation of tasks, the delivery of data between tasks, and recovery mechanisms. The amount of coding required to develop a workflow application is significantly reduced when using METEOR, as little must be written to implement the infrastructure of the workflow. Workflow applications built with METEOR can be distributed across multiple servers, as long as each server has a routable network address. As workflow is an active area of research and development, new features, such as improved error recovery and transactional semantics, are continually being implemented by WfMSs. As these capabilities appear, existing workflow applications can be upgraded to incorporate these new features.

A prototype workflow application to manage the generation of annotated clone sequence was developed using METEOR/WebWork. This prototype uses several systems at the University of Georgia to emulate the combined computing resources of a model geographically distributed genome project. This application is Web-based, meaning intertask communication and distribution of data are implemented using Web technology. Worklists and interaction with the workflow application are all done through a browser. Workflow tasks, such as Consed, which have their own GUI, require client machines to have an X windows server installed. A significant challenge to future, more complete versions of the application is the transport of very large data objects, such as sets of trace files, over a network. Addressing this problem may require increasing network bandwidth.

## REFERENCES

Altschul, S., Madden, T., Schaffer, A., Zhang, J., Zhang, Z., Miller, W., Lipman, D. (1997). Gapped BLAST and PSI-BLAST: A new generation of protein database search programs. Nucleic Acids Res 25, 3389–3402.

Anyanwu, K., Sheth, A., Miller, J, Kochut, K., Bhukhanwala, K. (1998). Healthcare enterprise process development and integration. Journal of Systems Integration, special issue.

Arnold, J., Cushion, M.T. (1997). Constructing a physical map of the *Pneumocystis* genome. J Eukaryot Microbiol 44, 8S.

Bennett, J.W. (1997). White paper: Genomics for filamentous fungi. Fungal Genet Biol 21, 3–7.

Casati, F., Castano, S., Fugini, M.G. (1998). Enforcing workflow authorization constraints using triggers. J Comp Security 6, 257–285.

Cichocki, A., Helal, A., Rusinkiewicz, M., Woelk, D. Workflow and Process Automation: Concepts and Technology. Boston: Kluwer Academic Publishers, 1998.

Collins, F.S., Patrinos, A., Jordan, E., Chakravarti, A., Gesteland, R., Walters, L. (1998) New goals for the U.S. human genome project: 1998-2003. Science 282, 682–689.

Cugola, G, Di Nitto, E., Fuggetta, A. Exploiting an event-based infrastructure to develop complex distributed systems. Proceedings of the 1998 International Conference on Software Engineering. IEEE Computer Society, 1998.

Ewing, B., Green, P. (1998). Base calling of automated sequencer traces using Phred II: Error probabilities. Genome Research 8, 186–194.

Goodman, N., Rozen, S., Stein, L.D. (2001). Data and workflow management for large scale biological research. In: Bioinformatics, Databases, and Systems. S.I. Letovsky, Editor. Boston: Kluwer Academic Publishers.

Goodman, N., Rozen, S., Stein, L.D. The LabFlow system for workflow management in large scale biology research laboratories. Proceedings of the 6th International Conference on Computational Biology: Intelligent Systems for Molecular Biology (ISMB '98). AAAI Press, 1998b, pp 69–77.

Goodman, N (1998a). The LabBase system for data management in large scale biology research laboratories. Bioinformatics 14, 562–574.

Gordon, D., Abajian, C., Green, P. (1998). Consed: A graphical tool for sequence finishing. Genome Res 8, 195–202.

Kochut K.J., Arnold, J., Miller, J.A., Potter, W.D. (1993). Design of an object-oriented database for reverse genetics. Proceedings of the 2nd International Conference on Computational Biology: Intelligent Systems for Molecular Biology (ISMB '93). AAAI Press, pp 234–242.

Kochut, K., Sheth, A., Miller, J.A. (1999). Optimizing workflows. Component Strategies 1, 45-57.

Kovacs, Z., Le Goff, J., McClatchey, R. (1998). Support for product data from design to production. Computer Integrated Manufacturing Systems 11, 285–290.

Krishnakumar, N., Sheth, A. (1995). Managing heterogeneous multisystem tasks to support enterprise-wide operations. Distributed and Parallel Databases 3, 155–186.

Miller, J., Sheth, A., Kochut, K., Wang, X. (1996). CORBA-based run-time architectures for workflow management systems. Journal of Database Management 7, 16–27.

Miller, J., Palaniswami, D., Sheth, A., Kochut, K., Singh, H. (1998). WebWork: METE-OR2's web-based workflow management system. Journal of Intelligent Information Systems 10, 185-215.

Natarajan, V. (1998). JFGDB: A Java based fungal genome database. Masters Thesis, University of Georgia, Athens, Georgia.

Prade, R.A., Griffith, J., Kochut, K., Arnold, J., Timberlake, W.E. 1997. In vitro reconstruction of the *Aspergillus nidulans* genome. Proc Natl Acad Sci USA 94, 14564-14569.

Schill, A., and Mittasch, C. (1997). CodAlf: A decentralized workflow management system on top of OSF DCE and DC++. Proceedings of the 3rd International Symposium on Autonomous Decentralized Systems (ISADS '97). Berlin: IEEE Computer Society, pp 205–212.

Sheth, A., Worah, D., Kochut, K., Miller, J., Zheng, K., Palaniswami, D., Das, S., (1997). The METEOR workflow management system and its use in prototyping significant healthcare applications. Proceedings of the Towards an Electronic Patient Record (TEPR '97), vol. 2, Nashville, TN, pp 267–278.

Sheth, A., Kochut, K., Miller, J., Worah, D., Das, S., Lin, C., Palaniswami, D., Lynch, J., Shevchenko, I. (1996). Supporting state-wide immunization tracking using multi-paradigm workflow technology. Proceedings of the 22nd International Conference on Very Large Data Bases, 263–273.

Smith, T.M., Abajian, C., Hood, L. (1997). Hopper: Software for automating data tracking and flow in DNA sequencing. CABIOS 13, 175–182.

Steele, T.W., Laugier, A., Falco, F. (1999). The impact of LIMS design and functionality on laboratory quality achievements. Accreditation and Quality Assurance 4, 102–106.

Wodtke, D., Weissenfels, J., Weikum, G., Dittrich, A.K., Muth, P. (1997). MENTOR workbench for enterprise-wide workflow management. Proceedings of the 1997 ACM SIGMOD International Conference on Management of Data. ACM Computer Society.

# Index